T0401713

312

Topics in Current Chemistry

Topics in Current Chemistry

Recently Published and Forthcoming Volumes

Unimolecular and Supramolecular Electronics I
Volume Editor: Robert M. Metzger
Vol. 312, 2012

Bismuth-Mediated Organic Reactions
Volume Editor: Thierry Ollevier
Vol. 311, 2012

Peptide-Based Materials
Volume Editor: Timothy Deming
Vol. 310, 2012

Alkaloid Synthesis
Volume Editor: Hans-Joachim Knölker
Vol. 309, 2012

Fluorous Chemistry
Volume Editor: István T. Horváth
Vol. 308, 2012

Multiscale Molecular Methods in Applied Chemistry
Volume Editors: Barbara Kirchner, Jadran Vrabec
Vol. 307, 2012

Solid State NMR
Volume Editor: Jerry C. C. Chan
Vol. 306, 2012

Prion Proteins
Volume Editor: Jörg Tatzelt
Vol. 305, 2011

Microfluidics: Technologies and Applications
Volume Editor: Bingcheng Lin
Vol. 304, 2011

Photocatalysis
Volume Editor: Carlo Alberto Bignozzi
Vol. 303, 2011

Computational Mechanisms of Au and Pt Catalyzed Reactions
Volume Editors: Elena Soriano, José Marco-Contelles
Vol. 302, 2011

Reactivity Tuning in Oligosaccharide Assembly
Volume Editors: Bert Fraser-Reid, J. Cristóbal López
Vol. 301, 2011

Luminescence Applied in Sensor Science
Volume Editors: Luca Prodi, Marco Montalti, Nelsi Zaccheroni
Vol. 300, 2011

Chemistry of Opioids
Volume Editor: Hiroshi Nagase
Vol. 299, 2011

Electronic and Magnetic Properties of Chiral Molecules and Supramolecular Architectures
Volume Editors: Ron Naaman, David N. Beratan, David H. Waldeck
Vol. 298, 2011

Natural Products via Enzymatic Reactions
Volume Editor: Jörn Piel
Vol. 297, 2010

Nucleic Acid Transfection
Volume Editors: Wolfgang Bielke, Christoph Erbacher
Vol. 296, 2010

Carbohydrates in Sustainable Development II
Volume Editors: Amélia P. Rauter, Pierre Vogel, Yves Queneau
Vol. 295, 2010

Carbohydrates in Sustainable Development I
Volume Editors: Amélia P. Rauter, Pierre Vogel, Yves Queneau
Vol. 294, 2010

Unimolecular and Supramolecular Electronics I

Chemistry and Physics Meet at Metal-Molecule Interfaces

Volume Editor: Robert M. Metzger

With Contributions by

D.L. Allara · H. Bässler · L. Echegoyen · C.R. Kagan · A. Köhler · M.M. Maitani · J.R. Pinzón · G. Saito · C.W. Schlenker · G. Szulczewski · M.E. Thompson · A. Villalta-Cerdas · Y. Yoshida

Springer

Editor
Prof. Robert M. Metzger
Department of Chemistry
The University of Alabama
Room 1088B, Shelby Hall
Tuscaloosa, AL 35487-0336
USA
rmetzger@ua.edu

ISSN 0340-1022 e-ISSN 1436-5049
ISBN 978-3-642-27283-7 e-ISBN 978-3-642-27284-4
DOI 10.1007/978-3-642-27284-4
Springer Heidelberg Dordrecht London New York

Library of Congress Control Number: 2011944817

Printed on acid-free paper

Springer is part of Springer Science+Business Media (www.springer.com)

Volume Editor

Prof. Robert M. Metzger

Department of Chemistry
The University of Alabama
Room 1088B, Shelby Hall
Tuscaloosa, AL 35487-0336
USA
rmetzger@ua.edu

Editorial Board

Topics in Current Chemistry Also Available Electronically

Topics in Current Chemistry is included in Springer's eBook package *Chemistry and Materials Science*. If a library does not opt for the whole package the book series may be bought on a subscription basis. Also, all back volumes are available electronically.

For all customers with a print standing order we offer free access to the electronic volumes of the series published in the current year.

If you do not have access, you can still view the table of contents of each volume and the abstract of each article by going to the SpringerLink homepage, clicking on "Chemistry and Materials Science," under Subject Collection, then "Book Series," under Content Type and finally by selecting *Topics in Current Chemistry*.

You will find information about the

- Editorial Board
- Aims and Scope
- Instructions for Authors
- Sample Contribution

at springer.com using the search function by typing in *Topics in Current Chemistry*.

Color figures are published in full color in the electronic version on SpringerLink.

Aims and Scope

The series *Topics in Current Chemistry* presents critical reviews of the present and future trends in modern chemical research. The scope includes all areas of chemical science, including the interfaces with related disciplines such as biology, medicine, and materials science.

The objective of each thematic volume is to give the non-specialist reader, whether at the university or in industry, a comprehensive overview of an area where new insights of interest to a larger scientific audience are emerging.

Thus each review within the volume critically surveys one aspect of that topic and places it within the context of the volume as a whole. The most significant developments of the last 5–10 years are presented, using selected examples to illustrate the principles discussed. A description of the laboratory procedures involved is often useful to the reader. The coverage is not exhaustive in data, but rather conceptual, concentrating on the methodological thinking that will allow the non-specialist reader to understand the information presented.

Discussion of possible future research directions in the area is welcome.

Review articles for the individual volumes are invited by the volume editors.

In references *Topics in Current Chemistry* is abbreviated *Top Curr Chem* and is cited as a journal.

Impact Factor 2010: 2.067; Section "Chemistry, Multidisciplinary": Rank 44 of 144

Preface

For these volumes in the Springer book review series *Topics in Current Chemistry*, it seemed natural to blend a mix of theory and experiment in chemistry, materials science, and physics. The content of this volume ranges from conducting polymers and charge-transfer conductors and superconductors, to single-molecule behavior and the more recent understanding in single-molecule electronic properties at the metal–molecule interface.

Molecule-based electronics evolved from several research areas:

1. A long Japanese tradition of studying the organic solid state (since the 1940s: school of Akamatsu).
2. Cyanocarbon syntheses by the E. I. Dupont de Nemours Co. (1950–1964), which yielded several interesting electrical semiconductors based on the electron acceptor 7,7,8,8-tetracyanoquinodimethan (TCNQ).
3. Little's proposal of excitonic superconductivity (1964).
4. The erroneous yet over-publicized claim of "almost superconductivity" in the salt TTF TCNQ (Heeger, 1973).
5. The first organic superconductor (Bechgard and Jérôme, 1980) with a critical temperature $T_c = 0.9$ K; other organic superconductors later reached T_c 13 K.
6. Electrically insulating films of polyacetylene, "doped" with iodine and sodium, became semiconductive (Shirakawa, MacDiarmid, Heeger, 1976).
7. The interest in TTF and TCNQ begat a seminal theoretical proposal on one-molecule rectification (Aviram and Ratner, 1974) which started unimolecular, or molecular-scale electronics.
8. The discovery of scanning tunneling microscopy (Binnig and Rohrer, 1982).
9. The vast improvement of electron-beam lithography.
10. The discovery of buckminsterfullerene (Kroto, Smalley, and Curl, 1985).
11. Improved chemisorption methods ("self-assembled monolayers") and physisorption methods (Langmuir–Blodgett films).
12. The growth of various nanoparticles, nanotubes, and nanorods, and most recently graphene.

All these advances have helped illuminate, inspire, and develop the world of single-molecule electronic behavior, and its extension into supramolecular assemblies.

These volumes bring together many of the leading practitioners of the art (in each case I mention only the main author). Bässler sets in order the theoretical understanding of electron transport in disordered (semi)-conducting polymers. Saito summarizes in fantastic detail the progress in understanding charge-transfer crystals and organic superconductivity. Echegoyen reviews the chemistry and electrochemistry of fullerenes and their chemical derivatives. Thompson reviews the progress made in organic photovoltaics, both polymeric and charge-transfer based. Ratner updates the current status of electron transfer theory, as is applies to measurements of currents through single molecules. Metzger summarizes unimolecular rectification and interfacial issues. Kagan discusses field-effect transistors with molecular films as the active semiconductor layer. Allara reminds us that making a "sandwich" of an organic monolayer between two metal electrodes often involves creep of metal atoms into the monolayer. Rampi shows how mercury drops and other techniques from solution electrochemistry can be used to fabricate these sandwiches. Wandlowski discusses how electrochemical measurements in solution can help enhance our understanding of metal–molecule interfaces. Hipps reviews inelastic electron tunneling spectroscopy and orbital-mediated tunneling. Joachim addresses fundamental issues for future molecular devices, and proposes that, in the best of possible worlds, all active electronic and logical functions must be predesigned into a single if vast molecular assembly. Szulczewski discusses the spin aspects of tunneling through molecules: this is the emerging area of molecular spintronics.

Many more areas could have been discussed and will undoubtedly evolve in the coming years. It is hoped that this volume will help foster new science and even new technology. I am grateful to all the coauthors for their diligence and Springer-Verlag for their hosting our efforts.

Tuscaloosa, Alabama, USA
Delft, The Netherlands
Dresden, Germany

Robert Melville Metzger

Contents

Top Curr Chem (2012) 312: 1–66
DOI: 10.1007/128_2011_218

Published online: 5 October 2011

Charge Transport in Organic Semiconductors

Heinz Bässler and Anna Köhler

Abstract Modern optoelectronic devices, such as light-emitting diodes, field-effect transistors and organic solar cells require well controlled motion of charges for their efficient operation. The understanding of the processes that determine charge transport is therefore of paramount importance for designing materials with improved structure-property relationships. Before discussing different regimes of charge transport in organic semiconductors, we present a brief introduction into the conceptual framework in which we interpret the relevant photophysical processes. That is, we compare a molecular picture of electronic excitations against the Su-Schrieffer-Heeger semiconductor band model. After a brief description of experimental techniques needed to measure charge mobilities, we then elaborate on the parameters controlling charge transport in technologically relevant materials. Thus, we consider the influences of electronic coupling between molecular units, disorder, polaronic effects and space charge. A particular focus is given to the recent progress made in understanding charge transport on short time scales and short length scales. The mechanism for charge injection is briefly addressed towards the end of this chapter.

Keywords Charge carrier mobility · Charge transport · Organic semiconductors · Molecular model · Gaussian disorder model · SSH model · Organic optoelectronics

Contents

H. Bässler (✉) and A. Köhler
Experimental Physics II, University of Bayreuth, Bayreuth, Germany
e-mail: baessler@staff.uni-marburg.de

1 Introduction

Charge transport in organic semiconductors is a timely subject. Today, organic semiconductors are already widely used commercially in xerography. For display and lighting applications they are employed as light emitting diodes (LEDs or OLEDs) or transistors, and they are making progress to enter the solar cell market [1–6]. As a result, interest in the science behind this novel class of materials has risen sharply. The optoelectronic properties of organic semiconductors differ from that of conventional inorganic crystalline semiconductors in many aspects and the knowledge of organic semiconductor physics is imperative to advance further with the associated semiconductor applications [7]. A central problem is the understanding of the mechanisms related to charge transport.

It may seem odd to write an article entitled "charge transport in organic semiconductors," notably polymers, when these materials are inherently insulators. This raises the question about the difference between a semiconductor and an insulator. The conductivity κ of the materials is the product of the elementary charge e, the mobility μ of charge carriers, and their concentration n, i.e., $\kappa = en\mu$. A material can be insulating either if there are no charges available or if they are immobilized. A prototypical example of the former case is quartz. Since the absorption edge of quartz is far in the ultraviolet region (at about 120 nm), the gap E_g between the valence and conduction band is about 10 eV [8]. This implies that, at ambient temperature, the concentration of free charge carriers is practically zero. However, if one generates charge carriers by high energy radiation, they would probably move with a mobility that is comparable to that of a conventional covalently bonded inorganic semiconductor such as silicon, i.e., 1,000 cm^2 V^{-1} s^{-1} or larger. Obviously, an inherent insulator can be converted into a semiconductor if free charge carriers are generated by either injection from the electrodes, by doping, or by optical excitation.

In traditional semiconductors such as silicon, germanium, or Ga_2As_3 the conductivity is between, say, 10^{-8} to $10^{-2}\ \Omega^{-1}\ cm^{-1}$. In an undoped solid, the concentration of free charge carriers is determined by $n = N_{eff}e^{-\frac{E_g}{2kT}}$ where N_{eff} is the effective density of valence or conduction band states and E_g is the band gap. For crystalline silicon, E_g is 1.1 eV and the charge carrier mobility is about 1,000 $cm^2\ V^{-1}\ s^{-1}$. This predicts an intrinsic conductivity of about $10^{-6}\ \Omega^{-1}\ cm^{-1}$ at room temperature. Note that a band gap of 1.1 eV translates into an absorption edge of 1,100 nm. In view of the relative dielectric constant as large as $\varepsilon = 11$, coulomb effects between electrons and holes are unimportant and electrons and holes are essentially free at room temperature. This implies that optical absorption is due to a transition from the valence band to a conduction band. The situation is fundamentally different in undoped molecular solids. Their absorption edge is usually larger than 2 eV and the dielectric constant is 3–4. In this case optical absorption generates coulomb bound electron-hole pairs with a binding energy of 0.5–1.0 eV. Even if one were to ignore the exciton binding energy and to identify incorrectly the optical absorption edge with a valence to conduction band transition, the resultant intrinsic conductivity would be much less than $10^{-12}\ \Omega^{-1}\ cm^{-1}$, assuming a charge carrier mobility of 1 $cm^2\ V^{-1}\ s^{-1}$, i.e., the materials are insulators. However, they can become semiconducting if charge carriers are generated extrinsically.

This chapter focuses on the electronic transport of organic semiconductors. The motivation is straightforward. Modern optoelectronic devices, such as light-emitting diodes, field effect transistors, and organic solar cells are based on charge transport. The understanding of the processes that control charge transport is therefore of paramount importance for designing materials with improved structure property relations. Research into this subject was essentially stimulated by studies on charge transport in molecularly doped polymers that are now commonly used in modern photocopying machines. It turns out that xerography is meanwhile a mature technology [1]. It is the only technology in which organic solids are used as active elements on a large industrial scale. An important step in the historic development of xerography was the recognition that one could profitably use aromatic molecules as a photoreceptor when they are embedded in a cheap inert flexible binding material such as polycarbonates. Meanwhile, most photocopiers and laser printers use this kind of receptors although few users will recognize that once they push the print button they start an experiment on transient photoconductivity in a polymeric photoreceptor. There is much hope that organic LEDs, FETs, and solar cells will be able to meet the competition from existing technology based upon inorganic materials and enter the market, similarly to xerography. OLEDs that are based on small molecules already constitute a substantial business.

Apart from the endeavor to optimize the structure property relations of materials used in modern optoelectronic devices there is the desire to understand the conceptual premises of charge transport in random organic solids. The use of amorphous, instead of crystalline, organic semiconductor materials is favored because they allow for a low cost of device fabrication and the use of flexible substrates, thus enabling mechanically flexible devices. The aim of this chapter is to introduce those new to this field to the already established understanding of charge transport in organic semiconductors, and to point those familiar with the field to current research activities where new insight emerges and to the challenges that remain.

2 Basic Concepts of Charge Transport in Organic Solids

2.1 Electronic Structure of Organic Solids

In order to understand charge transport in organic solids, we need to elaborate on the electronic structure of organic solids. Organic solids such as molecular crystals, amorphous molecular films, or polymeric films are made of molecular subunits. We shall therefore start from a molecular picture and consider any coupling between the molecular units afterwards. Organic semiconductors are hydrocarbon molecules with a backbone of carbon atoms. The strong bonds that form the molecular backbone arise from sp^2 hybridized atomic orbitals of adjacent carbon atoms that overlap yielding a bonding and antibonding molecular σ and σ^* orbitals. The remaining atomic p_z orbitals overlap to a lesser degree, so that the resulting molecular π and π^* orbitals are less binding or antibinding, thus forming the frontier orbitals of the molecule. In the ground state of the molecule, all bonding orbitals up to the highest occupied molecular orbital, the HOMO, are filled with two electrons of antiparallel spin while the antibonding orbitals, from the lowest unoccupied molecular orbital (LUMO) onwards, are empty. Neutral excited states can be formed for example by light absorption in a molecule, when an electron is promoted from the HOMO to the LUMO. In general, any configuration with an additional electron in an antibonding orbital and a missing electron in a bonding orbital, i.e., a hole, corresponds to a neutral excited state. Due to the low relative dielectric constant in organic semiconductors (on the order of $\varepsilon \approx 3$), coulomb attraction between electron and hole is strong, resulting in an exciton binding energy ranging from of 0.5 eV to more than 1 eV. Molecular orbital diagrams corresponding to the configurations in the ground or neutral excited states are shown in Fig. 1.

For charge transport in organic solids to take place, there must be a charge on the molecular unit. This may either be an additional electron that is accommodated in an antibonding orbital, or one that is removed from a bonding orbital. The molecule is then no longer in the ground state but rather in a charged excited state. The addition or removal of an electron from the molecule may be obtained in several ways:

1. Through injection or extraction of an electron at the interface between a metal electrode and the molecule, as is typically the case in the operation of a device such as light-emitting diodes (LED).

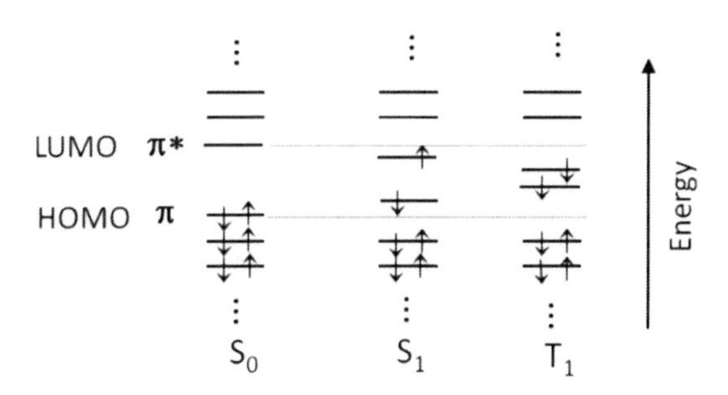

Fig. 1 Molecular orbital diagram showing the electronic configuration for the ground state (S_0), for the first spin-singlet excited state (S_1) and for the first spin-triplet excited state (T_1). The *arrows* indicate the electron spin, the *thin horizontal gray line* is a guide to the eye. In this representation, coulomb and exchange energies are explicitly included in the positions of the frontier orbitals

2. Through reduction or oxidation of the molecule by a dopant molecule. Atoms or molecules with high electron affinity, such as iodine, antimony pentafluoride ($SbCl_5$), or 2,3,5,6-tetrafluoro-7,7,8,8-tetracyanoquinodimethane (F_4-TCNQ), may oxidize a typical organic semiconductor such as poly(*p*-phenylene) derivatives, leaving them positively charged. Reduction, i.e., addition of an electron, may be obtained by doping with alkali metals.
3. Through exothermic dissociation of a neutral excited state in molecule by electron transfer to an adjacent molecule. This process leads to the generation of geminately bound electron-hole pairs as precursors of free positive and negative charges in an organic solar cell.

From electrochemical experiments it is well known that, after the removal of one electron from an individual molecule, more energy is required to remove a second electron. This implies that the relative positions of the molecular orbitals with respect to the vacuum level change upon removal or addition of an electron, as indicated in Fig. 2a in a qualitative fashion. Furthermore, when an electron is taken from a π-orbital or added to a π* orbital, this alters the spatial distribution of electrons in the more strongly bound σ-orbitals, resulting in different bond lengths of the molecule. The energy associated with this change in molecular geometry is known as the geometric reorganization energy, and the charge in combination with the geometric distortion of the molecule is referred to as a polaron. These effects due to electron–electron correlations and electron–phonon couplings are a manifestation of the low dielectric constant of organic semiconductors. They are absent in inorganic semiconductor crystals due to the strong dielectric screening with $\varepsilon \approx 11$.

A charged molecule may absorb light in the same fashion as does a neutral molecule, thereby promoting an electron from a lower to a higher molecular orbital. Possible optical transitions are indicated in Fig. 2a by arrows. These optical transitions can easily be observed in doped molecular films as well as in solution (see below). We note that, analogous to transitions in neutral molecules, absorption

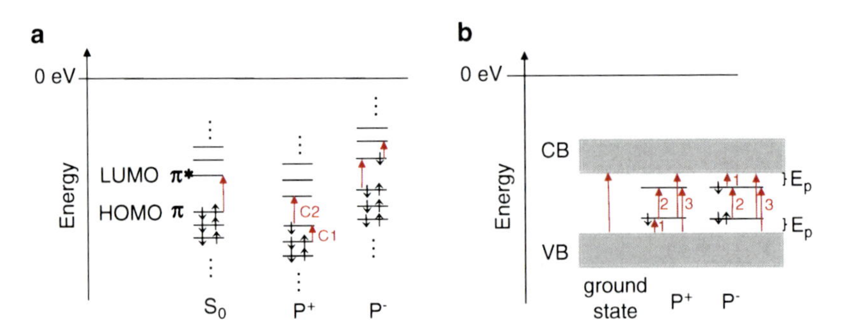

Fig. 2 (**a**) Molecular orbital diagram for a neutral molecule in the ground state (S_0), for a positively charged molecule (P^+), and for a negatively charged molecule (P^-). The shifts in the molecular orbital levels upon charging are only drawn in a qualitative fashion. Optical transitions are indicated by *red arrows*. C1 and C2 label the transitions seen in Fig. 4 further below. (**b**) Semiconductor band picture showing self-localized polaron energy levels within the band gap. The polaron binding energy E_p is also indicated. Predicted optical transitions involving the positive or negative polaron (P^+ or P^-, respectively) are indicated through *red arrows* and labeled by numbers

may cause a transition into different vibrational levels of the charged molecule, thus giving rise to vibrational structure in the polaron absorption spectra.

When molecules are not in a gas phase but in a solid, the absolute values of their energy levels shift with respect to the vacuum level due to the change in the polarization of their surroundings. If they are deposited, by spin-coating or evaporation, to form an amorphous film, the surrounding polarization varies spatially in a random fashion leading to a random distribution of the absolute values of the molecular energies. By the central limit theorem of statistics, this implies a Gaussian distribution of excited state energies [9] for both neutral and charged excited states, with a variance σ that is characteristic for the energetic disorder. Experimentally, this is observed as an inhomogeneous broadening of the optical spectra such as absorption, fluorescence, and phosphorescence spectra. Hole and electron transporting states are similarly disorder broadened although in this case state broadening is not directly amenable to direct absorption spectroscopy.

Such disorder is absent in a molecular crystal. In an inorganic semiconductor crystal, such as Si or Ge, atoms are bound by strong covalent bonds to form the crystal. Consequently, electronic interactions between the atomic orbitals are strong, and wide bands with bandwidths on the order of a few eV are formed that allow for charge transfer at high mobilities. In contrast, molecular crystals are kept together by weak van der Waals bonds. Consequently, electronic interactions between the molecular orbitals of adjacent lattice sites are weak and the resulting bands are narrow, with bandwidth below 500 meV [10]. In very pure molecular crystals of, say, naphthalene or perylene, band transport can therefore be observed from low temperatures up to room temperature [11–14]. At higher temperatures, intra- and intermolecular vibrations destroy the coherence between adjacent sites. A charge carrier is then scattered with a mean free path that approaches the distance between adjacent sites. As a result, band transport is no longer possible and charge carriers move by hopping.

On passing, we note that even though charge transport in pure molecular crystals takes place in a band, optical transitions in a molecular crystal do NOT take place between valence and conduction bands due to a lack of oscillator strength. This is an inherent consequence of the strong coulomb interaction present between charges in molecular crystals. While in inorganic crystals, the strong dielectric constant implies an effective shielding of coulomb forces, this is not the case in organic crystals due to their low dielectric constant. It implies that when an optical transition is to take place, in order for an electron to escape from its coulombically bound sibling, it had to overcome a coulomb capture radius which is about 20 nm. The electronic coupling among molecules that far apart is negligibly small, resulting in a negligible oscillator strength for such a "long distance charge-transfer type" transition. Therefore, a transition such that the electron is outside the coulomb capture radius of its sibling does not take place. Rather, absorption and emission in a crystal takes place between orbitals of an individual molecule on a particular lattice site, or between orbitals of immediately adjacent molecules, thus yielding strongly coulombically bound electron hole pairs, referred to as Frenkel excitons or charge transfer excitons, respectively. In a perfectly ordered crystal, the exciton, i.e., the two-particle excitation, is equally likely to be on any lattice site and thus couples electronically to neighboring sites. This results in the formation of an exciton band, i.e., a band for the two-particle excitation, within which the exciton moves in a delocalized fashion. Note that the exciton band describes the electronic coupling between an existing two-particle excitation on a molecule with its neighboring site (and thus the motion of an exciton), while the π or π^* bands describe the coupling of a one-particle molecular orbital with its neighbor. π or π^* bands are therefore suitable to portray the motion of a single charge carrier in a molecular crystal, yet, for the reasons just outlined, optical transitions between them do not occur.

Today's organic semiconductor devices such as LEDs, FETs, or solar cells may be made from amorphous molecular films, molecular crystals (in the case of some FETs), or from polymeric semiconductors. In polymers, molecular repeat units are coupled by covalent bonds allowing for electronic interaction between adjacent repeat units. As will be detailed in the next section, in a perfectly ordered polymer, such as crystalline polydiacetylene [15], this electronic interaction leads to the formation of a broad intra-chain exciton band as well as valence and conduction bands while inter-chain interactions are moderately weak and comparable with the situation of molecular crystals. In amorphous polymers, conformational disorder implies that coherence is only maintained over a few repeat units that thus form a chromophore [16]. We refer to this section of the polymer chain as the conjugation length. Naturally, the conjugation length in rigid, well ordered polymers such as MeLPPP is longer (on the range of 10–15 repeat units) than in polymers with a high degree of torsional disorder along the chain such as DOO-PPP [17, 18]. A charge carrier on a polymer chain may move coherently within the conjugation length, though hopping will take place between different conjugated segments [19, 20]. For the purpose of considering charge transport, it is therefore convenient to treat a conjugated segment of a polymer chain as a chromophore, i.e., analogous to a molecule.

So far we have outlined the conceptual framework in which we discuss charge transfer in organic semiconductors. It is based on a molecular picture where the molecular unit is considered central, with interactions between molecular units added afterwards. For amorphous molecular solids and for molecular crystals this approach is undisputed. In the case of semiconducting polymers, a conceptually different view has been proposed that starts from a one-dimensional (1D) semiconductor band picture, and that is generally known as the Su–Schrieffer–Heeger (SSH) model [21–24].

We feel the molecular approach we have taken gives an appropriate description of the underlying electronic structure. The conceptual framework one adopts however influences the interpretation of experimental results, for example when considering the absorption spectra of charge carriers. In order to place the discussion of charge transfer models for polymers into a larger context, it is beneficial to be aware of agreements and differences between a "molecular approach" and the SSH model. Therefore we shall digress here to a comparative discussion of the two approaches.

2.2 *Comparison of the Molecular Picture and the SSH Approach of Treating Charge Carriers in Semiconducting Conjugated Polymers*

The theory for a band picture of semiconducting polymers has been developed for a perfect, infinite, one-dimensional polymer chain. The simplest case to consider is polyacetylene, i.e., a chain of sp^2-hybridized carbon atoms. Early work on this "system" was carried out in the 1950s by Salem and Longuett-Higgins [25], who considered the electronic structure of a long sp^2-hybridized carbon chain with cyclic boundary conditions, i.e., forming a ring. The effect of a charge on such a system was later investigated by Su, Schrieffer, and Heeger [21], after synthesizing and doping polyacetylene. A similar theoretical "system" to consider is an infinite, planar chain of poly(*p*-phenylene) (PPP), which can be considered analogous to a one-dimensional "crystal" of phenyl units with strong coupling between the units. From an experimental point of view, a good realization of a perfect one-dimensional semiconducting polymer chain is given by crystalline polydiacetylene [15].

We will first sketch briefly how the electronic structure of a perfect one-dimensional polymer chain is perceived in a molecular picture before drawing the comparison to a semiconductor band picture. For our molecular based approach, we consider, say, a perfect PPP chain as a sequence of molecular repeat units such as phenylenes that are coupled by a covalent bond. As a result of the coupling, the molecular orbitals of adjacent units can interact and split. Due to the perfect order and symmetry, this process takes place across the entire chain leading to the

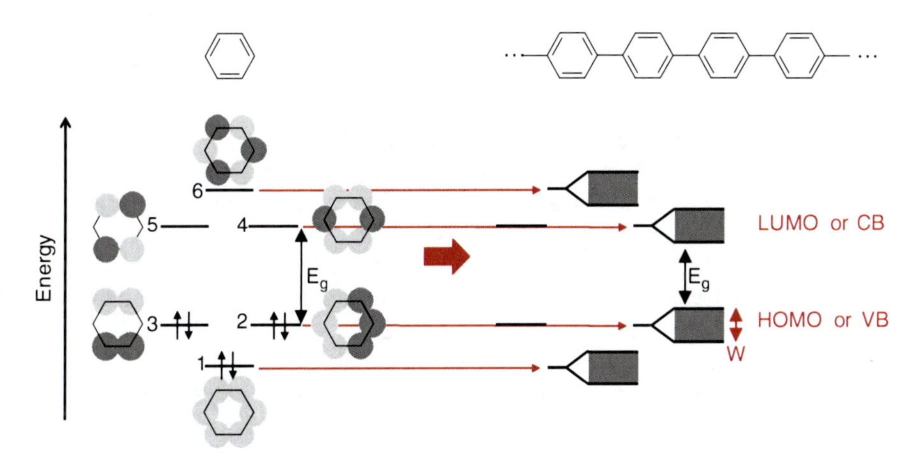

Fig. 3 Schematic, qualitatively illustrating the formation of bands from molecular orbitals when going from benzene to a perfectly ordered, infinite poly(*p*-phenylene) (PPP). (**a**) Energies and shapes of molecular orbitals for benzene in a simple Hückel-type picture. (**b**) Qualitative band structure resulting from electronic coupling between orbitals with electron density at the para-position. The frontier orbitals 2 and 4 in benzene can delocalize along the entire PPP chain, thus forming valence and conduction bands of width W. The lower and higher lying orbitals 1 and 6 in benzene can form corresponding lower and higher lying bands. Orbitals with nodes at the para-position such as 3 and 5 remain localized. See also [26]

formation of bands. For example, π and π* bands will arise from HOMO and LUMO orbitals, and they will take the role of a valence and conduction band. This is schematically illustrated in Fig. 3. In the molecular picture, coulomb interactions are considered to be strong, and consequently, for the same reasons as outlined in the case of a three-dimensional molecular crystal, optical excitations in a perfect polymer chain are assumed to result in the formation of strongly bound electron-hole pairs while direct transitions from a valence π-band to a conduction π* band are expected not to carry any oscillator strength. The π and π* bands in a perfect polymer in a perfect crystalline environment, and the energy gap separating them, owe their existence to the electronic coupling between repeat units. Their existence is independent of whether the system is aromatic or whether it has an alternation of single/double bonds. A critical quantity, however, is the relative size of the coupling energy between repeat units compared to the energetic variation of each unit (see Sect. 2.3 below). In amorphous polymer films, energetic disorder due to the polarization of the surroundings is strong, so that electronic coherence is only maintained over a few repeat units that are usually referred to as a conjugated segment.

In contrast, in the SSH model, the electrical bandgap E_g^{el} arises because of the alternation between single and double carbon–carbon bonds, a signature of the Peierls distortion in a 1D system. When a perfect 1D chain of equidistant carbon atoms is considered, the electronic structure resulting from the electronic coupling between the atomic p_z-orbitals is that of a half-filled π band, implying a metallic character. The introduction of an alternating bond length, however, leads to the formation of a filled π-band and an empty π* band, with a gap separating them, thus predicting semiconducting properties.

One of the key assumptions of the SSH model is that the electron–electron correlations and the coulomb attraction between electrons and holes are very small. As a direct consequence, the optical absorption is assigned to a valence band (VB) to conduction band (CB) transition as is in a conventional semiconductor rather than to the transition into a neutral excitonic state. The second key assumption in the SSH model relates to the magnitude of the electron–phonon coupling. Once a free electron–hole pair has been excited by an optically driven VB–CB transition, electrons and holes couple to phonons regardless if the associated chain distortions are conventional long wavelength phonons or rather more localized molecular vibrations. This type of coupling is inherent to both the molecular model and the semiconductor, i.e., SSH–model. It is a signature of the geometric reorganization a chain suffers when an electron is transferred from the HOMO to the LUMO. The reorganization energy is referred to as the polaron binding energy. The essential difference between the molecular and the SSH model relates to (1) the magnitude of the coupling and (2) the assignment of the sub-bandgap absorption features that show up when electrons and holes are excited. In the SSH model and the related Fesser – Bishop – Campbell model [23] a positively (negatively) charged self localized polaron P^+ (P^-) is created by removal (addition) of an electron with respect to the mid-gap Fermi-energy. As a result two energy levels form inside the band gap that are occupied with a total of one electron (three electrons). The polaron is associated with transitions among localized levels and non-localized band states (see Fig. 2). For example for P^+, the lowest transition is from the VB to a localized level (1), the second next lowest transition is between the localized levels (2), followed by two degenerate transitions (3). This implies that the lowest transition is a direct measure of the polaron binding energy E_p while the second next transition should occur at an energy of $E_g - 2E_p$, taking into account that the optical absorption edge is identified as a VB → CB transition. As a consequence of the neglect of the coulomb binding energy on the one hand and the assumed large electron–phonon coupling on the other, the collapse of two charges of the same kind should be an exothermic process leading to the formation of positively or negatively charged bipolarons. They are predicted to give rise to two sub-band optical absorption features.

Meanwhile there is overwhelming evidence that the basic assumptions of the SSH model are not applicable to π-bonded conjugated polymers. Coulombic and electron–electron correlation effects are large while electron–phonon coupling is moderately weak. As a consequence, the spectroscopic features in this class of materials are characteristic of molecular rather than of inorganic crystalline semiconductor systems. There are a number of key experimental and theoretical results that support this assignment:

1. A material that can be considered as a prototypical one-dimensional system consists of a poly-diacetylene (PDA) chain embedded in a perfect molecular precursor crystal at a concentration low enough that there is no inter-chain interaction. Such systems can be fabricated by controlled irradiation of a precursor crystal [15]. Some of the PDAs fluoresce. The absorption and fluorescence spectra are excitonic in character with resonant 0–0 transitions [15]. The Huang Rhys factor is small, indicating that coupling to molecular vibrations (and phonons) is weak. In conventional absorption spectroscopy the VB $\rightarrow$ CB transition is absent, although it shows up in electroabsorption spectroscopy. The energy difference of 0.55 eV between the exciton transition and the valence π-band $\rightarrow$ conduction π^* band transition is a direct measure of the exciton binding energy [27]. This value is supported by theory. In other π-conjugated polymers the magnitude of the exciton binding energy is similar [28]. In passing, we note that if the exciton binding energy was only about kT as implied by the SSH model there should be no efficient electroluminescence in organic LEDs, since in the absence of coulomb attraction electrons and holes would hardly find each other [29].
2. Level crossing between the two lowest singlet excited states was observed by the Kohler group through absorption and luminescence spectroscopy in oligoenes when the oligomer chain length increases. This can only be accounted for when electron–electron correlations are strong [30]. Another signature of the strong electron–electron interactions in π-bonded conjugated polymers is the observation of phosphorescence [31–34]. Phosphorescence spectra are separated from the fluorescence spectra by an exchange interaction energy of about $2J = 0.7$ eV (where J is the value of the exchange integral) [34, 35], implying a strong electron correlation effect.
3. The fact that the lowest charge induced absorption feature in π-conjugated polymers is near 0.5 eV is in disagreement with the notion that it is due to a transition involving a localized state and a band state, thus reflecting the magnitude of the polaron binding energy, which is half of the total reorganization energy $E_p = \frac{\lambda}{2}$. Even if one interpreted the temperature dependence of the hole mobility in the ladder type poly(*p*-phenylene) LPPP in terms of a disorder-free polaron transport (thus attributing all activation energy to polaronic effects) one would end up with a value of the polaron binding energy as low as 50 meV [36].
4. There is convincing evidence that the absorption spectra of charged π-conjugated oligomers and polymers are electronic transitions among different electronic levels of (monovalent) radical anions and cations rather than bipolarons (see, for example, Fig. 4) [37]. The spectra do not reflect the reorganization energy involved in ion formation but bear out vibronic splitting and follow the same relation on the reciprocal chain length dependence as do the absorption spectra of uncharged oligomers. However, in the experiments reported in [37] it has been observed that upon increasing the concentration of the oxidant/reductant the absorption features are shifted to higher energies. One could surmise that at high ion concentration bipolarons are indeed formed. Meanwhile it has been

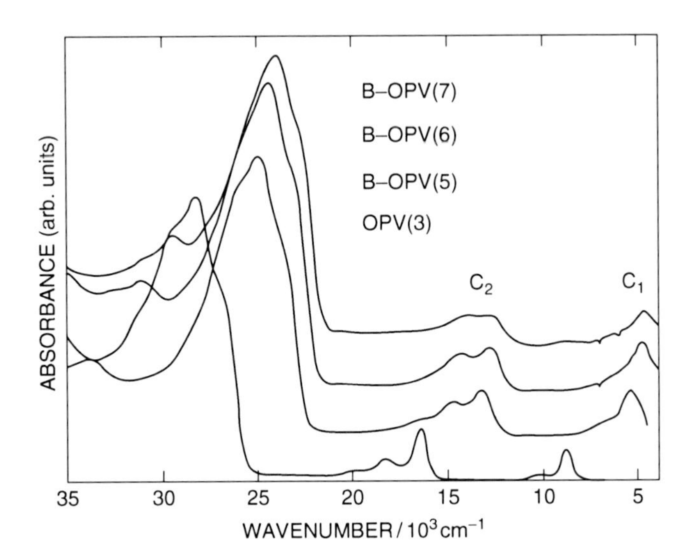

Fig. 4 Absorption spectra of radical cations of oligo-phenylenevinylenes OPV of different chain lengths in CH_2Cl_2 solution. C_1 and C_2 denote the transitions indicated in Fig. 2a. The radical ions are generated by adding $SbCl_5$ as an oxidant to the solution. From [37] with permission. Copyright (1993) by Elsevier

suggested, though, that the high energy features are due to the formation of pairs of monovalent polarons in which the radical ion state splits into a doublet in which the lower state is doubly occupied [38]. Related work has been performed on polyazulenes [39, 40]. Nöll et al. find that polyazulene can be doped up to a maximum number of one positive charge per three to four azulene units. At these high doping levels the charge carrying units are pairs of single-valent radical cations rather than bipolarons. At still higher doping levels the polymer starts decomposing. The energetic instability of bipolarons has further been proven by quantum chemical calculations on model systems consisting of a ring of thiophene units. The result is that, upon adding a second charge to the ring, both charges avoid each other rather than form a stable bipolaron [41]. More recent work indicates that a stable entity may only be formed when a pair of like charges is coupled with an oppositely charged moiety (a "trion") in which the coulomb repulsion is diminished [42]. Obviously the coulomb repulsion between a pair of like charges exceeds the gain in reorganization energy. Therefore bipolarons are unstable [43, 44]. By the way, it has never been questioned that the charge carrying species that is monitored in charge transport studies is a singly rather than a doubly charged entity.

This digression on the interpretation of the absorption from charged polymers illustrates the importance of the conceptual framework that is adopted. As already mentioned, for molecular glasses or crystals, a molecular picture has always been undisputed. For polymers, the debate conducted over the last two decades has eventually been largely settled on the same molecular view. Consequently, the discussion of the charge transfer models in this chapter is also based on a molecular picture throughout.

2.3 General Approach to Charge Transfer Mechanisms

There is quite a range of charge transfer models based on the molecular picture that are employed to describe charge transport in organic solids, such as models based on band transport, polaronic models, and models that focus on the effects of disorder. At the same time, organic solids are a broad class of materials, comprising crystals as well as molecular and polymeric glasses. It is therefore necessary to obtain some basic understanding on which parameters affect charge transport in order to assess which model may be suitable to describe a particular experimental situation.

In order to develop such a broader view and a general qualitative understanding of charge transport, it is beneficial to consider the general one-electron Hamiltonian shown in (1). In this approach we follow the outline taken in [45]. This Hamiltonian assumes a low carrier density, and effects due to electron correlation or coulomb interaction are not considered. Despite these limitations, the following general one-electron Hamiltonian is useful to illustrate different limiting cases:

$$H = H_0 + H_1 + H_2 + H_3 + H_4 \tag{1}$$

with

$$H_0 = \sum\nolimits_n \epsilon_n a_n^\dagger a_n + \sum\nolimits_\lambda \hbar\omega_\lambda \left(b_\lambda^\dagger b_\lambda + \frac{1}{2} \right)$$

being the electronic and vibrational excitation term,

$$H_1 = \sum\nolimits_{\substack{n,m \\ n \neq m}} J_{nm} a_n^\dagger a_m$$

being the electron transfer term,

$$H_2 = \sum\nolimits_\lambda \sum\nolimits_n g_{n\lambda}^2 \hbar\omega_\lambda a_n^\dagger a_n (b_\lambda + b_{-\lambda}^+)$$

being the dynamic diagonal disorder term,

$$H_3 = \sum\nolimits_{\substack{n,m \\ n \neq m}} \sum\nolimits_\lambda f_{nm\lambda}^2 \hbar\omega_\lambda a_n^\dagger a_m \left(b_\lambda + b_{-\lambda}^\dagger \right)$$

being the dynamic off-diagonal disorder term, and

$$H_4 = \sum\nolimits_n \delta\epsilon_n a_n^\dagger a_n + \sum\nolimits_{\substack{n,m \\ n \neq m}} \delta J_{nm} a_n^\dagger a_m$$

being the static diagonal and off-diagonal disorder term.
$a_n^\dagger$ (a_n) is the creation (destruction) operator for an excited electron in an orbital of energy ε_n at the molecular site n,
$b_n^\dagger$ (b_n) is the creation (destruction) operator for an vibrational mode of energy $\hbar\omega_\lambda$,
ε_n is the energy in a perfectly ordered lattice and $\delta\varepsilon_n$ is its variation due to static disorder,
J_{nm} is the electronic interaction between site m and n in a perfectly ordered lattice and δJ_{nm} is its variation due to static disorder, and
$g_{n\lambda}$ and $f_{nm\lambda}$ are dimensionless coupling constants for the electron–phonon coupling.

In (1), H_0 yields the total energy of system in which the molecules and the lattice are excited, yet there are no interactions between molecules and the lattice. The transfer of an electron from site m to site n is given by H_1. Polaronic effects, i.e., effects due to the interaction of the electronic excitation and the lattice, are given by H_2 and H_3. In H_2, the energy of the site is reduced by the interaction with the lattice vibration. In H_3, the lattice vibration alters the transition probability amplitude from site m to n. The term lattice vibration may refer to inter-molecular or intra-molecular vibrations. Static disorder effects are considered in H_4, which describes the changes to the site energy or transition probability amplitude by variations in the structure of the molecular solid.

The interactions considered in the polaronic terms H_2 and H_3 introduce "dynamic" disorder, since they are based on coupling of the electronic excitation to lattice vibrations. In contrast, the changes to site energy and transition rate in H_4 are independent of vibrations. They are merely due to variations in the morphological structure of the film or crystal, i.e., intermolecular distances and orientations, and they are thus referred to as "static" disorder. When (1) is written out in a matrix notation, the site energies appear on the diagonal position of the matrix, and thus energetic variations are sometimes called "diagonal disorder" while changes in the transition rate from site n to m are disguised by the term "off-diagonal disorder." In the Hamiltonian of (1), only linear coupling to lattice vibrations is considered. Throughout this chapter, the expression "disorder" usually refers to static disorder only, while we tend to employ the expression "polaronic effects" to discuss the effects due to the electron–phonon coupling expressed in H_2 and H_3.

Having clarified some of the terminology used, we can now turn to considering different modes of charge transfer. The nature of charge transfer is determined by the relative sizes of the interaction energy J_{nm}, the strength of the electron–phonon coupling expressed though the coupling constants in $g_{n\lambda}^2\hbar\omega_\lambda$ and $f_{nm\lambda}^2\hbar\omega_\lambda$, and the degree of static disorder present and expressed through $\delta\varepsilon$ and δJ_{nm}. Essentially, there are three limiting cases.

2.3.1 Band Transport

If the interaction energy with nearest neighbor, $J_{n,n+1}$,is large compared to any other energy present such as the effects of dynamic or static disorder, charge transport takes place through a band. The charge carrier delocalizes to form a propagating Bloch wave that may be scattered by lattice vibrations. Band transport can only occur if the bands are wider than the energetic uncertainty of the charge carrier. This requirement implies that by zero order reasoning [45] the charge carrier mobility must very roughly exceed $ea^2W/\hbar kT$, where e is the elementary charge, a is the lattice constant, and W is the bandwidth. For organic semiconductors, $W \approx 10\ kT$ and $a \approx 1$ nm so that band transport occurs if $\mu \approx 10\ \text{cm}^2\ \text{V}^{-1}\ \text{s}^{-1}$.

2.3.2 Polaronic Transport

If H_1 is small compared to H_2 and H_3, and if H_4 can be neglected, the transport is dominated by the coupling of the electronic excitation to intermolecular or intramolecular vibrations, and the charge carrier coupled to the lattice is termed a polaron. The interaction expressed in term H_2 causes a reduction of the site energy by the polaron binding energy. For charge transport, this needs to be overcome by thermal activation. The charge transfer itself takes place by an uncorrelated, phonon-assisted hopping process, and it is determined by H_3.

2.3.3 Disorder-Based Transport

If fluctuations in the intermolecular distances and orientations give rise to a large variation in the site energy and transition probability amplitude compared to the other terms, the static disorder dominates the charge transport. A charge carrier moves by uncorrelated hops in a broad density of states. Thermal activation is required to overcome the energy differences between different sites.

These different modes of transport result in a dissimilar temperature dependence of the charge carrier mobility, and this often provides a convenient means to investigate which transport regime may apply. In this chapter, due attention is therefore given to experimental approaches that allow for an investigation of the transport mechanism, and concomitantly of the underlying electronic structure.

In this chapter we start by considering charge transport for materials where the disorder aspect is dominant. This conceptual framework is then extended to include polaronic aspects. After discussing the effects of charge carrier density on charge transport in this disorder + polaronic dominated transport regime, we next consider how a stronger coupling between molecular units alters the mode of charge transport, finally arriving at the regime of band transport. Charge injection, which often precedes charge transport, is briefly addressed at the end of this chapter.

In the context of this chapter, we focus on the undoped or lightly doped π-conjugated systems that are commonly referred to as organic semiconductors. Conducting polymers, such as PEDOT:PSS, plexcore, polyaniline, polypyrrole, and others are not addressed here as their charge transfer mechanisms are rather different and would warrant an article in its own right.

3 Charge Transport at Low Carrier Density

The mobility of charge carrier is a key parameter for the understanding of electronic phenomena in organic semiconductors used, for instance, in electrophotography, and in modern devices such as organic light emitting diodes (OLEDs), field effect transistors (FETs), and photovoltaic (PV) cells. It determines both the device current and, concomitantly, the device efficiency as well as its response time. Devices of practical use are often layers of molecularly doped polymers, vapor deposited π-bonded oligomeric molecules, or π-conjugated main chain polymers. In such systems, disorder is a major issue for the structure–property relation. Since there is already a wealth of understanding of salient disorder phenomena pertinent to charge transport in such systems (see [46]), we shall only summarize earlier achievements and concentrate in more detail on more recent developments instead.

3.1 Experimental Approaches

The classic experiment to measure the mobility μ of charge carriers in a semiconductor is based upon the time of flight (ToF) technique. One creates a spatially narrow sheet of charge carriers next to the semitransparent top electrode in a sandwich-type sample by a short laser pulse and one records its arrival time (transit time) $t_{tr} = \mu/dF$ at the exit contact, d being the sample thickness and F being the electric field. Typically, one observes an initial spike followed by a plateau that falls off with a more or less pronounced kink. The initial spike reflects charge motion prior to the energetic relaxation in the DOS provided that the RC time constant of the device is short. Charges generated high in the density of states have a high hopping rate to neighboring sites since virtually all neighboring sites are at lower energy, and jumps down in energy are fast. This high hopping rate translates in a high current. Once in thermal equilibrium, the hopping rate is slower, reflected in a moderate and constant current. The initial spike is thus a genuine feature of a ToF signal in an amorphous film unless charge carriers are generated site-selectively at tail states of the DOS [47]. Experiments on molecularly doped polymers bear out this phenomenon consistently. It is not present in molecular crystals, where the mobility is time-independent. While the position of the kink in the current vs. time plot gives the transit time, the sharpness of the kink at the end of the plateau, i.e., the broadening of the signal, is a measure of the diffusion of the charge carriers while they are drifting under action of the applied field F. However, to observe ideal ToF signals requires that (1) the sample is free of charges without photoexcitation implying that the dielectric relaxation time $\varepsilon\varepsilon_0/\kappa$ is large compared to the transit time t_{tr}, (2) the RC-time constant is small compared to the transit time t_{tr}, (3) the thickness of the spatial spread of the packet of charge carriers is small compared to the film thickness d, (4) the concentration of charges is low enough that the charges do not interact, (5) there is no deep trapping, and (6) the mobility is time independent. Under intrinsic optical charge generation, condition (3) requires that the sample thickness is much larger than the penetration depth of light which is at least 100 nm or even larger. This implies a sample thickness of several micrometers. The problem can be circumvented if charges are photoinjected from a thin sensitizing dye layer [48]. This method has been applied to samples as thin as 300 nm [49]. Regarding condition (4), one usually assumes that it is fulfilled if the number of transported charges is less that 5% of the capacitor charge in order to prevent distortion of the ToF signal. For a field of 10^5 V/cm, a dielectric constant ε of 3 and a film thickness of $d = 2$ μm implies that the concentration of mobile charges inside the samples is smaller than 10^{15} cm^{-3} while a film thickness of 100 nm leads to a concentration of about 2×10^{16} cm^{-3}. These numbers suggest about 10^{-6} or, respectively, 2×10^{-5} charges per transport unit. The latter can be a molecule or a segment of a conjugated polymer. In order to overcome the problem of the RC time constant of the device exceeding the charge carrier transit time in thin samples, Klenkler et al. applied a transient electroluminescent technique to measure the electron mobility in Alq_3 [50]. The technique is fundamentally optical insofar that it decouples the carrier transit signal from the device charging signal, and it is free of RC time constant constraints. However, since it requires the fabrication of multilayer devices, it is applicable to polymer systems only if interlayer penetration can be avoided [51].

An alternative technique to measure the charge carrier mobility involves the injection of a space-charge-limited current from an ohmic electrode. In the absence of deep trapping the current is given by Child's law, i.e., $\varepsilon\varepsilon_0\mu F^2/d$. More recently, Juska et al. developed the technique of extracting charge carrier by linearly increasing voltage (CELIV) to measure μ [52]. In this technique one probes charge transport under steady state conditions. Therefore dispersion effects that are often important in ToF experiments are eliminated. However, the correct evaluation of the CELIV transients produced by photoexcitation nevertheless needs to be carried out with due care [53]. Dispersion effects are also eliminated when monitoring charge flow between coplanar source and drain electrodes in a field effect transistor. In an FET a variable gate voltage modulates a current injected from one of the electrodes. Since the number of charges is determined by the sample capacitance, the current is a direct measure of the carrier mobility. It turns out the mobility inferred from an FET-characteristic can exceed the value determined by a ToF experiment significantly. The reason is that the space charge existing in an FET fills up deep trapping states (see Sect. 4.1).

3.2 Conceptual Frameworks: Disorder-Based Models

A basic concept to analyze the charge carrier mobility in a disordered organic solid is the Gaussian disorder model (GDM) [46, 54] that describes hopping in a manifold of sites. In its original version the system is considered as an array of structureless point-like hopping sites with cubic symmetry whose energies feature a Gaussian-type density of energetically uncorrelated states distribution (DOS) with variance σ. The simplest ansatz for the hopping rate is that of Miller and Abrahams [55],

$$\nu_{ij} = \nu_0 \exp\left(-2\gamma a \frac{r_{ij}}{a}\right) \times \exp\left(-\left(\frac{\varepsilon_j - \varepsilon_i}{kT}\right)\right) \qquad \text{for } \varepsilon_j > \varepsilon_i, \tag{2a}$$

$$\nu_{ij} = \nu_0 \exp\left(-2\gamma a \frac{r_{ij}}{a}\right) \times 1 \qquad \text{for } \varepsilon_j \leq \varepsilon_i, \tag{2b}$$

where r_{ij}/a is the relative jump distance between sites i and j, a is the lattice constant, γ is the inverse localization radius related to the electronic coupling matrix element between adjacent sites, and ν_0 is a frequency factor. In a conventional ToF experiment or in Monte Carlo simulations one generates independent charge carriers at energetically arbitrary sites and one follows their motion under the action of an applied electric field. This implies that the charges execute a random walk. In its course they tend to relax energetically towards quasi equilibrium. Asymptotically, an occupational density of states distribution (ODOS) with the same variance yet displaced from the center of the original DOS by σ^2/kT is approached. Subsequent charge transport occurs by thermally activated jumps from the ODOS to a so-called transport energy somewhat below the center of the DOS [56]. This process is terminated when the charges arrive at the exit electrode. During the relaxation process the mean hopping rate, and thus the velocity of the packet of charges, decrease. This implies that the mobility decreases with time until a steady state condition is approached. Depending on the experimental parameters this relaxation process may not be completed before the charge carriers arrive at the exit contact. In a ToF experiment this results in a dispersive signal. In this case the ToF signal shows a featureless decay if plotted in a linear current vs time diagram. Only when displayed using logarithmic scales does a kink mark the arrival of the fastest carriers. However, the inferred "mobility" is no longer a material parameter. Rather it depends on experimental parameters such as sample thickness and electric field. However, even if the energetic relaxation of the charge carriers is completed before they reach the exit contact, the tail of the ToF signal is broader than expected for a hopping system devoid of disorder. The reason is that disorder gives rise to an anomalous spatial spreading of the packet of charges that increases with electric field and degree of disorder [57–61]. It turns out that this spread of the tail, defined as $w = \frac{(t_{1/2} - t_{\text{tr}})}{t_{1/2}}$, where $t_{1/2}$ is the time at which the current has decayed to half of the plateau value, is more or less universal, yielding $w = 0.4$–0.5 for a system in which σ is about 0.1 eV [62].

In an extended version of the hopping concept, positional ("off-diagonal") disorder in addition to energetic ("diagonal") disorder has been introduced [54, 63]. The simplest ansatz was to incorporate this by allowing the electronic overlap parameter $2\gamma a$ to vary statistically. Operationally, one splits this parameter into two site contributions, each taken from a Gaussian probability density, and defines a positional disorder parameter Σ, in addition to the energetic disorder parameter σ.

Monte Carlo simulations [54], analytical effective medium theory [64], and stochastic hopping theory [46] predict a dependence of the charge carrier mobility as a function of temperature and electric field given in (3):

$$\mu(\hat{\sigma}, F) = \mu_0 \exp\left(-\frac{4\hat{\sigma}^2}{9}\right) \exp\left\{C(\hat{\sigma}^2 - \Sigma^2)\sqrt{F}\right\} \quad \text{for } \Sigma \geq 1.5, \tag{3a}$$

$$\mu(\hat{\sigma}, F) = \mu_0 \exp\left(-\frac{4\hat{\sigma}^2}{9}\right) \exp\left\{C(\hat{\sigma}^2 - 2.25)\sqrt{F}\right\} \quad \text{for } \Sigma \leq 1.5, \tag{3b}$$

where $\hat{\sigma} = \sigma/kT$ and C is a numerical constant that depends on the site separation. If the lattice constant $a = 0.6$ nm, then C is 2.9×10^{-4} cmV$^{1/2}$. Equation (3) predicts a Poole Frenkel-like field dependence. It is important to note, though, that the Poole Frenkel-like field dependence is experimentally obeyed within a significantly larger field range than predicted by simple simulation. The reason is that the energies of the hopping sites are essentially determined by the van der Waals interaction between a charged site and its polarizable neighbor sites which may carry an additional static dipole moment. This implies that the site energies are correlated [65]. Dunlap et al. [66] pursued the idea further and proposed the following empirical relation for $\mu(\hat{\sigma}, F)$:

$$\mu(\hat{\sigma}_{\mathrm{d}}, F) = \mu_0 \exp\left(-\frac{9\hat{\sigma}_{\mathrm{d}}^2}{25}\right) \exp\left\{C_0\left(\hat{\sigma}_{\mathrm{d}}^{1.5} - \Gamma\right)\sqrt{\frac{eaF}{\sigma_{\mathrm{d}}}}\right\}, \tag{4}$$

where $C_0 = 0.78$, Γ describes the positional disorder, and σ_{d} is the width of the DOS caused by the electrostatic coupling of a charged site to neighboring dipoles. This correlated Gaussian disorder model (CGDM) explains the observed range of the $\ln \mu(F) \propto \sqrt{F}$ dependence and reproduces the $\ln \mu(F) \propto 1/T^2$ type of temperature dependence. Values for σ calculated by using (4) instead of using (3) turn out to be about 10% larger.

Equation (3) implies that the field dependence of the mobility can become negative if $\hat{\sigma} < \Sigma$ in (3a) or if $\Sigma > 1.5$ in (3b). This is a signature of positional disorder. The reason is the following. Suppose that a migrating charge carrier encounters a site from which the next jump in field direction is blocked because of poor electronic coupling. Under this condition the carrier may find it easier to circumvent that blockade. If the detour involves jumps against the field direction it will be blocked for higher electric fields. This process involves an interplay between energetic and positional disorder and it has been treated theoretically on an advanced level [63]. The treatment supports the conceptual premise and confirms that the effect is a genuine property of hopping within an energetically and positionally disordered system rather than a signature solely of diffusion of charge carriers at low electric fields [63, 67, 68]. At very high electric fields the velocity of charge carriers must saturate and, concomitantly, $\mu(F)$ must approach a $\mu \propto 1/F$ law because jumps in a backward direction are blocked and transport becomes entirely drift-controlled. Since this effect scales with eFa/σ, the onset of saturation should occur at lower fields if the energetic disorder decreases.

3.3 Conceptual Frameworks: Polaronic Contribution to Transport

So far we have disregarded polaronic effects. However, upon ionizing a molecule or a polymer chain by adding an extra electron there is a readjustment of bond lengths because the electron distribution changes. In optical transitions this effect is revealed by the coupling of the excitation to molecular vibrations. This effect can be quantified in terms of the Huang–Rhys factor. It determines the geometric relaxation energy between the initially generated vertical Franck Condon transition and relaxed electronic state. When transferring a charge between a pair of chromophores the concomitant relaxation energy has to be transferred as well, and this implies that transport is polaronic. Unfortunately, the relaxation energy associated with placing a charge on a chromophore is not amenable to direct probing. This lack of quantitative knowledge gave rise to a lively discussion in the literature on whether or not disorder effects or polaron effects control the temperature dependence of the charge carrier mobility [69]. Meanwhile it is generally agreed that an analysis of the temperature and field dependence of the mobility solely in terms of polaronic effects requires unrealistic parameters, notably an unacceptably large electronic overlap. Moreover, polaron effects cannot explain the observation of dispersive transport at lower temperatures.

An analytical theory based upon the effective medium approach (EMA) has been developed by Fishchuk et al. [70]. They consider the superposition of disorder and polaron effects and treat the elementary charge transfer process at moderate to high temperatures in terms of symmetric Marcus rates instead of Miller–Abrahams rates (see below). The predicted temperature and field dependence of the mobility is

$$\mu = \mu_0 \exp\left(-\frac{E_a}{kT} - \frac{\hat{\sigma}}{8q^2}\right) \exp\left(\frac{\hat{\sigma}^{1.5}}{2\sqrt{2}q^2}\sqrt{\frac{eaE}{\sigma}}\right) \exp\left(\frac{eaE}{4q^2kT}\right), \tag{5}$$

where E_a is half of the polaron binding energy, $q = \sqrt{1 - \sigma^2/8eakT}$. Note that in deriving this equation, site correlations have been included. Equation (5) agrees qualitatively with the empirical expression (6) derived from computer simulations [71]:

$$\mu = \mu_0 \exp\left(-\frac{E_\mathrm{a}}{kT} - 0.31\hat{\sigma}^2\right)\exp\left(0.78(\hat{\sigma}^{1.5} - 1.75)\sqrt{\frac{eaE}{\sigma}}\right). \tag{6}$$

These expressions have been successfully applied to polymeric systems of practical relevance, as detailed in the next section.

3.4 Survey of Representative Experimental Results

3.4.1 On the Origin of Energetic Disorder

Although most of the recent results on charge transport in organic solids have been obtained on π-conjugated polymers and oligomers used in organic OLEDs, FETs, and PV cells, it is appropriate to refer to a recent survey on charge transport in molecularly doped polymers by Schein and Tyutnev [72]. In fact, the prime intention to develop the Gaussian disorder model has been to understand charge transport in photoreceptors used in electrophotography. This survey elaborates on the origin of the energetic disorder parameter. It has been a straightforward assumption that the disorder parameter σ is a measure of the statistical spread of the electronic interaction of a charged transport molecule with induced dipole moments in the molecular environment, i.e., the van der Waals coupling, and of the interaction between permanent dipoles of both matrix and transport molecules. By measuring the temperature dependence of the charge mobility, it has been experimentally verified that in a sample in which hole transport is carried by 1,1-bis(di-4-tolylaminophenyl)cyclohexane (TAPC) molecules, whose dipole moment is small (about 1 D), the disorder parameter increases when the polarity of its surroundings increases. This occurs for example in the order of bulk film, TAPC blended with a polar polystyrene and TAPC blended with polycarbonate in which the carbonyl groups carry a high dipole moment [73]. This proves that the polarity of the matrix increases the energetic disorder. It is straightforward to conjecture that this increase of σ is of intermolecular origin and arises from the electrostatic coupling between the charged transport unit and the statistically oriented dipole moments of the carbonyl groups.

However, in that survey Schein and Tyutnev question the intermolecular origin of σ. They compared σ values derived from studies of hole transport in 1-phenyl-3-((diethylamino)styryl)-5-(*p*-(diethylamino)phenyl)pyrazoline (DEASP) molecules, derivatives of pyrazoline, whose dipole moment is 4.34 D, blended with either polystyrene or polycarbonate as function of concentration. They found that σ is independent of the matrix material and that σ remains constant when the concentration of DEASP increases from 10% to 70% while one would expect that σ increases as the concentration of the polar DEASP molecules increases. However, this expectation rests upon the assumption that the blend is homogeneous. It ignores aggregation effects that are particularly important for polar molecules. Since charge carriers will preferentially jump among nearest neighbor sites, dilution will only reduce the number of the transports paths between DEASP clusters rather than decreasing the ensemble averaged mean electronic coupling while the width of the DOS remains constant. Note, however, when the transport moieties are not rigid there can, in fact, be an intramolecular contribution to energetic disorder caused by a statistical distribution of conformations that translates into a spread of site energies [18].

In conjugated polymers there is an additional intra-chain contribution to the energetic disorder because the effective conjugation length of the entities that control the electronic properties is a statistical quantity. It turns out that the low energy tail of the absorption spectra as well as the high energy wing of the photoluminescence spectra can be fitted well to Gaussian envelope functions and their variances contain both intrachain and interchain contributions. Since the inhomogeneous line broadening of excitons and charge states has a common origin, it is a plausible assumption that the DOS of charge carriers in conjugated polymers is also a Gaussian, at least its low energy wing that is relevant for charge carrier hopping. Unfortunately, the DOS distribution for charge carriers is not amenable to absorption spectroscopy (see above). Indirect information can be inferred from that analysis of the temperature and field dependence of the charge carrier mobility and the shape of time of flight (ToF) signals. Note that if the DOS had an exponential rather than Gaussian tail a ToF signal would always be dispersive because charge carriers can never attain quasi equilibrium [74, 75].

3.4.2 Application of the Gaussian Disorder Model

A textbook example for the application of the uncorrelated GDM is the recent study by Gambino et al. on a light emitting dendrimer [49]. The system consists of a bis (fluorene) core, meta-linked biphenyl dendrons, and ethylhexyloxy surface units. ToF experiments shown in Fig. 5 were performed on 300 nm thick sandwich films

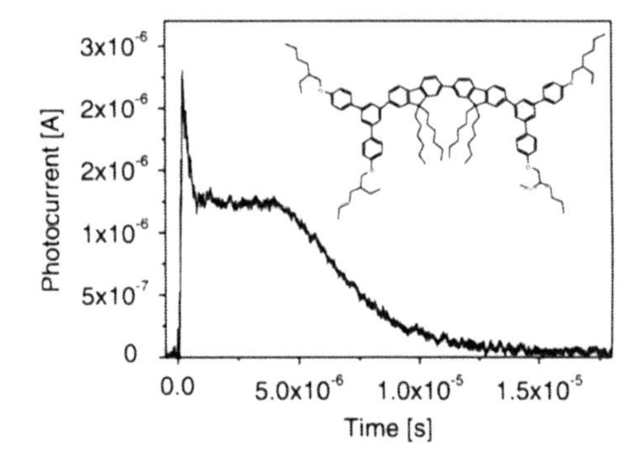

Fig. 5 Typical room temperature TOF hole transient for a first generation bis-fluorene dendrimer film of thickness 300 nm and an electric field of 1.6×10^5 V/cm. Also shown is the structure of the dendrimer. From [49] with permission. Copyright (2008) by Elsevier

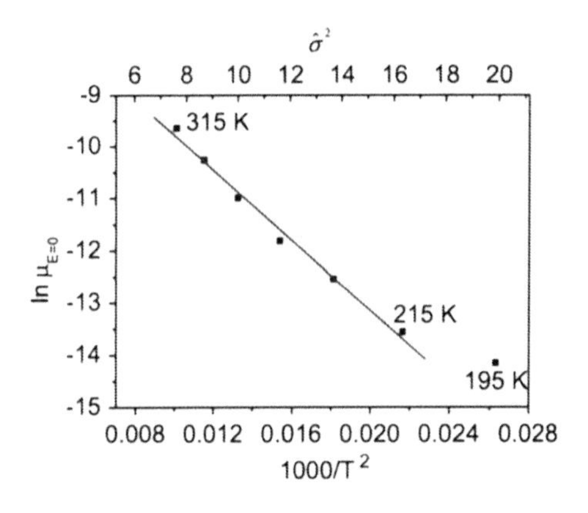

Fig. 6 Zero field hole mobility of the bis-fluorene dendrimer of Fig. 5 as a function of $1/T^2$. The deviation of the $\ln\mu(F)\propto 1/T^2$ dependence below 215 K is a signature of the onset of transit time dispersion. From [49] with permission. Copyright (2008) by Elsevier

within a temperature range between 315 and 195 K and within a field range between 1.5×10^4 and 3×10^5 V/cm using dye-sensitized injection (see Sect. 3.1). Data analysis yields an energetic disorder parameter $\sigma = 74 \pm 4$ meV, a positional disorder parameter $\Sigma = 2.6$ and $\mu_0 = 1.6 \times 10^{-3}$ cm^2 V^{-1} s^{-1}. Previous Monte Carlo simulations predicted that above a critical value of σ/kT ToF signals should become dispersive, indicating that charge carriers can no longer equilibrate energetically before they recombine with the electrode. For $\sigma = 74$ meV and a sample thickness of 300 nm the critical temperature is predicted to be 228 K. In fact, the experimental ToF signals lose their inflection points, i.e., become dispersive, at 215 K, as shown in Fig. 6. This a gratifying confirmation of the model.

Martens et al. inferred hole mobilities as a function of temperature and electric field in 100–300 nm thick films of four poly(p-phenylenevinylene) derivatives from space-charge-limited steady state currents injected from an ITO anode [76]. Within a dynamic range of two to three orders of magnitude the T-dependence of μ obeyed a ln μ vs T^{-2} dependence with σ values ranging from 93 meV (OC_1C_{10}-PPV) to 121 meV (partially conjugated OC_1C_{10}-PPV). In view of the extended range of the $\ln\mu$ vs $\sqrt{F}$ dependence, the data have been analyzed in terms of the correlated GMD model. Note that in their analysis the authors used a Poole–Frenkel-type of field dependence in Child's law for space-charge-limited current flow, $j_{\mathrm{Child}} = \frac{9}{8}\frac{\varepsilon\varepsilon_0\mu F^2}{d}$. In this approach, the authors do not consider the modification of the mobility due to filling of tail states in the DOS (see Sect. 4.1). However, this modification to Child's law is only justified if the field dependence of μ is weak since a field dependent mobility has a feedback on the spatial distribution of the space charge [77, 78]. Under these circumstances there is no explicit solution for $j_{\mathrm{Child}}(F)$ under space-charge-limited conditions [79]. However, the essential conclusion relates to the absolute value of the hole mobility and the verification of the predicted temperature dependence. The results confirm the notion that the molecular structure has an important bearing on charge transport. Broken conjugation limits transport, mainly due to the effective dilution of the fraction of the charge transporting moieties as evidenced by the low value of the prefactor to the mobility $\mu_0 = 4 \times 10^{-6}$ cm^2 V^{-1} s^{-1}. This prefactor is a measure of the electronic coupling among the

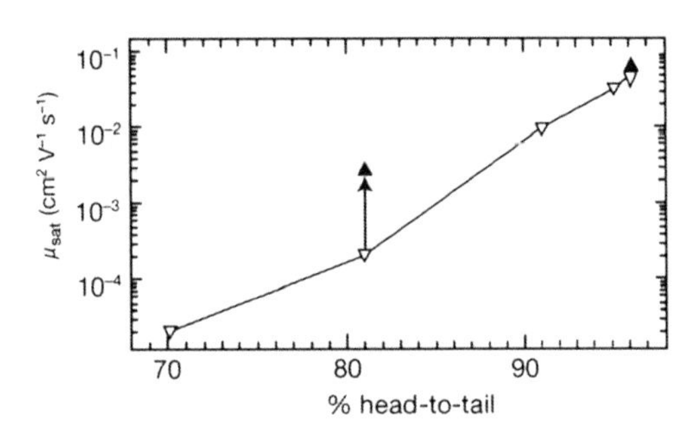

Fig. 7 Room temperature FET-mobility of P3HT in different microstructures. *Downward triangles*: spin-coated regioregular film, *upward triangles*: solution cast film. From [83] with permission. Copyright (1999) by Macmillan Publishers

transport sites. In this respect, bulky transport groups containing spiro-units are unfavorable [80, 81]. On the other hand, sterically demanding groups reduce charge trapping because they diminish the propensity of the sites for forming sandwich conformations that can act as charge carrier traps. Using a polymer with a high degree of regioregularity can significantly increase the mobility due to improved electronic interchain coupling and decreasing energetic disorder. Improved inter-chain ordering in substituted poly(3-hexylthiophene) (P3HT) can raise mobility up to 0.1 cm^2 V^{-1} s^{-1} [82, 83] as demonstrated in Fig. 7. The impact of this inter-chain ordering is also revealed in optical spectroscopy [84, 85]. This effect is profitably used in organic FETs and organic integrated circuits, employing, for instance ordered semiconducting self-assembled monolayers on polymeric surfaces. Such systems can be exploited in flexible monolayer electronics. Surprisingly, in the ladder-type poly-phenylene (MeLPPP), which is one of the least disordered of all π-conjugated polymers, the hole mobility is only about 3×10^{-3} cm^2 V^{-1} s^{-1} at room temperature [36]. Since the temperature dependence is low – because of low disorder – this has to be accounted for by weak inter-chain interactions. Obviously, the bulky substituents reduce the electronic coupling among the polymer chains.

Despite the success of the disorder model concerning the interpretation of data on the temperature and field dependence of the mobility, one has to recognize that the temperature regime available for data analysis is quite restricted. Therefore it is often difficult to decide if a ln μ vs T^{-2} or rather a ln μ vs T^{-1} representation is more appropriate. This ambiguity is an inherent conceptual problem because in organic semiconductors there is, inevitably, a superposition of disorder and polaron effects whose mutual contributions depend on the kind of material. A few representative studies may suffice to illustrate the intricacies involved when analyzing experimental results. They deal with polyfluorene copolymers, arylamine-containing polyfluorene copolymers, and σ-bonded polysilanes.

3.4.3 Polaronic Effects vs Disorder Effects

The most comprehensive study is that of Khan et al. [86]. They describe ToF experiments on sandwich-type samples with films of poly(9,9-dioctyl-fluorene)

(PFO), PFB, and a series of fluorene-triarylamine copolymers with different triarylamine content covering a broad temperature and field range. In all cases the field dependence of the hole mobility follows a $\ln \mu \propto \sqrt{F}$ dependence and a super-Arrhenius-type of temperature dependence. At lower temperatures the ToF signals are dispersive. When analyzing the experimental data the authors first checked whether or not the uncorrelated Gaussian disorder model (GDM) is appropriate. There are indeed reasonably good fits to the temperature and field dependence based upon (3). Recognizing, however, that experimentally observed $\ln \mu$ vs $\sqrt{F}$ dependence extends to lower fields than the GDM predicts, they went one step further and tested the correlated disorder model (CDM) in the empirical form of (4). Here the site separation enters as an explicit parameter. This analysis confirms the validity of the $\ln \mu \propto T^{-2}$ law except that the σ values turn out to be 10% larger because in the CDM the coefficient that enters the exponent in the temperature dependence is 3/5 instead of 2/3 in the GDM. The positional disorder parameters are comparable and the values for the site separation are realistic. Finally the authors took into account polaron effects by using the empirical expression (6). The difficulty is how to separate the polaron and disorder contributions to the T-dependence of μ. This can be done via an analysis of the field dependence of μ. Once σ is known the factor $\exp\left(-\frac{E_a}{kT}\right)$, that accounts for the polaron contribution, can be determined. The parameters inferred from the data fits are then compared by Khan and coworkers [86]. They find that by taking into account polaronic contributions, the σ value decreases while the prefactor to the mobilities increases by roughly one order of magnitude. The polaron binding energy $2E_a$ is significant and ranges between 0.25 eV and 0.40 eV: Nevertheless, energetic disorder plays a dominant role in hole transport. It is larger in the copolymers as compared to the homopolymers PFO and PFB.

A similar analysis has been carried out by Kreouzis et al. for hole transport in pristine and annealed polyfluorene films [87]. Consistent with the work of Khan et al. [86] on the copolymers, the results can best be rationalized in terms of the correlated disorder model including polaron effects. For different unannealed samples σ values are between 62 and 75 meV, the polaron activation energies are 180 meV, and the prefactor mobilities μ_0 are 0.4 and 0.9 $cm^2\ V^{-1}\ s^{-1}$. Annealing reduces the disorder parameters to 52 $\pm$ 1 meV and the prefactor to 0.3 $cm^2\ V^{-1}\ s^{-1}$. It is well known that PFO can exhibit different phases [88, 89]. Annealing an amorphous PFO film induces the formation of a fraction of the so-called β-phase, where chains are locked into a planar conformation resulting in a long conjugation length and low disorder. This lowers the geometric relaxation energy upon ionization, i.e., the polaron binding energy, for the β-phase.

However, one should be cautious about overinterpreting the field and temperature dependence of the mobility obtained from ToF measurements. For instance, in the analyses of the data in [86, 87], ToF signals have been considered that are dispersive. It is well known that data collected under dispersive transport conditions carry a weaker temperature dependence because the charge carriers have not yet reached quasi-equilibrium. This contributes to an apparent Arrhenius-type temperature dependence of μ that might erroneously be accounted for by polaron effects.

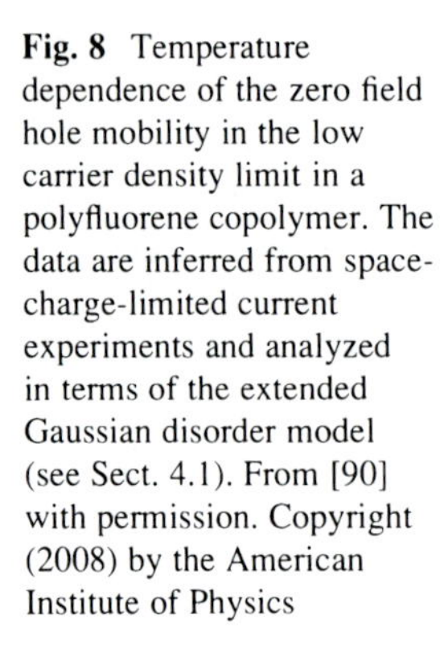

Fig. 8 Temperature dependence of the zero field hole mobility in the low carrier density limit in a polyfluorene copolymer. The data are inferred from space-charge-limited current experiments and analyzed in terms of the extended Gaussian disorder model (see Sect. 4.1). From [90] with permission. Copyright (2008) by the American Institute of Physics

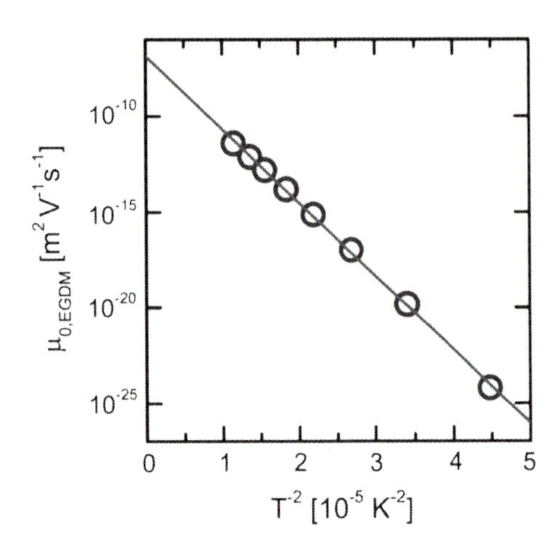

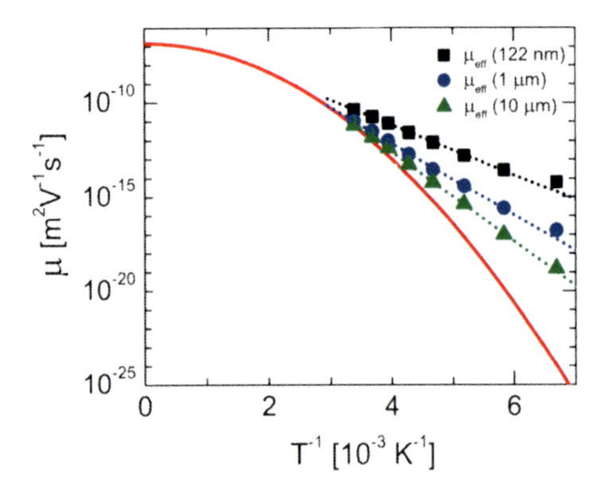

Fig. 9 Temperature dependence of the hole mobility of a polyfluorene copolymer inferred from space-charge-limited current measurements on samples of thicknesses 122 nm, 1 μm, and 10 μm. The *full curve* is an extrapolation to the low carrier density limit using the extended Gaussian disorder model. From [90] with permission. Copyright (2008) by the American Institute of Physics

In fact, in their recent work, Mensfoort et al. [90] conclude that in polyfluorene copolymers hole transport is entirely dominated by disorder. This is supported by a strictly linear $\ln \mu \propto T^{-2}$ dependence covering a dynamic range of 15 decades with a temperature range from 150 to 315 K (Fig. 8). Based upon stationary space-charge-limited current measurement, where the charge carriers are in quasi equilibrium so that dispersion effects are absent, the authors determine a width σ of the DOS for holes as large as 130 meV with negligible polaron contribution.

The work of Mensfoort et al. is a striking test of the importance of charge carrier density effects in space-charge-limited transport studies. For a given applied voltage the space charge concentration is inversely proportional to the device thickness. This explains why in Fig. 9 the deviation from the $\ln \mu \propto T^{-2}$

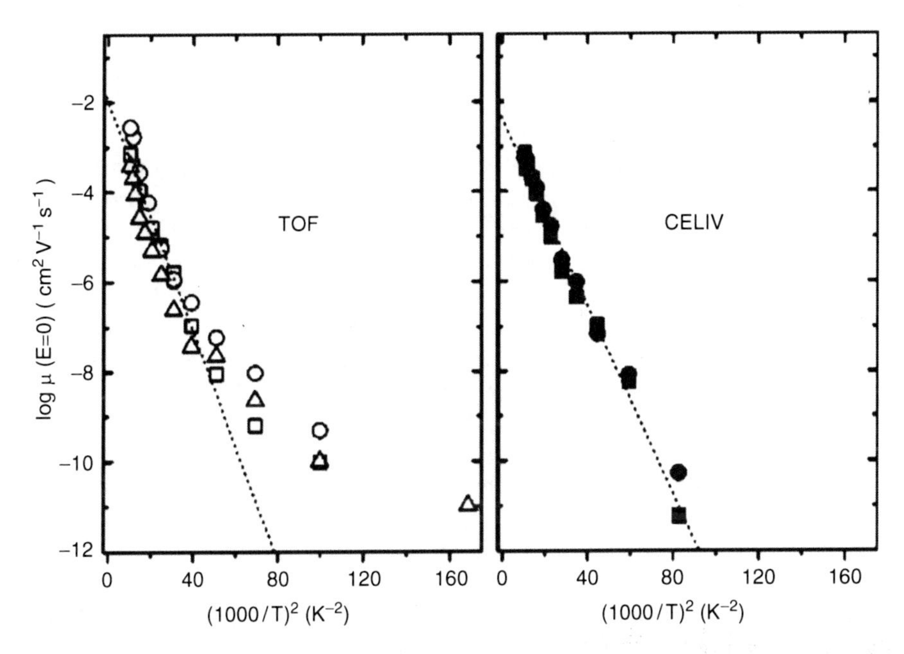

Fig. 10 Temperature dependence of the hole mobility in regioregular P3HT measured in TOF (*left*) and CELIV (*right*) configuration. *Different symbols* refer to different samples. From [91] with permission. Copyright (2005) by the American Institute of Physics

dependence of the hole mobility becomes more significant in thinner samples. This will be discussed in greater detail in Sect. 4.1.

In this context it is appropriate to recall the work of Mozer et al. [91] on hole transport in regio(3-hexylthiophene). These authors compared the field and temperature dependencies of the hole mobility measured via the ToF and CELIV methods. Quite remarkably, the temperature dependence deduced from ToF signals plotted on a ln μ vs T^{-2} scale deviate significantly from linearity while the CELIV data follow a $\ln\mu \propto T^{-2}$ law down to lowest temperatures (180 K) (see Fig. 10). The reason is that in a ToF experiment the charge carriers are generated randomly within the DOS and relax to quasi-equilibrium in their hopping motion while in a CELIV experiment relaxation is already completed. This indicates that a deviation from a $\ln\mu \propto T^{-2}$ form may well be a signature of the onset of dispersion rather than a process that is associated with an Arrhenius-type of temperature dependence such as polaron transport. Therefore the larger polaron binding energy that had been extracted from ToF data measured in the non-annealed PFO films should be considered with caution. Obviously, if one wants to distinguish between polaron and disorder effects based upon the temperature and field dependencies of the mobility one should ensure that dispersion effects are weak.

The conclusion that polaron effects contribute only weakly to the temperature dependence of the charge carrier mobility is supported by a theoretical study of polarons in several conjugated polymers. Meisel et al. [92] considered the electron–phonon interaction and calculated polaron formation in polythiophene, polyphenylenevinylene, and polyphenylene within an extended Holstein model. Minimization of the energy of the electronic state with respect to lattice degrees of freedom yields the polaron ground state. Input parameters of the Hamiltonian are obtained from ab initio calculations based on density-functional theory (DFT). The authors determined the size and the binding energies of the polarons as well as the lattice deformation as a function of the conjugation length. The binding energies decrease significantly with increasing conjugation length because the fractional change of bond lengths and angles decreases as the charges are more delocalized. The polaron extents are in the range of 6–11 nm for polythiophenes and polyphenylenevinylenes, and the associated polaron binding energies are 3 meV for holes and 7 meV for electrons. For polyphenylenes, the polaron size is about 2–2.5 nm and its binding energy is 30 meV for the hole and 60 meV for the electron. Although the calculations document that charge carriers are self-trapped, they indicate that polaron binding energies are much smaller than the typical width of the DOS of representative π-bonded conjugated polymers. This raises doubts on the conclusiveness of analyses of mobility data inferred from dispersive ToF signals.

Another cautionary remark relates to the field dependence of the charge carrier mobility. Ray Mohari et al. [93] measured the hole mobility in a blend of *N,N′*-diphenyl-*N,N′*-bis(3-methylphenyl)-(1,1′-biphenyl)-4,4′-diamine (TPD) and polystyrene in which the TPD molecules tend to aggregate. In the ordered regions the energetic disorder is significantly reduced relative to a system in which TPD is dispersed homogeneously. The experiments confirm that aggregation gives rise to a negative field dependence of the mobility. Associating that effect solely with positional disorder in a hypothetical homogenous system would yield a positional disorder parameter that is too large. These results demonstrate that changes of sample morphology can be of major impact on the field dependence of μ.

In the context of polaron effects we also mention the experimental work on hole transport in polysilanes that has been analyzed in terms of Fishchuk et al.'s analytical theory [70]. In this theory polaron effects are treated in Marcus terms instead of Miller–Abrahams jump rates, taking into account correlated energy disorder [see (5)]. The materials were poly(methyl(phenyl)silylene) (PMPSi) and poly(biphenyl(methyl)silylene) (PBPMSi) films. Polysilanes are preferred objects for research into polaron effects because when an electron is taken away from a σ-bonded, i.e., singly-bonded, polymer chain there ought to be a significant structural reorganization that gives rise to a comparatively large polaron binding energy. Representative plots for the temperature dependence of the hole mobility in PMPSi are shown in Fig. 11. Symbols show experimental data, full lines are theoretical fits. Considering that there is no arbitrary scaling parameter, those fits are an excellent confirmation of the theory. Note that the coupling element J that enters the Marcus rate has been inferred from the prefactor mobility. The data analysis also shows that the polaron binding energies in these materials are significant and depend on the pendant group.

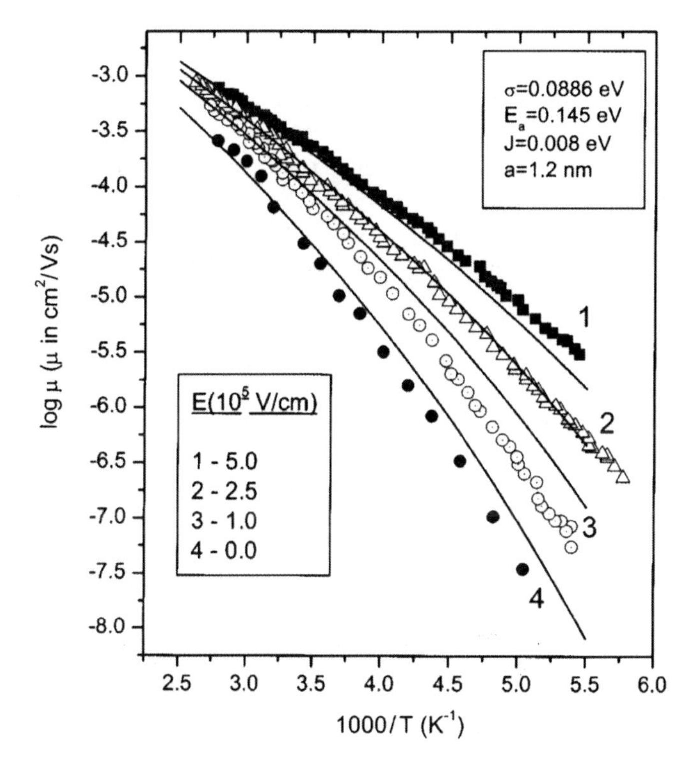

Fig. 11 Temperature dependence of the hole mobility in PMPSi at different electric fields. *Full curves* are calculated using the theory by Fishchuk et al. [70]. The fit parameters are the width σ of the density of states distribution, the activation energy E_a (which is $E_p/2$), the electronic exchange integral J, and the intersite separation a. From [70] with permission. Copyright (2003) by the American Institute of Physics

4 Charge Transport at High Carrier Density

4.1 Charge Transport in the Presence of Space Charge

The transport models discussed in Sect. 3 are premised on the condition that the interaction of the charge carriers is negligible. This is no longer granted if (1) a trapped space charge distorts the distribution of the electric field inside the dielectric, (2) ionized dopant molecules modify the DOS, or (3) the current flowing through the dielectric is sufficiently large so that a non-negligible fraction of tail states of the DOS is already occupied. The latter case is realized when either the current device is space-charge-limited (SCL) or the current is confined to a thin layer of the dielectric, for instance in a field effect transistor. It is conceptionally easy to understand that the temperature dependence of the charge carrier mobility must change when charge carriers fill up tail states of the DOS beyond the critical level defined by the condition of quasi-equilibrium. In this case the carrier statistics becomes Fermi–Dirac-like whereas it is Boltzmann-like if state filling is negligible. At low carrier density, a charge carrier in thermal equilibrium will relax to an energy $\varepsilon_\infty = \frac{\sigma^2}{kT}$ below the center of the DOS, provided it is given enough time to complete the relaxation process. Charge transport, however, requires a certain minimum energy to ensure there are enough neighboring sites that are energetically accessible [54]. To reach this so-called transport energy from the thermal equilibrium energy, an activation energy is needed. If, at higher carrier density, a quasi-Fermi level will be established that moves beyond ε_∞, the activation energy needed for a charge carrier to reach the transport level decreases and, concomitantly, the mobility increases (Fig. 12). This is associated not only with a weaker temperature dependence of μ but also with a gradual change from the $\ln\mu \propto T^{-2}$ dependence to an Arrhenius-type $\ln\mu \propto T^{-1}$ dependence because upward jumps of charge carriers no longer start from a temperature dependent occupational DOS but from the Fermi-level set by the applied voltage. The straightforward verification of this effect is the observation that the carrier mobilities measured under FET-conditions can be up to three orders of magnitude larger than the values inferred from ToF experiments [94]. Further, one observes a steeper increase of space-charge-limited current mobility with electric field than predicted by Child's law [76]. It is meanwhile recognized that this steeper increase is not due to a field dependence of the mobility under the premise of negligible concentration. Rather, as illustrated by Fig. 13, it is mostly an effect of the filling up of the DOS due to the increase of the

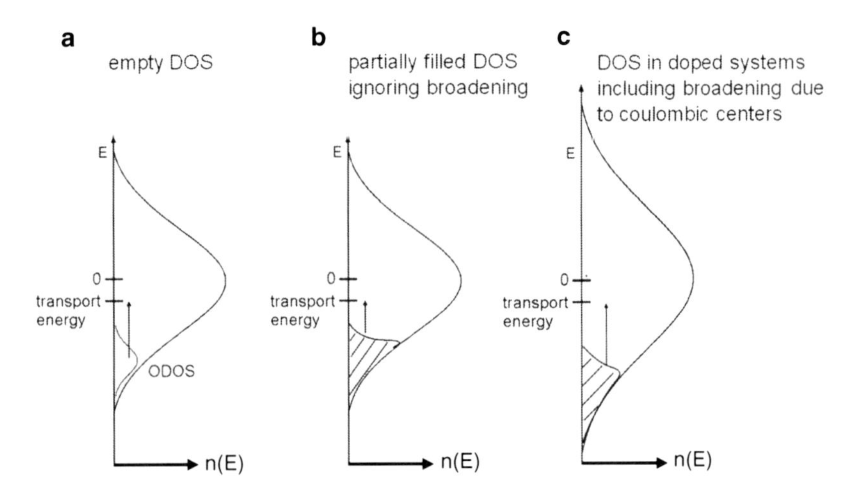

Fig. 12 Schematic view of the effect of state filling in the Gaussian distribution of the hopping states. (**a**) Charge carrier transport requires thermally activated transitions of a charge carrier from the occupational DOS (ODOS) to the transport energy E_{tr} in the low carrier limit. (**b**) Charge transport in the presence of a space charge obeying Fermi–Dirac statistics under the assumption that the space charge does not alter the DOS. (**c**) Charge transport in the presence of a space charge considering the broadening of the DOS due the countercharges generated, e.g., in the course of electrochemical doping. Note the larger width of the DOS

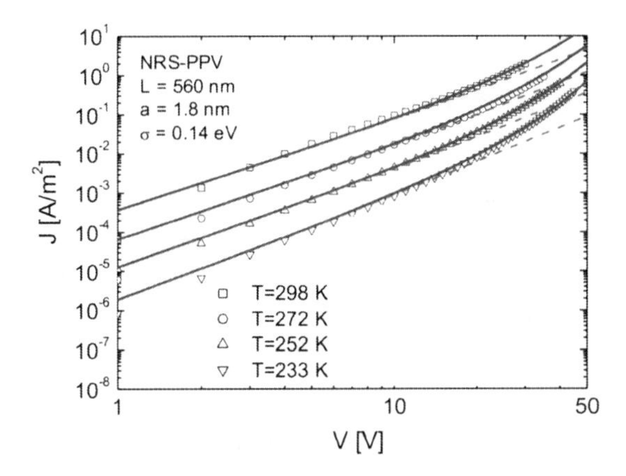

Fig. 13 Experimental (*symbols*) and theoretical (*lines*) data for the current-density as a function of applied voltage for a polymer film of a derivative of PPV under the condition of space-charge-limited current flow. *Full curves* are the solution of a transport equation that includes DOS filling (see text), *dashed lines* show the prediction of Child's law for space-charge-limited current flow assuming a constant charge carrier mobility. From [96] with permission. Copyright (2005) by the American Institute of Physics

charge carrier concentration [95]. Ignoring this effect in a data analysis would yield numerically incorrect results.

Among the first theoretical treatments of transport in the presence of a space charge is that of Arkhipov et al. [97]. These authors pointed out that in chemically doped materials and in the conduction channel of an FET the number of charge carriers occupying deep tail states of the Gaussian DOS can be significant relative to the total density of states. They developed a stochastic hopping theory based upon the variable range concept and incorporated the Fermi–Dirac distribution to describe the temperature dependence of the mobility. Currently the most frequently used formalism is that of Pasveer et al. [96]. It is based upon a numerical solution of the master equation representing charge carrier hopping in a lattice. Considering that a fraction of sites is already occupied, charge transport is considered as a thermally assisted tunneling process with Miller–Abrahams rates in a Gaussian manifold of states with variance σ, tacitly assuming that formation of a bipolaron, i.e., a pair of like charges on a given site, is prevented by coulomb repulsion. The results can be condensed into an analytical solution in factorized form,

$$\mu(T, F, n) = \mu_0(T) g_1(F, T) g_2(n), \tag{7}$$

where $\mu_0(T)$ is the temperature dependent mobility in the limit of $F = 0$, $g_1(F, T)$ is the mobility enhancement due to the electric field, and $g_2(n)$ is the enhancement factor due to state filling.

A more comprehensive theoretical treatment has been developed by Coehoorn et al. [98] in which the various approaches for charge carrier hopping in random organic systems have been compared. In subsequent work, Coehoorn [99] used two semi-analytical models to focus on charge transport in host guest systems, namely a relatively simple Mott-type model and a more advanced effective medium model. The latter model was generalized in order to be able to include the effect of different wave function extensions of host and guest molecules in a blend system.

At the same time Fishchuk et al. [100] developed an analytical theory based upon the effective medium approach to charge transport as a function of the charge carrier concentration within the DOS. In contrast to the work by Pasveer et al. they included polaron effects. It is obvious that how charges are transported, i.e., the trade-off between disorder and polaron effects, should have a major impact on the result. In the extreme case of vanishing disorder yet strong polaronic coupling, filling tail states of DOS by charge carriers should not have an effect on the transport except at very large charge concentration when coulomb repulsion becomes important. Filling-up tail states of the DOS will, however, become progressively important as the disorder contribution to charge transport increases. To incorporate polaron effects, Fishchuk et al. [100] replaced the Miller–Abrahams-type of hopping rate with a Marcus rate. Note, however, that the magnitude of the effect of DOS filling is solely determined by the ratio between the disorder energy σ and the polaron energy E_p and not by the analytical form of the hopping rate. The prize for doing analytical rather than numeric work is that it requires the solving of integrals of the hopping rate over the density of states in an effective medium approach, thus yielding a rather complicated expression. It is gratifying, though, that both the treatments of Arkhipov et al. and Pasveer et al. as well as Fishchuk et al.'s effective medium approach (EMA) are mutually consistent [96, 97, 100]. Figure 14 compares the results of the three models as a function of carrier concentration parametric in the disorder parameter σ. Figure 15 demonstrates that the effective medium approach provides a good fit to the concentration dependence of the carrier mobility in field effect transistors (FETs) using P3HT and OC_1C_{10}-PPV as active layers while Fig. 16 shows the temperature dependence of ln μ vs T^{-1} and ln μ vs T^{-2} parametric in the carrier concentration. It is obvious that there is a transition from the $\ln \mu \propto T^{-2}$ dependence at $n = 0$ to a $\ln \mu \propto T^{-1}$ law at higher concentrations. The Fishchuk et al. theory also confirms that, in a system in which polaron effects are dominant, $E_a/\sigma = 3$ implying $E_p/\sigma = 6$, i.e., the temperature dependence of the transport is dominated by polaron effects instead of disorder, so that the filling of the DOS with carrier density is unimportant. Unfortunately, the present EMA treatment does not allow encompassing of the parameter regime $E_a/\sigma < 3$.

Recently, Fishchuk et al. extended their effective medium approach to include a discussion of the so-called Meyer Neldel rule [102]. The Meyer Neldel rule is an empirical relation, originally derived from chemical kinetics. It describes the fact that enthalpy and entropy of a chemical reaction are functionally related to each other. More generally, it states that in a thermally activated rate process an increase in the activation energy needed is partially compensated by an increase in the prefactor. There are numerous examples, notably in semiconductor physics, that this rule is fulfilled for various reasons. Recently Emin advanced an adiabatic polaron hopping model that considers carrier-induced softening of the vibrations promoting charge carrier motion [103]. Fishchuk et al. were able to show that in

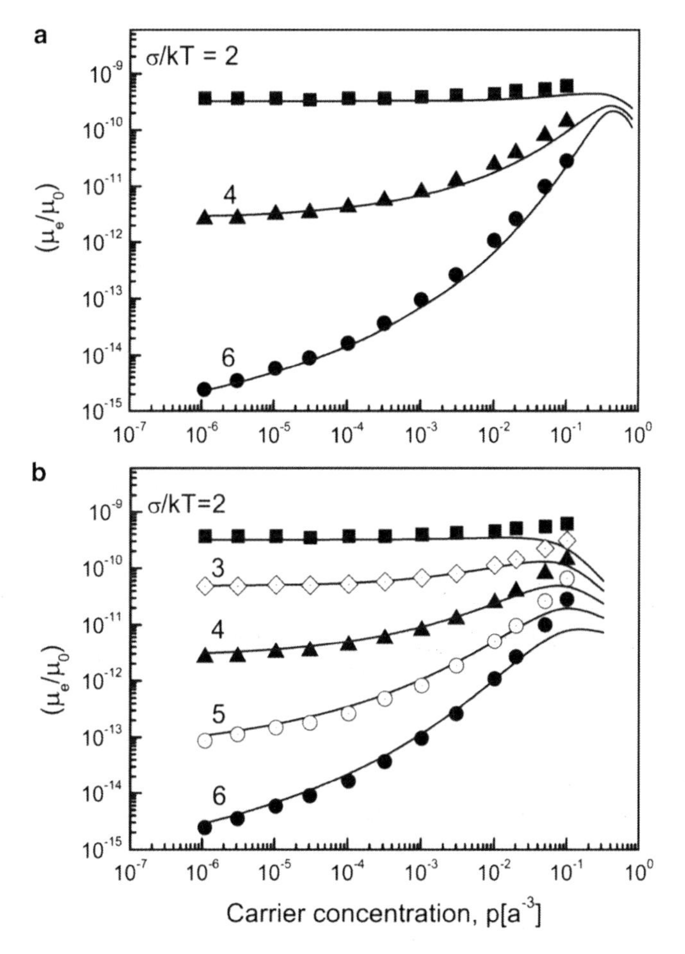

Fig. 14 A comparison of different approaches to describe the charge carrier mobility in a Gaussian-type hopping system as a function of the normalized concentration of the charge carriers. (**a**) *Full curves* are the result of effective medium calculations [100] while *symbols* are computer simulations [96]. (**b**) *Full curves* are calculated using the variable range hopping concept [101], *symbols* are the computer simulations. From [100] with permission. Copyright (2007) by the American Institute of Physics

a disordered system with a partially filled density of states distribution the Meyer–Nedel rule is indeed fulfilled but it is not related to polaronic transport. Instead, it is a genuine signature of hopping transport in a random system with Gaussian DOS distribution upon varying the charge carrier concentration as realized, e.g., in a field effect transistor at variable gate voltage. Figure 17 shows the temperature dependence of the hole mobility measured in a field effect transistor (FET) with pentacene (Fig. 17a) and P3HT (Fig. 17b) as active layers. This study also disproves the previous claim by Craciun et al. [104] that the temperature dependence of the charge carrier mobility in a variety of conjugated polymers

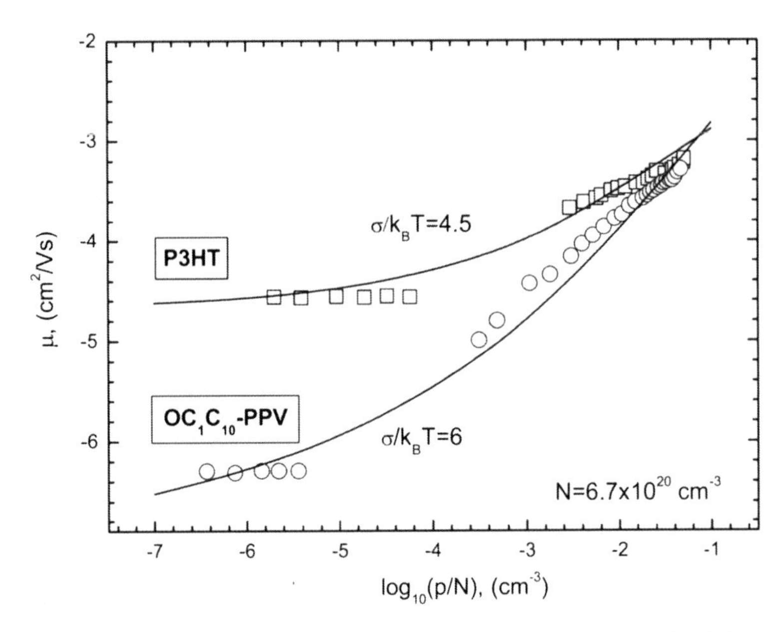

Fig. 15 Fits to experimental values of the hole FET-mobility using P3HT and OC_1C_{10}-PPV as active layers. From [100] with permission. Copyright (2007) by the American Institute of Physics

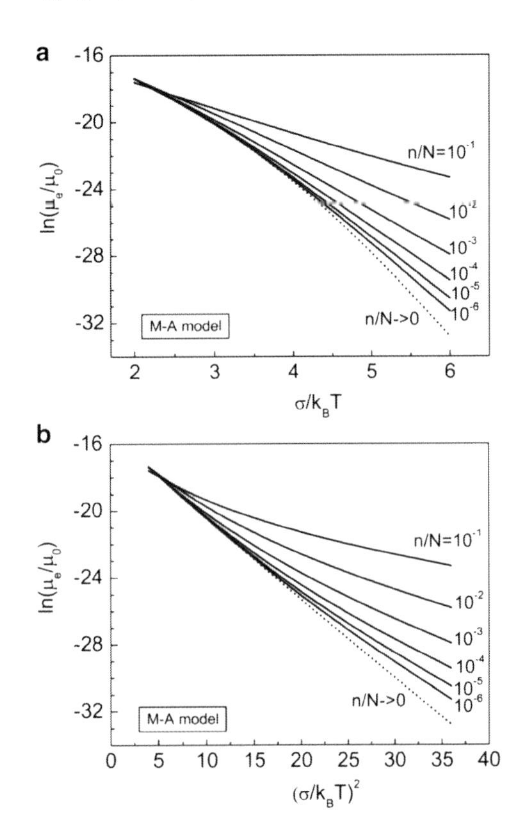

Fig. 16 Calculated charge carrier mobility in a Gaussian-type hopping system parametric in the charge carrier concentration and plotted (**a**) on a ln μ vs σ/kT scale and (**b**) on a ln μ vs $(\sigma/kT)^2$ scale. From [100] with permission. Copyright (2007) by the American Institute of Physics

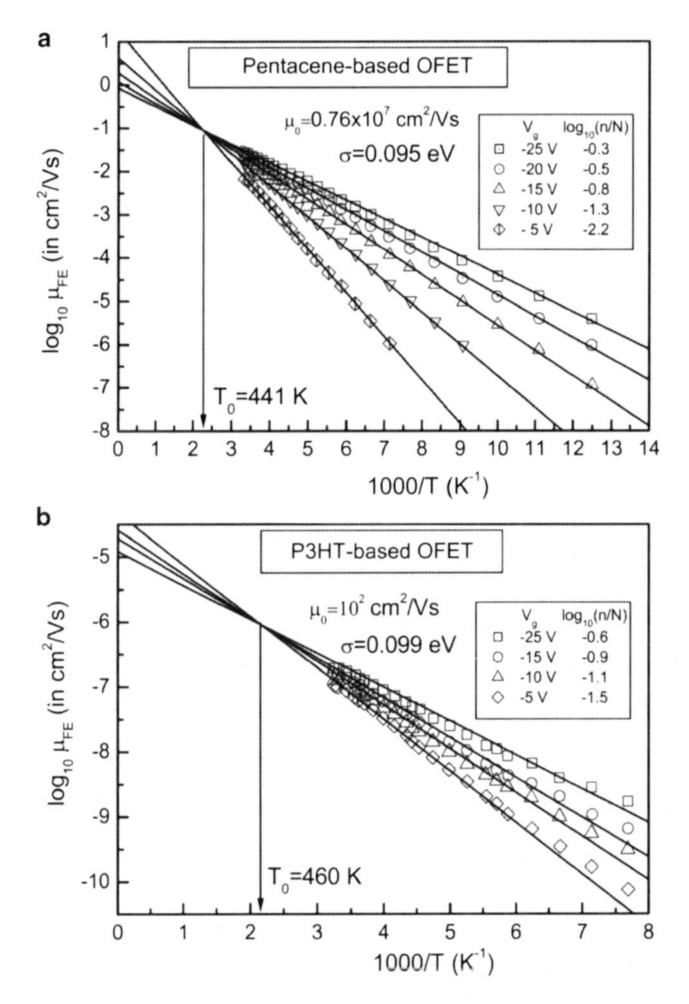

Fig. 17 Temperature dependence of the hole mobility measured in an FET with (**a**) pentacene and (**b**) P3HT as active layers. Parameter is the gate voltage. Data fitting using the Fishchuk et al. theory in [102] yields values for the mobility and the disorder potential extrapolated to zero electric field and zero carrier concentration. T_0 is the Meyer–Nedel temperature (see text). From [102] with permission. Copyright (2010) by the American Institute of Physics

extrapolates to an universal value of 30 $cm^2\ V^{-1}\ s^{-1}$ in the $T \to \infty$ limit. This has been corroborated by the work of Mensfoort et al. (see Fig. 9).

Meanwhile the Pasveer et al. formalism [96] has been termed as "extended Gaussian disorder model (EDGM)" and "extended correlated disorder model (ECDM)" depending on whether the correlation of the site energies is included or not. The EGDM has recently been applied to analyze the SCL current injected from an ohmic PEDOT:PSS anode into a polyfluorene based light emitting polymer layer with different layer thicknesses [105]. It is instructive to compare the experimental current–voltage characteristics to curves calculated by using the classic Child's law combined with a Poole–Frenkel-type field dependence of the mobility to the predictions of the EGDM model in which DOS filling has been taken into account. The comparison indicates that (1) the EGDM model can reproduce the experimental results with remarkable accuracy, yet (2) high precision regarding data quality is required to distinguish among the various models, and (3) the influence of DOS filling diminishes in thicker samples. The latter effect results from the inverse decay of the carrier concentration with cell thickness at constant applied electric field. This may explain why Agrawal et al. [106] could successfully explain their SCL current flow in copper phthalocyanine (CuPc) layers with thicknesses ranging from 100 to 400 nm. The same reasoning applies to the work of Mensfoort et al. [90] on SCL current flow in the polyfluorene diode mentioned above. The authors demonstrate that in a thick sample the temperature dependence of the mobility strictly follows a $\ln \mu \propto T^{-2}$ law yet it acquires a $\ln \mu \propto T^{-1}$ branch at lower temperatures with an activation energy that decreases with decreasing layer thickness. At lower temperatures, a Fermi level is formed at an energetic position that depends on the carrier concentration and thus on the film thickness. Their quantitative data analysis confirms the notion that in this material polaron effects are unimportant. This is consistent with the work of Meisel et al. [92].

4.2 *Transport in Doped Semiconductors*

The conventional way to increase the conductivity of a semiconductor is to introduce dopants that can act as electron donors or/and acceptors. In fact, significant progress with inorganic semiconductors could only be obtained once carrier transport was no longer determined by impurities but could be controlled and tuned by doping. Controlled and stable doping can be accomplished easily in inorganic semiconductors yet it imposes serious problems for organic semiconductors. The problem is related to the level spectrum of transport states. P-type doping requires the transfer of an electron from the filled HOMO of the host to the LUMO of the dopant at no or only little energy expense as illustrated in Fig. 18.

Correspondingly, the HOMO of the dopant has to be close to the LUMO of the host in order to promote n-type doping. This puts serious constraints on the mutual energy levels. In most organic materials the HOMO is around −5 to −6 eV below the vacuum level. Assuming an electrical bandgap of 2.5 eV, a p-type dopant therefore has to act as a very strong electron acceptor. N-type dopants had to have a HOMO level near −2.5 to −3.5 eV. Clearly, this is difficult to achieve. On the other hand, it would be of considerable advantage to be able to raise the concentration of mobile charge carriers significantly. In addition to yielding a higher carrier mobility, as outlined above, a higher charge carrier density should also reduce ohmic losses at internal interfaces. Note that in organic semiconductors the intrinsic conductivity is low. Therefore a device behaves like a dielectric medium, and an applied voltage drops across the bulk of the sample rather than at

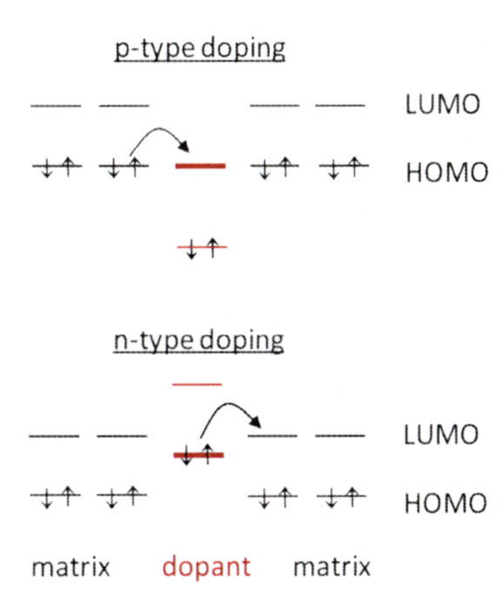

Fig. 18 Doping mechanisms for molecular p-type doping (*top*) and for n-type doping (*bottom*). P-type (n-type) doping is achieved when the molecular dopant acts as acceptor (donor). After [107]

an active interfacial layer between, say, donor and acceptor layers. If one were to increase the conductivity of the material, in which the charge carriers are transported towards the interfacial layer by doping, one would ensure that a larger fraction of the applied voltage drops at the active layer. Therefore doping should minimize ohmic losses.

Attempts to dope organic semiconductors have been made very early in the field, motivated by the prospect of possibly reaching metallic conductivities [108, 109]. These "synthetic metals," however, have not been realized. While p-type doping could be obtained, for example, with iodine gases for poly-*p*-phenylene vinylene (PPV) derivatives, and n-type doping was demonstrated with sodium for a cyano-derivative of PPV, the doping levels obtained were not stable with time. The dopant molecules readily diffused into the organic semiconductor, yet also out of it. Due to the lack of stability, these approaches were not suitable for commercial applications.

Pioneering work on *stable* doping in organic LEDs has been carried out by the group of K. Leo in Dresden and has been reviewed by Walzer [107]. It is now clear that F_4-TCNQ can act as a dopant because its electron affinity is close to 5 eV [110, 111], which is close to the ionization potential of triphenylamine derivatives and to some phthalocyanines (Pc) [107]. It turns out that doping of ZnPc by 2% of F_4-TCNQ raises the conductivity to a level of $10^{-3}\ \Omega^{-1}\ \mathrm{cm}^{-1}$. When using TCNQ instead of F_4-TCNQ the conductivity is only $10^{-6}\ \Omega^{-1}\ \mathrm{cm}^{-1}$. This illustrates the importance of the fluoro-substituents that raise the electron affinity. After all, the key parameter for efficient p-type doping is the difference between the LUMO of the dopant and the HOMO of the host [112, 113], although some level mismatch can be compensated by the effect of disorder broadening of the distribution of transport states. One can expect that disorder ameliorates a fatal level mismatch in a similar way as it is the case of thermally activated charge injection from an electrode (see Sect. 6).

The concentration of free holes generated by the p-dopant can be calculated under the assumption that doping does not alter the hole mobility. Doing so, Zhang et al. [114] derived a concentration of free holes of $4 \times 10^{16}\ \mathrm{cm}^{-3}$ upon doping a PPV film by F_4-TCNQ at a doping ratio of 1:600. At such a doping level processable films are still formed. By relating the charge concentration to the mass ratio the authors estimated that only 1% of the dopant molecules are ionized. This is supported by impedance spectroscopy on Schottky diodes. However, the assumption of a constant mobility is not trivial. There is a superposition of several conceivably compensating effects including level filling (see below), DOS broadening (see below), charge percolation, and changes of the wavefunction overlap parameter that controls charge carrier hopping [115].

In contrast to p-type doping, n-type doping is intrinsically much more difficult to achieve because dopants with high lying HOMO levels can easily and inadvertently be reduced by oxygen. One way towards n-type doping is the use of alkali metals such as lithium or cesium [107]. They are frequently employed to improve electron injection from the anode of an OLED [116]. In small molecule OLED devices, fabricated by evaporation, the dopant metal can be coevaporated. Replacing lithium by cesium has the advantage that Cs^+ ions have a lower diffusivity compared to the small Li^+ ions. This makes the devices less sensitive to temperature and helps keep the dopant away from the charge recombination layer. This is important because ion diffusion into the recombination zone causes quenching of the electroluminescence.

A first study of controlled molecular n-type doping in molecular organic semiconductors was presented by Nollau et al. [117]. They doped naphthalenetetracarboxylic dianhydride (NTCDA) by co-sublimation with the donor molecule bis(ethylenedithio)-tetrathiafulvalene. It was shown that the Fermi level shifts towards the transport level and that the conductivity was increases. However, the conductivities achieved were rather low and only one to two orders of magnitude above the background conductivity of nominally undoped NTCDA. More successful was doping of hexadecafluorophthalocyaninatozinc (F_{16}ZnPC) by tetrathianaphthacene (TTN). The dependence of the UPS spectra on doping suggested that TTN acts as an efficient donor. Essentially no doping was found with Alq_3 as a host material because its LUMO is too high to allow for efficient electron injection from TTN [118]. Another pair of host and guest materials, investigated by the Kahn group [119], is an electron transporting tris(thieno)hexaazatriphenylene derivative doped with bis(cyclopentadienyl)-cobalt(II) that has an unusually low ionization energy of 4 eV in the condensed phase. By UV, X-ray, and inverse photoemission studies the authors concluded that the dopant shifts the Fermi level 0.56 eV towards the unoccupied states of the host. A three orders of magnitude increase in the current was demonstrated. The result indicates that the electron is still quite localized at the dopant site and requires thermal activation for complete ionization.

Another way to accomplish n-type doping is by using cationic dyes [120, 121]. The cationic doping method has been applied successfully for solar cells in which materials with low lying LUMOs are used for electron transporting purposes. The straightforward message is that one has to employ strong electron acceptors as hole transporters, since the HOMO of a stable n-type dopant cannot be higher than, say,

4 eV. Of course, there is a price to pay for this because the lower the LUMO of the hole acceptor the more likely it is that an impurity acts as a hole trap and vice versa.

In the case of doping by organic salts one should be aware of the effects caused by the inevitable presence of counter ions. They act as coulomb wells and modify the DOS of the transport states. By fabricating a gated electrochemical cell with PPV as a transporting film and Au/Pt as source-drain electrodes immersed in an electrolyte (0.1 M tetrabutylammonium perchlorate or hexafluorophosphate or $LiClO_4$ in acetonitrile) Hulea et al. were able to quantify this effect [122]. When the applied potential is increased, holes are injected into PPV. Their charge is counterbalanced by ClO_4^- or PF_6^- ions that enter the film from the electrolyte. The number of holes, inferred from the differential capacitance, allows the energy spectrum of transport states (DOS) to be mapped out. The resulting DOS carries a Gaussian core centered at 5.55 ± 0.02 eV with a variance of 0.19 ± 0.01 eV followed by an exponential tail. However, at very low doping, 10^{-4} (states/eV) per monomer, the DOS is a single Gaussian with variance of 0.11 eV. This is consistent with charge transport studies on an undoped film [123]. Obviously, the presence of random coulomb centers broadens the Gaussian distribution and creates deep tail states, in corroboration of earlier Monte Carlo simulations [124] and more recent analytical theory [125]. The broadening of the tail states is illustrated in Fig. 19. The net effect of this tail broadening is a decreasing charge carrier mobility at low to moderate doping levels. At higher doping levels the coulomb traps overlap spatially and smooth the energy landscape. A striking documentation of this effect

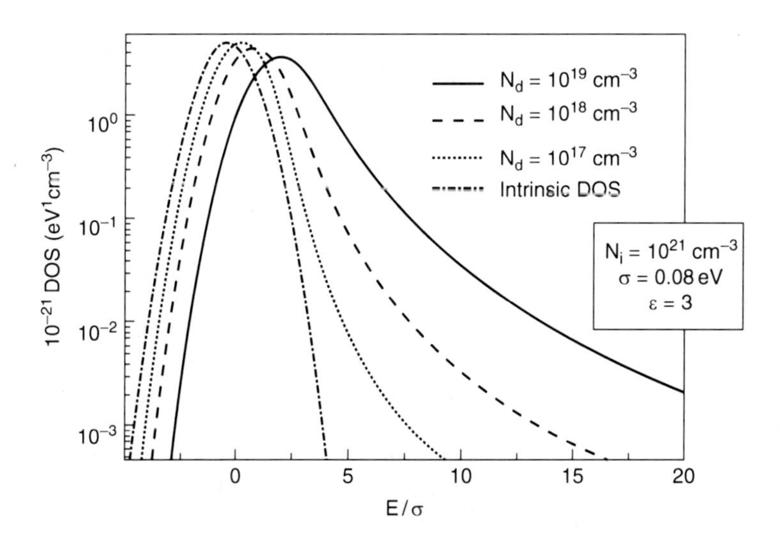

Fig. 19 The effect of doping on the density of states distribution in a disordered organic semiconductor at variable concentration of charged dopants. The energy scale is normalized to the width of the DOS, expressed through σ, of the undoped sample. The parameters are the intrinsic site concentration N_i and the dopant concentration N_d. From [125] with permission. Copyright (2005) by the American Institute of Physics

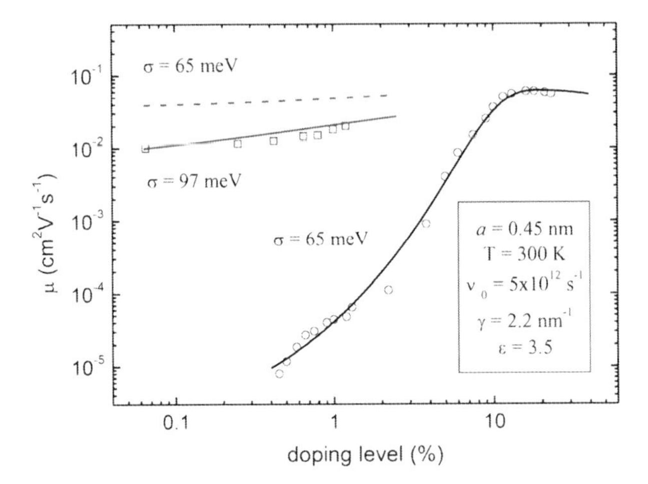

Fig. 20 Charge carrier mobility in P3HT as a function of the charge carrier concentration. *Squares* refer to an experiment performed on a field effect transistor while *circles* refer to experiments done on an electrochemically doped sample. In the latter case the mobility is inferred from the steady state current at a given doping level. *Solid* and *dashed lines* have been fitted using the theory of [101]. The fit parameters are the site separation a, the prefactor ν_0 in the Miller–Abrahams-type hopping rate, the inverse wavefunction decay parameter γ and the dielectric constant ε. From [101] with permission. Copyright (2005) by the American Institute of Physics

is presented in Fig. 20. It shows the dependence of the hole mobility in a P3HT film as a function of concentration of holes generated either by electrochemical doping or by injection from electrodes in an FET structure. In the former case the counter charges are ions incorporated within the film while in the latter they are weakly bound electrons inside the gate electrode. More recent Monte Carlo simulations confirm the interplay between an increase of disorder and a concomitant decrease of the carrier mobility in the presence of a moderate concentration of extra charges and a smoothing of the energy landscape at higher concentrations [126]. By the way, broadening of the DOS distribution by counter charges also effects the Langevin-type electron hole recombination process, as was demonstrated by Monte-Carlo simulations by van der Holst and coworkers [127].

The preceding discussion of doping effects pertain to systems with low to moderate doping levels. Under this premise charge transport tends to remain of the hopping type though the shape of the DOS may be altered by the dopant. This no longer true in the case of highly doped materials that are metallic or quasi-metallic such as PEDOT [128, 129], poly-aniline [130–132], and poly-pyrrole [133, 134]. While there are calls for new concepts, notably because highly doped systems can no longer be considered as being homogeneous [128] and because there are issues concerning a possible semiconductor–metal transition, some of the experimental results can be rationalized in terms of the variable range, i.e., a modified, hopping concept [134, 135]. This is a topic in its own right and shall, therefore, not be discussed here.

5 Charge Transport in the Strong Coupling Regime

5.1 *Intra-Chain Transport at Short Time Scales*

In amorphous molecular systems, even the fastest hopping process is determined by the strength of inter-molecular coupling among the adjacent transport sites. A crude measure for the jump time can be obtained from the mobility extrapolated to infinite temperatures, $\mu(T \to \infty) = \mu_0$, in a $\ln \mu$ vs T^2 plot. For this rough estimation we use Einstein's ratio $eD = \mu kT$ and assume isotropic hopping with a diffusion constant $D = \frac{1}{6}av^2$ where $a = 1$ nm is the inter-site separation and v is the inter-hop frequency. Taking $\mu_0 = 10^{-2}$ cm^2 V^{-1} s^{-1}s as a representative value for the prefactor mobility, we obtain $v = 10^{11}$ s^{-1}, equivalent to a minimum jump time of 10 ps for iso-energetic jumps.

The situation is different for systems in which the hopping sites are extended as realized in conjugated polymers. Here, one expects a hierarchy of transport processes, i.e., there should be fast on-chain transport followed by slower inter-site transport as visualized in Fig. 21. Calculations of the effective mass of charge carriers in conjugated polymers [28] predict a charge carrier mobility as high as 1,000 cm^2 V^{-1} s^{-1}, i.e., six to nine orders of magnitude larger than measured in a conventional ToF experiment. The obvious reason for this discrepancy is related to the spatial scale over which transport is measured. It is reasonable to assume that, in principle, transport is fast as long it is not affected by intrinsic or extrinsic scattering and/or localization.

The ideal systems to test this conjecture are single-crystalline poly-diacetylenes (PDAs), notably perfect PDA chains embedded in a crystalline lattice [15]. Employing the electroreflection method, Weiser and Möller [136] analyzed the

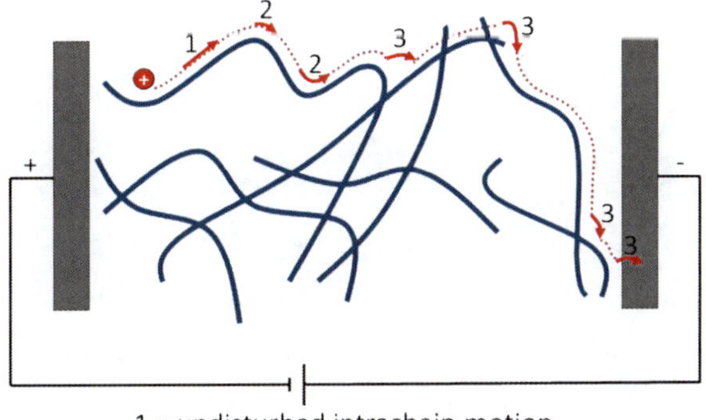

1 = undisturbed intrachain motion
2 = disturbed intrachain motion
3 = interchain motion

Fig. 21 A schematic view of the hierarchy of charge carrier hopping in a network of disordered conjugated polymer chains. *1* depicts ultra-fast motion within an ordered segment of the chain while *2* and *3* illustrate intra-and interchain hopping processes

reflection feature occurring approximately 0.5 eV above the excitonic absorption edge in terms of the Franz–Keldysh effect acting on the otherwise hidden valence to conduction band transition. They came up with an effective mass of approximately 0.05 m_e, in agreement with theoretical calculations [28]. This would translate into an electron mobility of approximately 1,000 $cm^2/V^{-1}s^{-1}$. So far, nobody has confirmed this prediction. Employing various methods, there is consensus that the macroscopic mobility of, presumably, electrons in crystalline PDA is greater than 1 $cm^2/V^{-1}s^{-1}$ along the chain and a factor 10^{-2} to 10^{-3} lower perpendicular to it [20, 137]. However, this value refers to macroscopic samples rather than to individual chains.

In order to obtain information pertaining to the on-chain motion, one has to rely on analogous spectroscopic studies on excitons. The Schott group studied the photoluminescence of single PDA chains in an unreacted crystalline matrix. They found that at low temperatures the excited state is an exciton that moves coherently within a distance of tens of micrometers with a coherence time of up to 1 ps. However, in a non-crystalline conjugated polymer such as MEH-PPV the phase coherence time of an optical excitation is as short as 100 fs [138, 139]. This strongly suggests that dephasing is due to scattering at static or dynamic chain imperfections rather than a generic property of conjugated polymers in general. It is straightforward to conjecture that the scattering time of charge carriers is similarly short and that their motion on macroscopic dimension is disorder controlled.

It is instructive to estimate the displacement of a charge carrier after generation. Using Einstein ratio $eD = \mu kT$, one arrives at a mean square displacement of a charge carrier, $\sqrt{\langle \Delta x^2 \rangle}$, of 20 nm (60 nm) if scattering occurred at 100 fs (1 ps) after excitation and if the initial mobility were 1,000 $cm^2/V^{-1}s^{-1}$ (Fig. 22). This has to be compared to the electronic correlation length of a chain, the so-called effective conjugation length. Usually, although not well founded, it is inferred from the dependence of the energy of a singlet excitation as a function of the lengths of oligomers. A conservative estimate is a value less than 10 nm. This indicates that

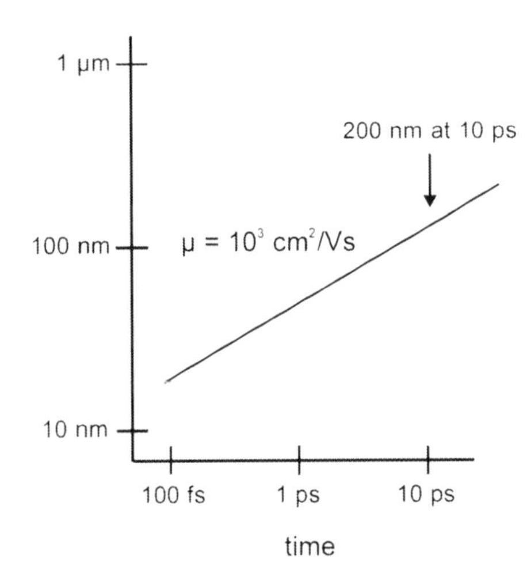

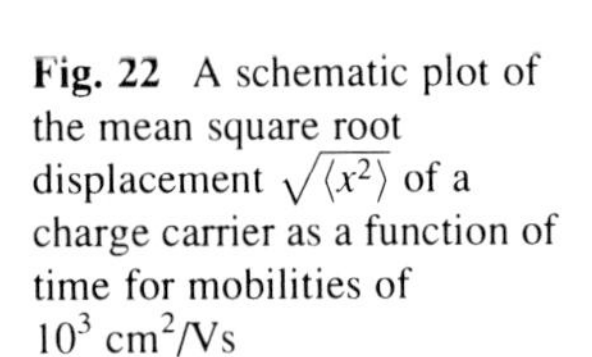
Fig. 22 A schematic plot of the mean square root displacement $\sqrt{\langle x^2 \rangle}$ of a charge carrier as a function of time for mobilities of 10^3 cm^2/Vs

one has to probe charge carrier motion on a very short length and on an ultra-fast time scale in order to be able to monitor the intrinsic, i.e., defect-free mobility.

Pioneering work on the time-resolved probing of charge carrier motion along the chain of a conjugated polymer has been performed by the Delft group employing the time resolved microwave conductivity (TRMC) technique [140]. It is an electrode-less technique. Charge carriers are generated homogeneously inside the sample by irradiation with 5–20 ns pulses of 3 MeV electrons from a van der Graaff electron accelerator. One probes their oscillatory motion by a microwave field on a short length scale yet within a nanosecond to microsecond time scale, i.e., long compared to the inverse probing microwave frequency. Typical frequencies are either 10 GHz or 34 GHz. The peak amplitude of the microwave field is of the order of 100 V/cm, i.e., low as compared to the field strengths commonly used in ToF experiments. The radiation induced conductivity is inferred from the decrease of the microwave power reflected from the cell. It is given by the product of the charge carrier concentration and their mobility. The charge concentration can be estimated from the irradiation dose and the known ionization efficiency. In order to probe charge motion on isolated polymer chains, the polymer is dissolved in benzene solution. However, the technique can also be used to study solid state effects in bulk samples. The high energy electrons absorbed inside the sample scatter on the solvent molecules and produce a virtually uniform distribution of excess electrons and benzene cations with a known concentration. Benzene cations diffuse towards the polymer chains and form polymer cations. As this diffusion-controlled reaction proceeds, an increase of the transient conductivity is observed on a time scale of hundreds of nanoseconds. This increase indicates that the positive charge on the chains are more mobile than the benzene cation in solution whose mobility is 1.2×10^{-3} cm^2/V^{-1}s^{-1} at room temperature. Monitoring the motion of electrons on the polymer chain requires thorough degassing and doping with a strong hole scavenger such as tetramethyl-*p*-phenylenediamine (TMPD). In order to study selectively the motion of holes the solution can be saturated by oxygen that is an efficient scavenger for excess electrons.

A representative example for the information extracted from a TRMC experiment is the work of Prins et al. [141] on the electron and hole dynamics on isolated chains of solution-processable poly(thienylenevinylene) (PTV) derivatives in dilute solution. The mobility of both electrons and holes as well as the kinetics of their bimolecular recombination have been monitored by a 34-GHz microwave field. It was found that at room temperature both electrons and holes have high intra-chain mobilities of $\mu_- = 0.23 \pm 0.04$ cm^2/V^{-1}s^{-1} and $\mu_+ = 0.38 \pm 0.02$ cm^2/V^{-1}s^{-1}. The electrons become trapped at defects or impurities within 4 μs while no trapping was observed for holes. The essential results are (1) that the trap-free mobilities of electrons and holes are comparable and (2) that the intra-chain hole mobility in PTV is about three orders of magnitude larger than the macroscopic hole mobility measured in PTV devices [142]. This proves that the mobilities inferred from ToF and FET experiments are limited by inter-chain hopping, in addition to possible trapping events. It also confirms the notion that there is no reason why electron and hole mobilities should be principally different. The fact that electron mobilities observed in devices are usually much lower than the hole mobilities is exclusively due to trapping. Note that in a polymer with high lying HOMO, the LUMO level is also high, implying that inadvertent impurities such as oxidation products can act as electron traps. Analogous reasoning applies to holes. Similar experiments were performed to measure the mobility of holes along isolated chains of polyphenylenevinylene and polythiophene backbones. The values are 0.43 $cm^2/V^{-1}s^{-1}$ for MEH-PPV and 0.02 $cm^2/V^{-1}s^{-1}$ for P3HT [143].

The TRMC technique has also been applied to solid samples. It delineates the effects of sample morphology. Among the materials investigated were crystalline poly-diacetylenes, $\pi-\pi$ stacked columnar discotic liquid crystals, and conjugated polymers [144]. The largest values, on the order of 10 $cm^2/V^{-1}s^{-1}$, were found for single-crystal poly-diacetylenes. Much lower values covering the range from 0.009 to 0.125 $cm^2/V^{-1}s^{-1}$ were obtained for solution synthesized conjugated polymers with six different backbone structures. This is attributed mainly to their complex morphology and the resulting static disorder in the backbone structure. The highest mobilities for this class of materials, ca. 0.1 $cm^2/V^{-1}s^{-1}$, were found for liquid crystalline derivatives of polyfluorene and poly(phenylenevinylene). Even larger values, close to 1 $cm^2/V^{-1}s^{-1}$, were measured with discotic materials in crystalline and liquid crystalline phases. This is a signature of their self-organizing nature and hence their degree of structural order, which compensates for the weaker electronic coupling between monomeric units in the discotics as compared with covalently bonded conjugated polymers [145].

The TRMC technique has successfully been used to answer open questions regarding the relation between morphology and charge carrier mobility in layers of P3HT with different molecular weight [146]. In agreement with intuition, the TRMC experiment probes the local mobility within ordered grains in which the chains are fully elongated. It depends only weakly on molecular weight. In contrast, the macroscopic mobility of medium molecular weight layers is two orders of magnitude lower than the local mobility and *decreases* with increasing temperature. The rate limiting process is transport through disordered material surrounding ordered grains [147]. A clue for the unusual temperature dependence is provided by temperature dependent UV–Vis absorption spectra. They suggest that the aggregates undergo a "premelting" significantly below the actual melting temperature. As a result, the aggregate width decreases. The concomitant increase in width of the interlamellar zones, as well as the likely increase of disorder in these amorphous regions, is presumably the main reason for the drop in the macroscopic mobility of short chain deuterated P3HT upon heating. In contrast, long chains may interconnect the crystalline domains in high molecular weight deuterated P3HT, thus bypassing the disordered interlamellar regions and rendering the macroscopic charge transport in this material less susceptible to changes of the sample heterogeneity.

Although the above mobilities in isolated chains of conjugated polymers and in solid phases are much larger than values inferred from ToF experiments, in no case do they come close to the value of 1,000 $cm^2/V^{-1}s^{-1}$ expected for a perfect one-dimensional π-bonded polymer chain. Moreover, they are morphology-sensitive.

Improved ordering enhances the mobility. It is straightforward to conjecture that even the on-chain mobility is limited by static and dynamic disorder. A simple estimate will illustrate this. Suppose that an initially generated charge carrier had an ultra-high mobility of 1,000 $cm^2/V^{-1}s^{-1}$ and, accordingly, a diffusion coefficient of 25 cm^2/s at 295 K. Within 10 ps, which is the characteristic time scale of the microwave field, the diffusional spread of a packet of charge carriers should be 200 nm. This value is much larger that the effective conjugation length of conjugated polymers except for polydiactylenes in crystalline matrix. Therefore the carrier transport is mediated by scattering events and even blockades due to chain defects and chain ends. This notion is experimentally verified by probing the carrier motion at different microwave frequencies. If the frequency is increased their motion is confined to a smaller length scale, i.e., a carrier experiences fewer stopping events between the energy barriers. This, in turn, leads to a higher mobility upon increasing frequency. Obviously the TRMC technique does not yield a well-defined mobility value but depends on the experimental conditions. This is an inherent feature of the technique and can be exploited to extract information regarding carrier motion inside of a polymer chain. i.e., the scattering mechanism. One can even extrapolate on the carrier motion prior to scattering and blocking at a chain defect/end.

To extract pertinent information from experimental results requires a theoretical model. Based upon earlier work by Grozema et al. [143], Prins et al. [148] developed a framework to understand the effects that static disorder as well as chain dynamics have on the charge motion inside a π-conjugated chain. The basic idea is that, in conjugated oligomers and polymers, torsions, static conjugation breaks, or chain ends can all act as barriers to charge transport. The presence of these barriers leads to an increase of the charge carrier mobility with increasing microwave frequency because at higher frequency the carrier motion is probed on a smaller length scale, i.e., between these barriers. A major source of time-dependent disorder is presented by thermally driven torsional motion between repeat units of, say, a poly-phenylenevinylene or a poly-thiophene chain. It results in a variation of the electronic coupling between the repeat units. In the theory the electronic coupling was calculated by density functional theory. The results confirm that the carrier mobility probed by the TRMC technique increases with increasing length of the chain and with conjugation length. Experiments on PV oligomers with varying length and polymers containing a variable fraction of chemical conjugation breaks are in good agreement with theory. The concept of time-dependent disorder also explains why the hole mobility in MEH-PPV probed by 34-GHz radiation is 0.46 $cm^2/V^{-1}s^{-1}$ while it is only 0.02 $cm^2/V^{-1}s^{-1}$ in P3HT. The reason is the larger deviation of coplanar alignment of the structural units of P3TH as compared to MEH-PPV.

It was straightforward to apply the TRMC technique to study on-chain charge transport to ladder-type poly-phenylene (LPPP) systems because covalent bridging between the phenylene rings planarizes the chain skeleton, eliminates ring torsions, and reduces static disorder. One can conjecture that in these systems intra-chain motion should be mostly limited by static disorder and chain ends. To confirm this notion, Prins et al. [149] measured the complex microwave conductivity in solid samples of either phenyl- or methyl-substituted LPPP with different chain lengths, i.e., of Ph-LPPP with $\langle n \rangle = 13, 16, 35$, and of MeLPPP with $\langle n \rangle = 54$, $\langle n \rangle$ being the average number of repeat units. In such an experiment, the real part of the conductivity reflects the field-induced barrier-less in-phase drift velocity of charges undergoing conventional Gaussian diffusion along an ordered, infinitely long polymer chain while the imaginary part reflects the carrier motion hindered by barriers such as chain ends. The observation that the mobility extracted from the imaginary part of the conductivity increases from 0.056 to 0.08 $cm^2/V^{-1}s^{-1}$ and to 0.14 $cm^2/V^{-1}s^{-1}$ when $\langle n \rangle$ increases from 13 to 16 and to 35 proves that the carrier motion is limited by chain ends. In order to obtain the mobility in an infinitely long chain the authors developed a model for one-dimensional diffusional motion between infinitely high reflection barriers as a function of the chain length. By extrapolation, the authors arrived at an infinite chain mobility of 30 $cm^2/V^{-1}s^{-1}$. Analogous experiments and analyzing procedures on MeLPPP in dilute solution yield a spectacular value of 600 $cm^2/V^{-1}s^{-1}$, comparable to the hypothetical value for a perfect chain [150]. The fact that the infinite chain mobility in solid LPPP-type systems is a factor of 30 less than in solution testifies to the role of inter-chain disorder in bulk systems. Although these extrapolated infinite-chain mobilities should be viewed with some caution, they provide an idea of how charge carriers move between scattering barriers in a conjugated polymer in a device.

When interpreting the results derived from TRMC experiments one should keep in mind that a GHz microwave field interrogates the charge carrier dynamics on a time scale ranging from nano- to microseconds. This implies that one does not probe nascent charge carriers but, rather, carriers that have had enough time to relax to energetically more favorable sections of a polymer chain. To reveal the carrier dynamics on an ultra-fast time scale requires all-optical probing. The technique of choice is time resolved terahertz (THz) spectroscopy that offers picosecond time resolution and low probing fields (kV/cm) [151–153]. It yields the far-infrared conductivity of charges generated by femtosecond light pulses. One can monitor the evolution of charge motion as a function of delay time between generation and probe pulses. The price to pay for shifting the observation window to picosecond and to sub-picosecond time scales is (1) the difficulty to retrieve the pertinent information from the raw data and (2) that the measured property is a conductivity, i.e., the product of the carrier yield and their mobility. The Sundström group [153] applied this technique to study holes on low-bandgap conjugated polymers consisting of an alternating sequence of a low bandgap unit as an electron accepting group and a dialkoxy-phenylene unit. The copolymer was blended with the fullerene derivative PCBM which is often used as an acceptor in organic solar cells. Upon photoexcitation the electron is transferred to the PCBM. The hole remains on the polymer chain and is responsible for the ultra-fast conductivity because the electron motion among the PCBM units is much slower. From the intricate analysis of the experimental results combined with simulation the authors conclude an intrinsic carrier mobility of 40 $cm^2/V^{-1}s^{-1}$ within unperturbed chain segments comprising about four repeat units. Like the electron, the hole generated by the dissociation process carries an initial excess of kinetic energy that allows it to pass easily over potential barriers with an intrinsic mobility of about 2 $cm^2/V^{-1}s^{-1}$. Subsequently, the hole cools down at an initial relaxation rate of 1/180 fs^{-1} and gets trapped at an initial rate of 1/860 fs^{-1}. This is manifested by a steep drop of the conductivity. The results are consistent with the notion of an ultra-fast mobility of charges on a π-conjugated chain between scattering events. They also indicate that the time averaged mobility derived from a TRMC experiment is that of charge carriers that already suffered some relaxation.

An independent and complementary method to measure the charge carrier mobility on ultra-fast time scales has recently been introduced by Devizis et al. [154]. It is based on time resolved electric field induced second harmonic generation (SHG). This method is commonly used to determine molecular hyperpolarizabilities. Any process that changes the electric field distribution in the material will affect the temporal SHG signal. In turn, the SHG intensity can be taken as a probe of changes of the electric field due to charge motion. Upon generating charge carriers in a charged capacitor by a short laser flash the moving charge partially shields the electric field and the SHG efficiency decreases. Measuring the decrease of the SHG signal as a function of time after the laser pulse yields the time dependence of the carrier motion up to a detection time of 3 ns. Integrated photocurrents measured within a time window of 10 ns to 10 μs complement the information on the time dependence of the mobility. The technique was applied to poly-spiro-bifluorene-*co*-benzothiadiazol (PSF-BT). It turns out that the sum of the electron and hole mobilities probed at 1 ps after excitation is about 0.1 $cm^2/V^{-1}s^{-1}$. It decreases to 10^{-6} $cm^2/V^{-1}s^{-1}$ at 1 μs featuring an $\ln \mu$ vs $t^{-\beta}$ law with β ranging from 0.84 at lower fields (about 4×10^5 V/cm) to 0.75 (at about 1.5×10^6 V/cm) (see Fig. 23). These results prove unambiguously that within a time range of 6 decades transport is dispersive and, concomitantly, controlled by disorder.

5.2 *Band Transport*

In this chapter we so far focussed on charge transport in organic solids that are used as active materials in modern opto-electronic devices such as OLEDs, solar cells, field effect transistors, and photocopiers. Dictated by the need for cost-efficient device manufacturing and the realization of optimized structure-property relations for special applications, the active device elements are usually non-crystalline, if not amorphous. This implies that charge transport is of the hopping type. It turned out, though, that crystalline organic semiconductors may profitably be employed as active layers in field effect transistors where band-like carrier transport prevails. For this reason we shall very briefly address recent developments in this area without attempting to cover the growing field of organic field effect transistors in greater detail.

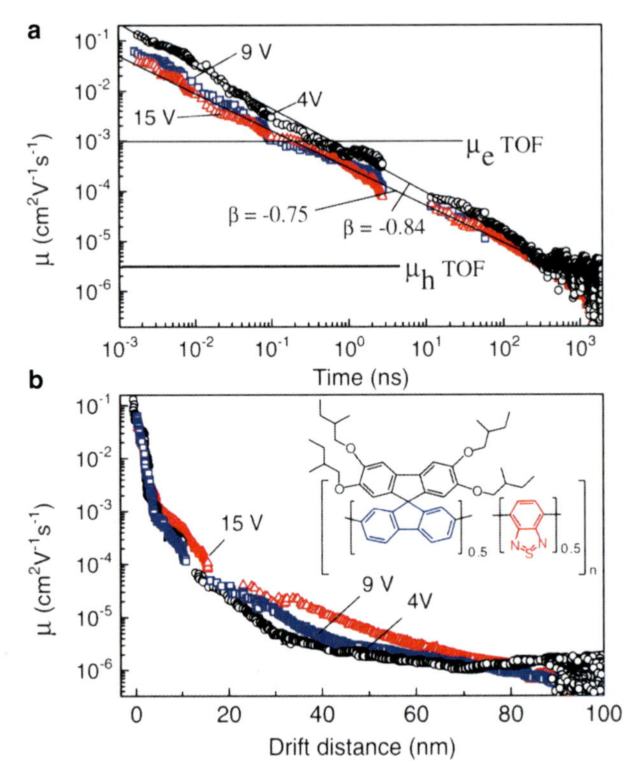

Fig. 23 (**a**) Dependence of the hole mobility in a film of poly-spiro-bifluorene-*co*-benzothiazole (PSF-BT) as function of the time elapsed after charge carrier generation by a 130 fs laser pulse at different applied voltages. The *horizontal lines* represent the electron and hole mobilities inferred from ToF experiments. (**b**) Momentary mobility as a function of the averaged distance that a carrier travelled after a given time. The *inset* depicts the chemical structure of PSF-BT. From [154] with permission. Copyright (2009) by the American Institute of Physics

Band-like charge transport in molecular crystals was investigated experimentally in the 1980s and 1990s. The pioneering work of Karl and his group in Stuttgart showed that close to and below room temperature, and dependent on crystallographic direction, the mobility of both electrons and holes feature a T^{-n} law with $n = 2$–3, provided that they can be purified efficiently (see [155]). For less pure crystals, the mobility is temperature activated with an activation energy that is given by the difference of the HOMO (LUMO) levels between host and guest. This proves that transport is trap limited. The temperature at which the transition from band to trap limited transport occurs depends on the trap concentration and the trap depth. In view of the narrow width of valence and conduction bands in molecular crystals the charge carrier mean free paths are only a few lattice sites at most. This implies that transport is on the borderline between being coherent and incoherent. This problem has been discussed intensively in the books of Pope and Swenberg [45] and of Silinsh and Capek [156] and recently by Fratini and Chiuchi [157]. A comprehensive review of more recent advances has been published by Coropceanu et al. [10]. Based upon quantum chemical treatments the authors developed a consistent theoretical framework of the electronic coupling and the electron–phonon interaction as a function of the lattice structure. The electronic transfer integral is the matrix element that couples the wavefunctions of two charge-localized states via an electronic Hamiltonian of the system. It depends on the mutual orientation of the molecules and thus on the crystallographic direction. Typical values range between about 10 and 83 meV for the (100) direction of a rubrene crystal. Since the widths of the transport bands is four times the transfer integral, this results in bandwidths of the order of some 10–100 meV. Although these calculations refer to molecular crystals they are also relevant for disorder systems because the charge carrier mobility extrapolated to infinite temperature (μ_0) depends on the electronic coupling among the structural building blocks.

Organic crystalline materials that may be used in FETs are rubrene and pentacene because the relevant electronic transfer integrals are comparatively large. Accordingly, both materials have comparatively high hole mobilities of 10 $cm^2/V^{-1}s^{-1}$ or higher at room temperature if measured in FET configuration (see [158]). Podzorov et al. [159] measured a value of 30 $cm^2/V^{-1}s^{-1}$ at 200 K. In later work Zeis et al. [160] reported a maximal mobility of 13 $cm^2/V^{-1}s^{-1}$ with strong anisotropy. A decrease of μ at lower temperatures is a signature of charge carrier trapping. Applying a hot wall deposition method, a hole mobility of only 2.4 $cm^2/V^{-1}s^{-1}$ in rubrene has been measured. Obviously sample preparation and purification have a profound effect on the crystal properties, particularly if the sample is polycrystalline instead of single crystalline [161]. For pentacene, an extrapolated value as high as 50 $cm^2/V^{-1}s^{-1}$ for the in-plane mobility has been inferred from space-charge-limited current measurements performed in surface configuration [162]. A high hole mobility is consistent with a valence bandwidth along the 100 direction of 240 meV at 120 K and 190 meV at 295 K determined from photoemission spectroscopy [163] and with DFT calculations for the charge transfer integral that yield values of 70–75 meV [10]. However, the way the sample is prepared has a major effect on the mobility. This is illustrated by the work of Minarin et al. [164]. These authors measured the hole mobility in a single-grain pentacene FET within a temperature range between 300 K and 5.8 K and found a room temperature mobility of about 1 $cm^2/V^{-1}s^{-1}$ and weakly activated transport below with an activation energy of 4.6 meV. In a polycrystalline sample the room temperature mobility is about 0.3 $cm^2/V^{-1}s^{-1}$ and the activation energy is 55 meV. Obviously, grain boundaries act as charge carrier traps [165, 166].

In an attempt to combine band-like charge carrier motion realized in an – inevitably fragile – crystalline FET structure with structural robustness and flexibility, Sakanoue and Sirringhaus [167] prepared FETs using spin coated films of 6,13-bis(triisopropylsilylethynyl)(TIPS)-pentacene films in contact with a perfluorinated, low dielectric-constant polymer gate electrode. The (linear) hole mobility at room temperature is 0.8 $cm^2/V^{-1}s^{-1}$ with tendency of an apparent "band-like" negative temperature coefficient of the mobility ($d\mu/dT < 0$).

The authors use optical spectroscopy of gate-induced charge carriers to show that, at low temperature and small lateral electric field, charges become localized onto individual molecules in shallow trap states, but that at moderate temperatures an electric field is able to detrap them, resulting in transport that is not temperature-activated. This work demonstrates that transport in such systems can be interpreted in terms of classical semiconductor physics and there is no need to invoke one-dimensional Luttinger liquid physics [168].

6 Charge Injection

6.1 Mechanism of Charge Carrier Injection

In most cases, injection from the electrodes is the process by which charge carriers are generated in OLEDs and FETs. Usually it is limited by an energy barrier between the Fermi-level of the electrode and the transport levels of the dielectric. In conventional crystalline inorganic semiconductors, the relevant processes are either Richardson–Schottky emission or Fowler–Nordheim tunneling (see Fig. 24). The former process implies that a thermally excited electron from the Fermi level travels across the maximum of the electrostatic potential modified by the coulomb potential of the image charge and the applied electric field without being scattered. It gives rise to an Arrhenius-type of temperature dependence ($\ln j \propto T$), and a Poole–Frenkel-type of field dependence, i.e., $\ln j \propto \sqrt{F}$. In the classic Fowler–Nordheim case, one ignores the image potential and one assumes that an electron at the Fermi level of the metal tunnels through a triangular potential barrier set by the interfacial energy barrier and the applied potential. For both mechanisms, the crucial condition is that there is strong electronic coupling among the constituting lattice elements that leads to wide valence and conduction bands. This implies that the scattering length of charge carriers is much larger than the interatomic separation. In organic solids this condition is violated because electronic coupling between molecules is of van der Waals type and thus weak. Accordingly, transport is incoherent and of the hopping type. Therefore, the condition of collision-free charge injection across the maximum of the electrostatic potential, implied by Schottky theory, is violated. An experimental signature of this failure of the classic injection models is that the temperature dependence of the injection current is (1) weaker than expected based upon the estimated energy barriers and (2) sub-linear on an Arrhenius scale. This excludes a classic Richardson–Schottky emission process, yet this also eliminates Fowler–Nordheim tunneling as a mechanism since these observations hold even within a field range at which tunneling has got to be inefficient. Thus, theoretical reasoning and experimental observation imply that the classical inorganic semiconductor mechanisms for charge injection do not apply. Therefore, alternative, more suited approaches are needed.

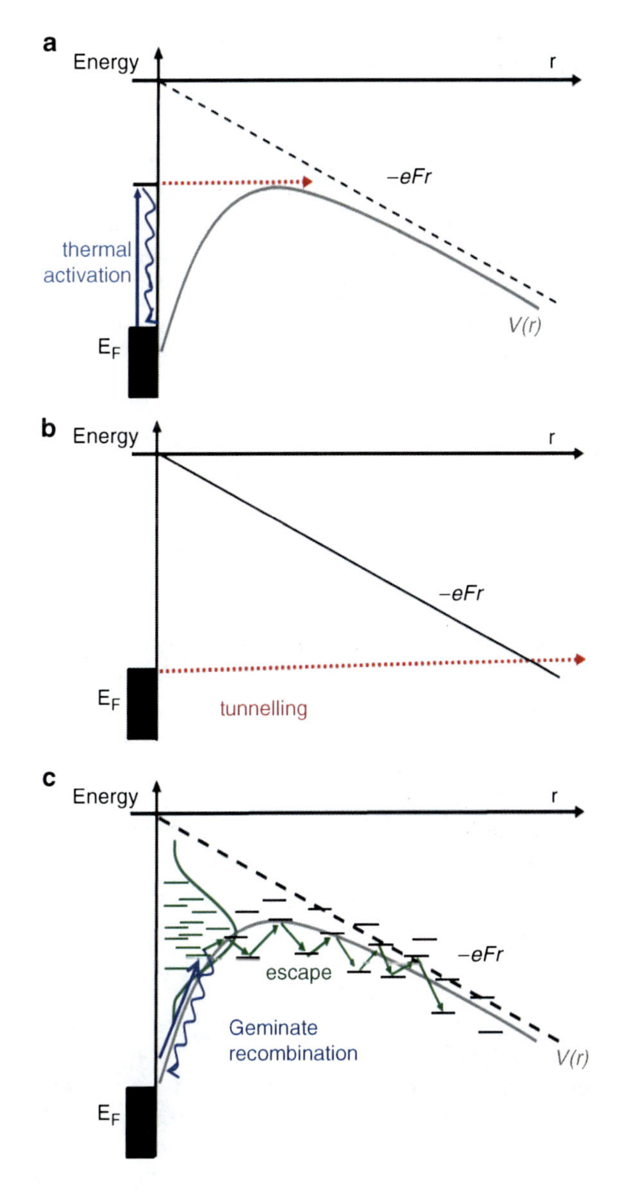

Fig. 24 Schematic representation of electron injection from a metallic electrode into a semiconductor (**a**) via Schottky emission, (**b**) via Fowler–Nordheim tunneling, and (**c**) via hopping in a disordered organic solid. F is the applied electric field, r denotes the distance from the electrode

Based upon simulation [169] and analytical theory [170] a model has meanwhile been developed that takes into account (1) the existence of the image charge at the electrode, (2) the hopping-type of charge transport, and (3) the presence of disorder existing in a non-crystalline system. The underlying idea, originally introduced by Gartstein and Conwell [171], is that a thermally activated jump raises an electron from the Fermi level of the electrode to a tail state of the density of states distribution of transport sites of the dielectric medium subject to the condition that this site has at least one hopping neighbor at equal or even lower energy. It is fulfilled by optimizing a transport parameter with regard to both jump distance and jump energy. This condition ensures that the primarily injected carrier can continue its motion away from the interface rather than recombine with its image charge in the electrode. Subsequently, the carrier is considered to execute a diffusive random walk in the combined coulomb potential of the image charge and the externally applied potential, and this is described by using a one-dimensional Onsager's theory. Simulation and analytical theory for injection barriers ranging from 0.2 to 0.7 eV and for temperatures between 300 and 200 K are mutually consistent. Apparently, the neglect of the stochastic nature of the carrier motion in the vicinity of the electrode as well as the neglect of disorder on the one-dimensional Onsager-type escape process is uncritical in the considered parameter regime. The field dependence of the injection efficiency follows a Poole–Frenkel-type law and bears out a sub-linear temperature dependence on an Arrhenius scale with a mean activation energy that is significantly lower than one might suspect based upon the literature values of the work-function of the electrode and the electron affinity/ionization energy of the dielectric. The qualitative explanation of this ubiquitously observed phenomenon [172] is the following. Upon lowering the temperature, the transport energy, and therefore also the critical site energy from which a charge carrier can start its diffusive motion within the DOS distribution, decreases. Consequently the activation energy needed for injection is not constant but decreases with decreasing temperature.

A textbook example for the successful application of the model of Arkhipov et al. is the work of van Woudenbergh et al. [173]. More recently, Agrawal et al. [106] compared injection limited currents and space-charge-limited currents in a copper-phthalocyanine sandwich cell with ITO and Al electrodes. An analysis of experimental data yields consistent values for the width of the DOS distribution as well as for inter-site separation [174]. These studies support the model of thermally activated injection into a Gaussian DOS distribution of hopping sites and confirm the notion that disorder facilitates injection because it lowers the injection barrier, although the transport velocity decreases with increasing disorder.

In 2000, A. Burin and M. Ratner [175] presented an injection model based upon the idea that injection and transport occur through one-dimensional straight paths, effectively lowering the injection barrier. Meanwhile, van der Holst et al. [127] developed a three-dimensional master equation model that includes the disorder in the transport and avoids the shortcomings of the one-dimensional model, in which carrier blockade events can be crucial (Fig. 25). It is fair to say, though, that for devices operating close to room temperature, the simpler treatment of Arkhipov et al. is sufficient. It is worth pointing out, however, that all these treatments rely on the notion that the initial injection event is largely controlled by the existence of tail states of the DOS at the interface and that it is defined by an energy vs hopping range optimization procedure. Since in a bulk system the low energy sites are spatially fixed, the injection process is NOT spatially homogeneous, but filamentary as pointed out by van der Holst et al.

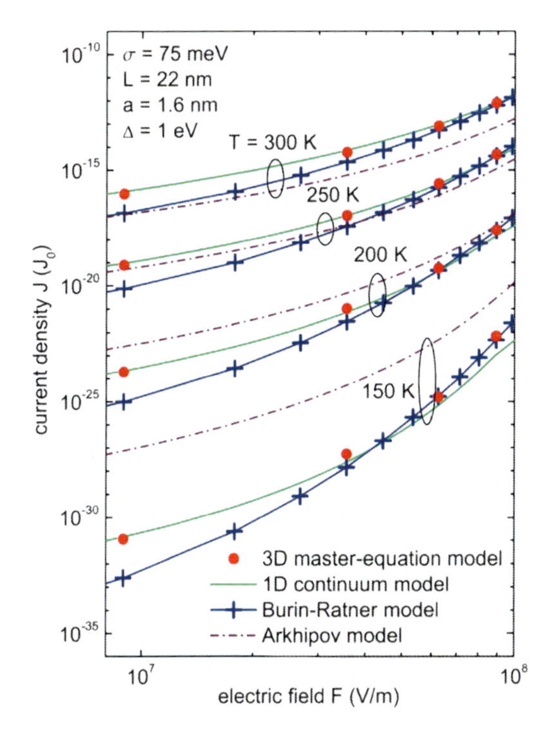

Fig. 25 Comparison of the predictions of various models for current injection from a metal electrode into a hopping system featuring a Gaussian DOS of variance $\sigma = 75$ meV as a function of the electric field at different temperatures. The 1D continuum and the 3D master equation model have been developed by van der Holst et al. [127]. The calculations based upon the Burin-Ratner and the Arkhipov et al. models are taken from [175] and [170] respectively. Parameters are the sample length L, the intersite separation a and the injection barrierΔ. From [127] with permission. Copyright (2009) by the American Institute of Physics

The magnitude of the injection barrier is open to conjecture. Meanwhile there is consensus that energy barriers can deviate significantly from the values estimated from vacuum values of the work-function of the electrode and from the center of the hole and electron transporting states, respectively. The reason is related to the possible formation of interfacial dipole layers that are specific for the kind of material. Photoelectron spectroscopy indicates that injection barriers can differ by more than 1 eV from values that assume vacuum level alignment [176, 177]. Photoemission studies can also delineate band bending close to the interface [178].

An example of the difficulties encountered when trying to fabricate an ohmic electrode, able to sustain a space-charge-limited current, is the recent work of the Neher group [179]. The authors deposited barium as an electron injection cathode on top of an electron transporting polymer based on a naphthalene diimide core whose LUMO is as low as -4 eV below vacuum level. Although the Fermi level of barium should be above the LUMO of the polymer, the electron current is, unexpectedly, contact-limited rather than space-charge-limited. It appears that evaporation of reactive metals onto layers of conjugated polymers introduces injection barriers through the formation of oxides and chemical defects. While this effect can be masked by the low bulk currents in the majority of n-type polymers with low electron mobilities, it is important in the case of this naphthalene diimide polymer that has an unusually high electron mobility of 3.5×10^{-3} cm^2/ $V^{-1}s^{-1}$ at 295 K.

6.2 *Ohmic Injection*

At low injection barriers the surface charge injected into a layer next to the electrode can become comparable with the capacitor charge. If this condition is met at arbitrary applied voltage, such a contact serves as an "ohmic" electrode from which charge injection occurs. In this case the current is solely determined by transport within the bulk of the dielectric rather than by the injection process, i.e., by the charge carrier mobility, and it is considered to be space-charge-limited (SCL). Establishment of this condition depends on both the magnitude of the injection barrier and the charge carrier mobility. For typical OLED parameters, the critical energy barrier is about 0.3 eV [180, 181]. Interfacial doping, for instance by electron acceptors such as C_{60} [182, 183] and the more electronegative tetrafluoronocyanoquinodimethane (F_4-TCNQ) [184], can lower the injection barrier. Another way to improve injection is to insert a thin interfacial layer, such as a self-assembled monolayer [185]. The effect of such a layer can be due to interfacial tunneling [186] or to changes of the sample morphology and/or the electronic structure at the interface.

It is obvious that in an OLED efficient charge injection is crucial. A simple estimate will illustrate this. Suppose that a concentration n of charge carriers equivalent to the capacitor charge CV, where C is the capacitance per unit area and V is the applied voltage, is distributed homogeneously within a dielectric layer of thickness d. We find $n = \dfrac{\varepsilon\varepsilon_0 F}{ed}$. The bimolecular recombination time with a counter charge is $\frac{1}{n\gamma}$, where γ is the bimolecular recombination coefficient. Chemical kinetics predicts $\gamma = 4\pi\langle R\rangle(D_+ + D_-)$, where $\langle R\rangle$ is the mean distance at which charges recombine, and D_+ and D_- are the diffusion constants of the charge carriers. It is taken to be the coulomb capture radius $\langle R\rangle = \frac{e^2}{4\pi\varepsilon_0\varepsilon\, kT}$. Using the Einstein ratio $eD = \mu kT$ between the sum of the diffusion coefficients D and the sum of the mobilities μ, one obtains $\frac{1}{n\gamma} = \frac{d}{\mu F}$, i.e., $\frac{1}{n\gamma}$ is equal to the transit time of the charge carrier. This shows that in an OLED the concentration of positive and negative charges must at least be comparable to the capacitor charge in order to ensure that charges recombine bimolecularly rather than wastefully at the electrodes. A crucial test of this condition is to check whether or not the unipolar current injected from a supposedly ohmic electrode obeys Childs's law for SCL current flow, $j = \frac{9}{8}\varepsilon\,\varepsilon_0\mu\frac{F^2}{d}$, i.e., at a given electric field the current must follow a 1/d dependence on sample thickness [106]. There should also be consistency regarding the value of the mobility inferred from a ToF experiment performed under weak injection conditions and the SCL current [187]. However, even if one would expect an electrode to be ohmic based upon the estimate for the injection barrier, ohmicity is not always granted. An example is the work on poly(9, 9′-dioctyl-fluorenyl-2,7-diyl)-*co*-(4,4′-N-(4-*sec*-butyl))diphenylamine (TBB) with ITO/PEDOT:PSS serving as a hole injecting electrode. Due to appropriate synthesis and purification procedures, the mobility can be raised to 10^{-2} $cm^2/V^{-1}s^{-1}$. Unexpectedly, the hole current is not space-charge-limited, most likely because the sheet resistance of the ITO/PEDOT:PSS electrode can become the limiting parameter relative to the bulk resistance if the mobility is high [188]. Obviously the design of an ohmic electrode with low sheet resistance, able to sustain a large SCL current, is still a challenge.

7 Summary and Conclusions

In this chapter we have tried to summarize the current understanding of charge transport in organic semiconductors. It includes a brief description of the methods to measure charge transport, an overview of pertinent experimental results, and an outline of theoretical concepts. These conceptual frameworks had been developed in the course of a most fruitful interaction between theoreticians and experimentalists. This scientific progress can also be expected to bring about technological advances. For example, in the beginning of electrophotography it was not at all clear whether or not an organic material like a polymer could ever replace inorganic photoreceptors such as selenium and arsenic selenide due to the omnipresent charge carrier trapping in the organic material. Polymers can never be purified up to the level attained in silicon or selenium. However, it was the intuition of organic chemists and electrochemists who realized that one can overcome the problem of charge carrier trapping in an organic photoreceptor used in xerography by using a transport material with low oxidation potential, so that most impurities are inactive. Nowadays, photocopying machines with organic semiconductors are present in virtually every office worldwide.

In a similar way, the role of disorder, that inevitably is present in a polymer or a molecularly doped polymer, can now be quantified due to the significant advances in our theoretical understanding that have been outlined in this chapter. Ultimately, this will allow for the design of transport materials with sufficiently large carrier mobilities. Until we arrive there, however, it is worth recalling what we do understand and which challenges remain. Today, we have a reasonably well-founded understanding of the macroscopic, i.e., ensemble-averaged process of charge transport in disordered organic solids. This applies to single-component materials as well as to multi-component materials such as doped organic semiconductors. The corresponding models were presented in Sects. 3 and 4. However, our understanding on the microscopic scale, in particular on very short time scales and over very short distances, is still incomplete. The present advances towards understanding this regime have been outlined in Sect. 5. These short-scale, short-range processes are of particular importance to solar cells.

It is a generally accepted notion that the most efficient organic cells utilize donor-acceptor blends forming internal heterojunctions as active materials. In these cells the primary step is the absorption of a photon in (usually) the donor phase. The generated exciton diffuses towards the internal interface and transfers an electron to the acceptor, thus forming an electron-hole pair (synonymously called a charge transfer (CT) state or geminate pair). The pair has to dissociate and both electron and hole must drift under the action of the built-in potential and be collected at the electrodes without suffering further geminate and non-geminate recombination. Meanwhile there is consensus that the crucial step is the escape of the electron-hole pair from its mutual coulomb potential. It is an open question, though, why in efficient organic cells the geminate recombination of the initially generated pair is greatly reduced. (1) Is there a disorder related blockade of the non-radiative decay of the pair [61], (2) is the formation of the geminate pair an ultra-fast process during which the carriers move in a high mobility state (see Sect. 5) so that the initially generated pair is only loosely bound, (3) is there a shielding of the coulomb potential due to dark dipoles existing at their internal interface [189], (4) is the dissociation process disorder assisted [190–192], and (5) does the conjugation length of the donor or acceptor play a role [193]? At the moment there are no unique answers to these questions and there is likely to continue to be no unique answers because polymers and small molecule devices should behave differently. It is highly probable, however, that microscopic charge transport, controlled by morphology, plays a crucial role, and current research points this way [194–197].

The progress reported here has been achieved through simultaneous advances in theoretical methodology and experimental techniques and, importantly, through good scientific communication between theoretically and experimentally-minded physicists and chemists. Our experimental abilities as well as computing powers and theoretical methodologies are constantly expanding. Optical experiments can be conducted in the femtosecond range, techniques such as the use of terahertz spectroscopy allow for the probing of carrier mobility at ultrafast time scales, and sophisticated computer simulations can be based on realistic models of the organic film morphology. It will be interesting to see which concepts will have been established in the future concerning charge transport on short time scales and distances, and to which advanced optoelectronic applications our established knowledge on macroscopic carrier transport has led us from its simple beginnings in the photocopying process.

References

1. Borsenberger PM, Weiss DS (1998) Organic photoreceptors for xerography. Marcel Dekker, New York
2. Friend RH, Gymer RW, Holmes AB, Burroughes JH, Marks RN, Taliani C, Bradley DDC, Dos Santos DA, Bredas JL, Logdlund M, Salaneck WR (1999) Electroluminescence in conjugated polymers. Nature 397:121
3. Forrest SR (2004) The path to ubiquitous and low-cost organic electronic appliances on plastic. Nature 428:911
4. All Articles in (2007) Special issue on organic electronics and optoelectronics. Chem Rev 107:923
5. Müllen K, Scherf U (2006) Organic light emitting devices: synthesis, properties and applications. Wiley-VCH, Weinheim
6. Yersin H (2007) Highly efficient OLEDs with phosphorescent materials. Wiley-VCH, Weinheim
7. Hertel D, Bässler H (2008) Photoconduction in amorphous organic solids. Chemphyschem 9:666
8. Chang EK, Rohlfing M, Louie SG (2000) Excitons and optical properties of alpha-quartz. Phys Rev Lett 85:2613
9. Kador L (1991) Stochastic-theory of inhomogeneous spectroscopic line-shapes reinvestigated. J Chem Phys 95:5574
10. Coropceanu V, Cornil J, da Silva DA, Olivier Y, Silbey R, Bredas JL (2007) Charge transport in organic semiconductors. Chem Rev 107:926
11. Warta W, Stehle R, Karl N (1985) Ultrapure, high mobility organic photoconductors. Appl Phys A 36:163
12. Karl N (2000) In: Madelung O, Schulz M, Weiss H (eds) Semiconductors (Landolt-Boernstein (New Series), Group III). Springer, Heidelberg, p 106
13. Karl N (2001) In: Farchioni R, Grosso G (eds) Organic electronic materials. Springer-Verlag, Berlin
14. Schwoerer M, Wolf HC (2007) Organic molecular solids. Wiley-VCH, Weinheim
15. Schott M (2006) In: Lanzani G (ed) Photophysics of molecular materials: from single molecules to single crystals. Wiley-VCH, Weinheim, p 49
16. Collini E, Scholes RD (2009) Coherent intrachain energy migration in a conjugated polymer at room temperature. Science 323:369
17. Barford W, Trembath D (2009) Exciton localization in polymers with static disorder. Phys Rev B 80:165418
18. Hoffmann ST, Bässler H, Köhler A (2010) What determines inhomogeneous broadening of electronic transitions in conjugated polymers. J Phys Chem B 114:17037
19. Bässler H (1985) In: Bloor D, Chance RR (eds) Polydiacetylenes. Martinus Nijhof, Dordrecht, The Netherlands, p 135
20. Blum T, Bässler H (1988) Reinvestigation of generation and transport of charge-carriers in crystalline polydiacetylenes. Chem Phys 123:431
21. Su WP, Schrieffer JR, Heeger AJ (1979) Solitons in polyacetylene. Phys Rev Lett 42:1698
22. Heeger AJ, Kivelson S, Schrieffer JR, Su WP (1988) Solitons in conducting polymers. Rev Mod Phys 60:781
23. Fesser K, Bishop AR, Campbell DK (1983) Optical-absorption from polarons in a model of polyacetylene. Phys Rev B 27:4804
24. Sariciftci NS (1997) Primary photoexcitations in conjugated polymers: molecular exciton versus semiconductor band model. Word Scientific, Singapore
25. Longuet-Higgins HC, Salem L (1959) The alternation of bond lengths in long conjugated chain molecules. Proc Royal Soc London A 251:172

26. Salaneck WR, Staftström S, Bredas J-L (1996) Conjugated polymer surfaces and interfaces. Electronic and chemical structure of interfaces for polymer light emitting devices. Cambridge University Press, Cambridge
27. Weiser G (1992) Stark-effect of one-dimensional Wannier excitons in polydiacetylene single-crystals. Phys Rev B 45:14076
28. Van der Horst JW, Bobbert PA, Michels MAJ, Bässler H (2001) Calculation of excitonic properties of conjugated polymers using the Bethe–Salpeter equation. J Chem Phys 114:6950
29. Albrecht U, Bässler H (1995) Efficiency of charge recombination in organic light-emitting-diodes. Chem Phys 199:207
30. Kohler BE, Woehl JC (1995) A simple-model for conjugation lengths in long polyene chains. J Chem Phys 103:6253
31. Romanovskii YV, Gerhard A, Schweitzer B, Personov RI, Bässler H (1999) Delayed luminescence of the ladder-type methyl-poly(para-phenylene). Chem Phys 249:29
32. Monkman AP, Burrows HD, Hamblett I, Navaratnam S, Scherf U, Schmitt C (2000) The triplet state of the ladder-type methyl-poly(p-phenylene) as seen by pulse radiolysis-energy transfer. Chem Phys Lett 327:111
33. Hertel D, Setayesh S, Nothofer HG, Scherf U, Müllen K, Bässler H (2001) Phosphorescence in conjugated poly(para-phenylene)-derivatives. Adv Mater 13:65
34. Köhler A, Wilson JS, Friend RH, Al-Suti MK, Khan MS, Gerhard A, Bässler H (2002) The singlet-triplet energy gap in organic and Pt-containing phenylene ethynylene polymers and monomers. J Chem Phys 116:9457
35. Köhler A, Beljonne D (2004) The singlet-triplet exchange energy in conjugated polymers. Adv Funct Mater 14:11
36. Hertel D, Bässler H, Scherf U, Hörhold HH (1999) Charge carrier transport in conjugated polymers. J Chem Phys 110:9214
37. Deussen M, Bässler H (1993) Anion and cation absorption-spectra of conjugated oligomers and polymer. Synth Met 54:49
38. Bäuerle P, Segelbacher U, Maier A, Mehring M (1993) Electronic-structure of monomeric and dimeric cation radicals in end-capped oligothiophenes. J Am Chem Soc 115:10217
39. Nöll G, Lambert C, Lynch M, Porsch M, Daub J (2008) Electronic structure and properties of poly- and oligoazulenes. J Phys Chem C 112:2156
40. Osterholm A, Petr A, Kvarnstrm C, Ivaska A, Dunsch L (2008) The nature of the charge carriers in polyazulene as studied by in situ electron spin resonance UV–visible–near-infrared spectroscopy. J Phys Chem B 112:14149
41. van Haare JAEH, Havinga EE, van Dongen JLJ, Janssen RAJ, Cornil J, Bredas JL (1998) Redox states of long oligothiophenes: two polarons on a single chain. Chem Eur J 4:1509
42. Kadashchuk A, Arkhipov VI, Kim C-H, Shinar J, Lee D-W, Hong Y-R, Jin J-I, Heremans P, Bässler H (2007) Localized trions in conjugated polymers. Phys Rev B 76:235205
43. Furukawa Y (1996) Electronic absorption and vibrational spectroscopies of conjugated conducting polymers. J Phys Chem 100:15644
44. Sakamoto A, Nakamura O, Tasumi M (2008) Picosecond time-resolved polarized infrared spectroscopic study of photoexcited states and their dynamics in oriented poly(p-phenylene-vinylene). J Phys Chem B 112:16437
45. Pope M, Swenberg CE (1999) Electronic processes in organic crystals and polymers. Oxford University Press, Oxford
46. Arkhipov VI, Fishchuk II, Kadashchuk A, Bässler H (2007) In: Hadziioannou G, Malliaras GG (eds) Semiconducting polymers: chemistry, physics, engineering, vol 1. Wiley-VCH, Weinheim
47. Borsenberger PM, Pautmeier L, Bässler H (1991) Hole transport in bis(4-N, N-diethylamino-2-methylphenyl)-4-methylphenylmethane. J Chem Phys 95:1258
48. Markham JPJ, Anthopoulos TD, Samuel IDW, Richards GJ, Burn PL, Im C, Bässler H (2002) Nondispersive hole transport in a spin-coated dendrimer film measured by the charge-generation-layer time-of-flight method. Appl Phys Lett 81:3266

49. Gambino S, Samuel IDW, Barcena H, Burn PL (2008) Electric field and temperature dependence of the hole mobility in a bis-fluorene cored dendrimer. Org Electron 9:220
50. Klenkler RA, Xu G, Aziz H, Popovic ZD (2006) Charge-carrier mobility in an organic semiconductor thin film measured by photoinduced electroluminescence. Appl Phys Lett 88:242101
51. Bange S, Kuksov A, Neher D (2007) Sensing electron transport in a blue-emitting copolymer by transient electroluminescence. Appl Phys Lett 91:143516
52. Juska G, Arlauskas K, Viliunas M, Kocka J (2000) Extraction current transients: new method of study of charge transport in microcrystalline silicon. Phys Rev Lett 84:4946
53. Bange S, Schubert M, Neher D (2010) Charge mobility determination by current extraction under linear increasing voltages: case of nonequilibrium charges and field-dependent mobilities. Phys Rev B 81:035209
54. Bässler H (1993) Charge transport in disordered organic photoconductors – a Monte-Carlo simulation study. Phys Status Solidi B 175:15
55. Miller A, Abrahams E (1960) Impurity conduction at low concentrations. Phys Rev 120:745
56. Arkhipov VI, Emelianova EV, Adriaenssens GJ (2001) Effective transport energy versus the energy of most probable jumps in disordered hopping systems. Phys Rev B 6412:125125
57. Pautmeier L, Richert R, Bässler H (1991) Anomalous time-independent diffusion of charge-carriers in a random potential under a bias field. Phil Mag B 63:587
58. Richert R, Pautmeier L, Bässler H (1989) Diffusion and drift of charge-carriers in a random potential – deviation from Einstein law. Phys Rev Lett 63:547
59. Roichman Y, Tessler N (2002) Generalized Einstein relation for disordered semiconductors – implications for device performance. Appl Phys Lett 80:1948
60. Tal O, Epstein I, Snir O, Roichman Y, Ganot Y, Chan CK, Kahn A, Tessler N, Rosenwaks Y (2008) Measurements of the Einstein relation in doped and undoped molecular thin films. Phys Rev B 77:201201
61. Tessler N, Preezant Y, Rappaport N, Roichman Y (2009) Charge transport in disordered organic materials and its relevance to thin-film devices: a tutorial review. Adv Mater 21:2741
62. Borsenberger PM, Pautmeier LT, Bässler H (1993) Scaling behavior of nondispersive charge-transport in disordered molecular-solids. Phys Rev B 48:3066
63. Fishchuk II, Kadashchuk A, Bässler H, Abkowitz M (2004) Low-field charge-carrier hopping transport in energetically and positionally disordered organic materials. Phys Rev B 70:245212
64. Movaghar B, Grünewald M, Ries B, Bässler H, Wurtz D (1986) Diffusion and relaxation of energy in disordered organic and inorganic materials. Phys Rev B 33:5545
65. Gartstein YN, Conwell EM (1994) High-field hopping mobility of polarons in disordered molecular-solids – a Monte-Carlo study. Chem Phys Lett 217:41
66. Dunlap DH, Parris PE, Kenkre VM (1996) Charge-dipole model for the universal field dependence of mobilities in molecularly doped polymers. Phys Rev Lett 77:542
67. Cordes H, Baranovskii SD, Kohary K, Thomas P, Yamasaki S, Hensel F, Wendorff JH (2001) One-dimensional hopping transport in disordered organic solids. I. Analytic calculations. Phys Rev B 63:094201
68. Hirao A, Nishizawa H, Sugiuchi M (1995) Diffusion and drift of charge carriers in molecularly doped polymers. Phys Rev Lett 75:1787
69. Schein LB, Glatz D, Scott JC (1990) Observation of the transition from adiabatic to nonadiabatic small polaron hopping in a molecularly doped polymer. Phys Rev Lett 65:472
70. Fishchuk II, Kadashchuk A, Bässler H, Nespurek S (2003) Nondispersive polaron transport in disordered organic solids. Phys Rev B 67:224303
71. Parris PE, Kenkre VM, Dunlap DH (2001) Nature of charge carriers in disordered molecular solids: are polarons compatible with observations? Phys Rev Lett 87:126601
72. Schein LB, Tyutnev A (2008) The contribution of energetic disorder to charge transport in molecularly doped polymers. J Phys Chem C 112:7295

73. Borsenberger PM, Bässler H (1991) Concerning the role of dipolar disorder on charge transport in molecularly doped polymers. J Chem Phys 95:5327
74. Scher H, Montroll EW (1975) Anomalous transit-time dispersion in amorphous solids. Phys Rev B 12:2455
75. Arkhipov VI, Iovu MS, Rudenko AI, Shutov SD (1979) Analysis of the dispersive charge transport in vitreous 0.55 As_2S_3-0.45 Sb_2S_3. Physica Status Solidi (a) 54:67
76. Martens HCF, Blom PWM, Schoo HFM (2000) Comparative study of hole transport in poly(*p*-phenylene vinylene) derivatives. Phys Rev B 61:7489
77. Young RH (1994) Trap-free space-charge-limited current – analytical solution for an arbitrary mobility law. Phil Mag Lett 70:331
78. Abkowitz M, Pai DM (1986) Comparison of the drift mobility measured under transient and steady-state conditions in a prototypical hopping system. Phil Mag B 53:193
79. Young RH (1994) A law of corresponding states for hopping transport in disordered materials. Phil Mag B 69:577
80. Laquai F, Wegner G, Im C, Bässler H, Heun S (2006) Nondispersive hole transport in carbazole- and anthracene-containing polyspirobifluorene copolymers studied by the charge-generation layer time-of-flight technique. J Appl Phys 99:033710
81. Laquai F, Wegner G, Im C, Bässler H, Heun S (2006) Comparative study of hole transport in polyspirobifluorene polymers measured by the charge-generation layer time-of-flight technique. J Appl Phys 99:023712
82. Bao Z, Dodabalapur A, Lovinger AJ (1996) Soluble and processable regioregular poly (3-hexylthiophene) for thin film field-effect transistor applications with high mobility. Appl Phys Lett 69:4108
83. Sirringhaus H, Brown PJ, Friend RH, Nielsen MM, Bechgaard K, Langeveld-Voss BMW, Spiering AJH, Janssen RAJ, Meijer EW, Herwig P, de Leeuw DM (1999) Two-dimensional charge transport in self-organized, high-mobility conjugated polymers. Nature 401:685
84. Brown PJ, Sirringhaus H, Harrison M, Shkunov M, Friend RH (2001) Optical spectroscopy of field-induced charge in self-organized high mobility poly(3-hexylthiophene). Phys Rev B 63:125204
85. Brown PJ, Thomas DS, Köhler A, Wilson JS, Kim J-S, Ramsdale CM, Sirringhaus H, Friend RH (2003) Effect of interchain interactions on the absorption and emission of poly (3-hexylthiophene). Phys Rev B 67:064203
86. Khan RUA, Poplavskyy D, Kreouzis T, Bradley DDC (2007) Hole mobility within arylamine-containing polyfluorene copolymers: a time-of-flight transient-photocurrent study. Phys Rev B 75:035215
87. Kreouzis T, Poplavskyy D, Tuladhar SM, Campoy-Quiles M, Nelson J, Campbell AJ, Bradley DDC (2006) Temperature and field dependence of hole mobility in poly (9,9-dioctylfluorene). Phys Rev B 73:235201
88. Scherf U, List EJW (2002) Semiconducting polyfluorenes – towards reliable structure-–property relationships. Adv Mater 14:477
89. Khan ALT, Sreearunothai P, Herz LM, Banach MJ, Köhler A (2004) Morphology-dependent energy transfer within polyfluorene thin films. Phys Rev B 69:085201
90. Van Mensfoort SLM, Vulto SIE, Janssen RAJ, Coehoorn R (2008) Hole transport in polyfluorene-based sandwich-type devices: quantitative analysis of the role of energetic disorder. Phys Rev B 78:085208
91. Mozer AJ, Sariciftci NS, Pivrikas A, Österbacka R, Juska G, Brassat L, Bässler H (2005) Charge carrier mobility in regioregular poly(3-hexylthiophene) probed by transient conductivity techniques: a comparative study. Phys Rev B 71:035214
92. Meisel KD, Vocks H, Bobbert PA (2005) Polarons in semiconducting polymers: study within an extended Holstein model. Phys Rev B 71:205206
93. Mohan SR, Joshi MP, Singh MP (2009) Negative electric field dependence of mobility in TPD doped polystyrene. Chem Phys Lett 470:279

94. Tanase C, Wildeman J, Blom PWM, Mena Benito ME, de Leeuw DM, van Breemen AJJM, Herwig PT, Chlon CHT, Sweelssen J, Schoo HFM (2005) Optimization of the charge transport in poly(phenylene vinylene) derivatives by processing and chemical modification. J Appl Phys 97:123703
95. Van Mensfoort SLM, Coehoorn R (2008) Effect of Gaussian disorder on the voltage dependence of the current density in sandwich-type devices based on organic semiconductors. Phys Rev B 78:085207
96. Pasveer WF, Cottaar J, Tanase C, Coehoorn R, Bobbert PA, Blom PWM, de Leeuw DM, Michels MAJ (2005) Unified description of charge-carrier mobilities in disordered semiconducting polymers. Phys Rev Lett 94:206601
97. Arkhipov VI, Heremans P, Emelianova EV, Adriaenssens GJ, Bässler H (2002) Weak-field carrier hopping in disordered organic semiconductors: the effects of deep traps and partly filled density-of-states distribution. J Phys Condens Matter 14:9899
98. Coehoorn R, Pasveer WF, Bobbert PA, Michels MAJ (2005) Charge-carrier concentration dependence of the hopping mobility in organic materials with Gaussian disorder. Phys Rev B 72:155206
99. Coehoorn R (2007) Hopping mobility of charge carriers in disordered organic host-guest systems: dependence on the charge-carrier concentration. Phys Rev B 75:155203
100. Fishchuk II, Arkhipov VI, Kadashchuk A, Heremans P, Bässler H (2007) Analytic model of hopping mobility at large charge carrier concentrations in disordered organic semiconductors: polarons versus bare charge carriers. Phys Rev B 76:045210
101. Arkhipov VI, Emelianova EV, Heremans P, Bässler H (2005) Analytic model of carrier mobility in doped disordered organic semiconductors. Phys Rev B 72:235202
102. Fishchuk II, Kadashchuk AK, Genoe J, Ullah M, Sitter H, Singh TB, Sariciftci NS, Bässler H (2010) Temperature dependence of the charge carrier mobility in disordered organic semiconductors at large carrier concentrations. Phys Rev B 81:045202
103. Emin D (2008) Generalized adiabatic polaron hopping: Meyer-Neldel compensation and Poole-Frenkel behavior. Phys Rev Lett 100:166602
104. Craciun NI, Wildeman J, Blom PWM (2008) Universal Arrhenius temperature activated charge transport in diodes from disordered organic semiconductors. Phys Rev Lett 100:056601
105. Blakesley JC, Clubb HS, Greenham NC (2010) Temperature-dependent electron and hole transport in disordered semiconducting polymers: analysis of energetic disorder. Phys Rev B 81:045210
106. Agrawal R, Kumar P, Ghosh S, Mahapatro AK (2008) Thickness dependence of space charge limited current and injection limited current in organic molecular semiconductors. Appl Phys Lett 93:073311
107. Walzer K, Maennig B, Pfeiffer M, Leo K (2007) Highly efficient organic devices based on electrically doped transport layers. Chem Rev 107:1233
108. Chiang CK, Fincher CR, Park JYW, Heeger AJ, Shirakawa H, Louis EJ, Gau SC, MacDiarmid AG (1977) Electrical conductivity in doped polyacetylene. Phys Rev Lett 39:1098
109. Heeger AJ (1989) Charge transfer in conducting polymers. Striving toward intrinsic properties. Faraday Discuss Chem Soc 88:203
110. Pfeiffer M, Fritz T, Blochwitz J, Nollau A, Plönnigs B, Beyer A, Leo K (1999) Controlled doping of molecular organic layers: physics and device prospects. Adv Sol State Phys 39:77
111. Gao W, Kahn A (2001) Controlled p-doping of zinc phthalocyanine by coevaporation with tetrafluorotetracyanoquinodimethane: a direct and inverse photoemission study. Appl Phys Lett 79:4040
112. Matsushima T, Adachi C (2008) Enhancing hole transports and generating hole traps by doping organic hole-transport layers with p-type molecules of 2,3,5,6-tetrafluoro-7,7,8,8-tetracyanoquinodimethane. Thin Solid Films 517:874

113. Lee J-H, Leem D-S, Kim J-J (2010) Effect of host organic semiconductors on electrical doping. Org Electron 11:486
114. Zhang Y, de Boer B, Blom PWM (2009) Controllable molecular doping and charge transport in solution-processed polymer semiconducting layers. Adv Funct Mater 19:1901
115. Maennig B, Pfeiffer M, Nollau A, Zhou X, Leo K, Simon P (2001) Controlled p-type doping of polycrystalline and amorphous organic layers: self-consistent description of conductivity and field-effect mobility by a microscopic percolation model. Phys Rev B 64:195208
116. Kido J, Nagai K, Okamoto Y (1993) Bright organic electroluminescent devices with double-layer cathode. IEEE Trans Electron Devices 40:1342
117. Nollau A, Pfeiffer M, Fritz T, Leo K (2000) Controlled n-type doping of a molecular organic semiconductor: naphthalenetetracarboxylic dianhydride (NTCDA) doped with bis (ethylenedithio)-tetrathiafulvalene (BEDT-TTF). J Appl Phys 87:4340
118. Tanaka S, Kanai K, Kawabe E, Iwahashi T, Nishi T, Ouchi Y, Seki K (2005) Doping effect of tetrathianaphthacene molecule in organic semiconductors on their interfacial electronic structures studied by UV photoemission spectroscopy. Jpn J Appl Phys 44:3760
119. Chan CK, Amy F, Zhang Q, Barlow S, Marder S, Kahn A (2006) N-Type doping of an electron-transport material by controlled gas-phase incorporation of cobaltocene. Chem Phys Lett 431:67
120. Werner AG, Li F, Harada K, Pfeiffer M, Fritz T, Leo K (2003) Pyronin B as a donor for n-type doping of organic thin films. Appl Phys Lett 82:4495
121. Werner A, Li F, Harada K, Pfeiffer M, Fritz T, Leo K, Machill S (2004) n-Type doping of organic thin films using cationic dyes. Adv Funct Mater 14:255
122. Hulea IN, Brom HB, Houtepen AJ, Vanmaekelbergh D, Kelly JJ, Meulenkamp EA (2004) Wide energy-window view on the density of states and hole mobility in poly(p-phenylene vinylene). Phys Rev Lett 93:166601
123. Tanase C, Meijer EJ, Blom PWM, de Leeuw DM (2003) Unification of the hole transport in polymeric field-effect transistors and light-emitting diodes. Phys Rev Lett 91:216601
124. Silver M, Pautmeier L, Bässler H (1989) On the origin of exponential band tails in amorphous-semiconductors. Solid State Commun 72:177
125. Arkhipov VI, Heremans P, Emelianova EV, Bässler H (2005) Effect of doping on the density-of-states distribution and carrier hopping in disordered organic semiconductors. Phys Rev B 71:045214
126. Zhou J, Zhou YC, Zhao JM, Wu CQ, Ding XM, Hou XY (2007) Carrier density dependence of mobility in organic solids: a Monte Carlo simulation. Phys Rev B 75:153201
127. Van der Holst JJM, van Oost FWA, Coehoorn R, Bobbert PA (2009) Electron-hole recombination in disordered organic semiconductors: validity of the Langevin formula. Phys Rev B 80:235202
128. Prigodin VN, Hsu FC, Park JH, Waldmann O, Epstein AJ (2008) Electron-ion interaction in doped conducting polymers. Phys Rev B 78:035203
129. Lee HJ, Lee J, Park S-M (2010) Electrochemistry of conductive polymers. 45. Nanoscale conductivity of PEDOT and PEDOT:PSS composite films studied by current-sensing AFM. J Phys Chem B 114:2660
130. Jung JW, Lee JU, Jo WH (2009) High-efficiency polymer solar cells with water-soluble and self-doped conducting polyaniline graft copolymer as hole transport layer. J Phys Chem C 114:633
131. Banerjee S, Kumar A (2010) Dielectric behavior and charge transport in polyaniline nanofiber reinforced PMMA composites. J Phys Chem Solids 71:381
132. Li D, Huang J, Kaner RB (2009) Polyaniline nanofibers: a unique polymer nanostructure for versatile applications. Acc Chem Res 42:135
133. Lee K, Miller EK, Aleshin AN, Menon R, Heeger AJ, Kim JH, Yoon CO, Lee H (1998) Nature of the metallic state in conducting polypyrrole. Adv Mater 10:456
134. Hulea IN, Brom HB, Mukherjee AK, Menon R (2005) Doping, density of states, and conductivity in polypyrrole and poly(p-phenylene vinylene). Phys Rev B 72:054208

135. Mott NF (1969) Conduction in non-crystalline materials. 3. Localized states in a pseudogap and near extremities of conduction and valence bands. Phil Mag 19:835
136. Weiser G, Möller S (2002) Directional dispersion of the optical resonance of π-π* transitions of alpha-sexithiophene single crystals. Phys Rev B 65:045203
137. Zuilhof H, Barentsen HM, van Dijk M, Sudhölter EJR, Hoofman RJOM, Siebbeles LDA, de Haas MP, Warman JM (2001) In: Nalwa HS (ed) Supramolecular photosensitive and electroactive materials. Elsevier, San Diego
138. Milota F, Sperling J, Szöcs V, Tortschanoff A, Kauffmann HF (2004) Correlation of femtosecond wave packets and fluorescence interference in a conjugated polymer: towards the measurement of site homogeneous dephasing. J Chem Phys 120:9870
139. Dykstra TE, Kovalevskij V, Yang X, Scholes GD (2005) Excited state dynamics of a conformationally disordered conjugated polymer: a comparison of solutions and film. Chem Phys 318:21
140. Gelinck GH, Warman JM (1996) Charge carrier dynamics in pulse-irradiated polyphenylenevinylenes: effects of broken conjugation, temperature, and accumulated dose. J Phys Chem 100:20035
141. Prins P, Candeias LP, van Breemen AJJM, Sweelssen J, Herwig PT, Schoo HFM, Siebbeles LDA (2005) Electron and hole dynamics on isolated chains of a solution-processable poly (thienylenevinylene) derivative in dilute solution. Adv Mater 17:718
142. Huitema HEA, Gelinck GH, van der Putten JBPH, Kuijk KE, Hart CM, Cantatore E, de Leeuw DM (2002) Active-matrix displays driven by solution-processed polymeric transistors. Adv Mater 14:1201
143. Grozema FC, van Duijnen PT, Berlin YA, Ratner MA, Siebbeles LDA (2002) Intramolecular charge transport along isolated chains of conjugated polymers: effect of torsional disorder and polymerization defects. J Phys Chem B 106:7791
144. Warman JM, de Haas MP, Dicker G, Grozema FC, Piris J, Debije MG (2004) Charge mobilities in organic semiconducting materials determined by pulse-radiolysis time-resolved microwave conductivity: π-bond-conjugated polymers versus π−π-stacked discotics. Chem Mater 16:4600
145. Adam D, Schuhmacher P, Simmerer J, Häussling L, Siemensmeyer K, Etzbach KH, Ringsdorf H, Haarer D (1994) Fast photoconduction in the highly ordered columnar phase of a discotic liquid-crystal. Nature 371:141
146. Pingel P, Zen A, Abellon RD, Grozema FC, Siebbeles LDA, Neher D (2010) Temperature-resolved local and macroscopic charge carrier transport in thin P3HT layers. Adv Funct Mater 20:2286
147. Joshi S, Grigorian S, Pietsch U, Pingel P, Zen A, Neher D, Scherf U (2008) Thickness dependence of the crystalline structure and hole mobility in thin films of low molecular weight poly(3-hexylthiophene). Macromolecules 41:6800
148. Prins P, Grozema FC, Siebbeles LDA (2006) Efficient charge transport along phenylene-vinylene molecular wires. J Phys Chem B 110:14659
149. Prins P, Grozema FC, Schins JM, Savenije TJ, Patil S, Scherf U, Siebbeles LDA (2006) Effect of intermolecular disorder on the intrachain charge transport in ladder-type poly (p-phenylenes). Phys Rev B 73:045204
150. Prins P, Grozema FC, Schins JM, Patil S, Scherf U, Siebbeles LDA (2006) High intrachain hole mobility on molecular wires of ladder-type poly(*p*-phenylenes). Phys Rev Lett 96:146601
151. Parkinson P, Joyce HJ, Gao Q, Tan HH, Zhang X, Zou J, Jagadish C, Herz LM, Johnston MB (2009) Carrier lifetime and mobility enhancement in nearly defect-free core-shell nanowires measured using time-resolved terahertz spectroscopy. Nano Lett 9:3349
152. Parkinson P, Lloyd-Hughes J, Johnston MB, Herz LM (2008) Efficient generation of charges via below-gap photoexcitation of polymer-fullerene blend films investigated by terahertz spectroscopy. Phys Rev B 78:115321

153. Němec H, Nienhuys H-K, Perzon E, Zhang F, Inganäs O, Kužel P, Sundström V (2009) Ultrafast conductivity in a low-band-gap polyphenylene and fullerene blend studied by terahertz spectroscopy. Phys Rev B 79:245326
154. Devižis A, Serbenta A, Meerholz K, Hertel D, Gulbinas V (2009) Ultrafast dynamics of carrier mobility in a conjugated polymer probed at molecular and microscopic length scales. Phys Rev Lett 103:027404
155. Warta W, Karl N (1985) Hot holes in naphthalene: high, electric-field-dependent mobilities. Phys Rev B 32:1172
156. Silinsh EA, Capek V (1994) Organic molecular crystals. Interaction, localization and transport properties. American Institute of Physics, New York
157. Fratini S, Ciuchi S (2009) Bandlike motion and mobility saturation in organic molecular semiconductors. Phys Rev Lett 103:266601
158. Hasegawa T, Takeya J (2009) Organic field-effect transistors using single crystals. Science Tech Adv Mater 10:024314
159. Podzorov V, Menard E, Borissov A, Kiryukhin V, Rogers JA, Gershenson ME (2004) Intrinsic charge transport on the surface of organic semiconductors. Phys Rev Lett 93:086602
160. Zeis R, Besnard C, Siegrist T, Schlockermann C, Chi X, Kloc C (2006) Field effect studies on rubrene and impurities of rubrene. Chem Mater 18:244
161. Wang L, Fine D, Basu D, Dodabalapur A (2007) Electric-field-dependent charge transport in organic thin-film transistors. J Appl Phys 101:054515
162. Jurchescu OD, Baas J, Palstra TTM (2004) Effect of impurities on the mobility of single crystal pentacene. Appl Phys Lett 84:3061
163. Koch N, Vollmer A, Salzmann I, Nickel B, Weiss H, Rabe JP (2006) Evidence for temperature-dependent electron band dispersion in pentacene. Phys Rev Lett 96:156803
164. Minari T, Nemoto T, Isoda S (2006) Temperature and electric-field dependence of the mobility of a single-grain pentacene field-effect transistor. J Appl Phys 99:034506
165. Hamadani BH, Richter CA, Gundlach DJ, Kline RJ, McCulloch I, Heeney M (2007) Influence of source-drain electric field on mobility and charge transport in organic field-effect transistors. J Appl Phys 102:044503
166. Hallam T, Lee M, Zhao N, Nandhakumar I, Kemerink M, Heeney M, McCulloch I, Sirringhaus H (2009) Local charge trapping in conjugated polymers resolved by scanning Kelvin probe microscopy. Phys Rev Lett 103:256803
167. Sakanoue T, Sirringhaus H (2010) Band-like temperature dependence of mobility in a solution-processed organic semiconductor. Nat Mater 9:736
168. Yuen JD, Menon R, Coates NE, Namdas EB, Cho S, Hannahs ST, Moses D, Heeger AJ (2009) Nonlinear transport in semiconducting polymers at high carrier densities. Nat Mater 8:572
169. Wolf U, Arkhipov VI, Bässler H (1999) Current injection from a metal to a disordered hopping system. I. Monte Carlo simulation. Phys Rev B 59:7507
170. Arkhipov VI, Wolf U, Bässler H (1999) Current injection from a metal to a disordered hopping system. II. Comparison between analytic theory and simulation. Phys Rev B 59:7514
171. Gartstein YN, Conwell EM (1995) High-field hopping mobility in molecular systems with spatially correlated energetic disorder. Chem Phys Lett 245:351
172. Akuetey G, Hirsch J (1991) Contact-injected currents in polyvinylcarbazole. Phil Mag B 63:389
173. Van Woudenbergh T, Blom PWM, Vissenberg MCJM, Huiberts JN (2001) Temperature dependence of the charge injection in poly-dialkoxy-*p*-phenylene vinylene. Appl Phys Lett 79:1697
174. Hosseini AR, Wong MH, Shen Y, Malliaras GG (2005) Charge injection in doped organic semiconductors. J Appl Phys 97:023705
175. Burin AL, Ratner MA (2000) Temperature and field dependence of the charge injection from metal electrodes into random organic media. J Chem Phys 113:3941

176. Ishii H, Sugiyama K, Ito E, Seki K (1999) Energy level alignment and interfacial electronic structures at organic/metal and organic/organic interfaces. Adv Mater 11:605
177. Koch N, Elschner A, Johnson RL, Rabe JP (2005) Energy level alignment at interfaces with pentacene: metals versus conducting polymers. Appl Surf Sci 244:593
178. Blakesley JC, Greenham NC (2009) Charge transfer at polymer-electrode interfaces: the effect of energetic disorder and thermal injection on band bending and open-circuit voltage. J Appl Phys 106:034507
179. Steyrleuthner R, Schubert M, Jaiser F, Blakesley JC, Chen Z, Facchetti A, Neher D (2010) Bulk electron transport and charge injection in a high mobility n-type semiconducting polymer. Adv Mater 22:2799
180. Davids PS, Campbell IH, Smith DL (1997) Device model for single carrier organic diodes. J Appl Phys 82:6319
181. Wolf U, Barth S, Bässler H (1999) Electrode versus space-charge-limited conduction in organic light-emitting diodes. Appl Phys Lett 75:2035
182. Koo Y-M, Choi S-J, Chu T-Y, Song O-K, Shin W-J, Lee J-Y, Kim JC, Yoon T-H (2008) Ohmic contact probed by dark injection space-charge-limited current measurements. J Appl Phys 104:123707
183. Wang ZB, Helander MG, Greiner MT, Qiu J, Lu ZH (2009) Energy-level alignment and charge injection at metal/C_{60}/organic interfaces. Appl Phys Lett 95:043302
184. Fehse K, Olthof S, Walzer K, Leo K, Johnson RL, Glowatzki H, Broker B, Koch N (2007) Energy level alignment of electrically doped hole transport layers with transparent and conductive indium tin oxide and polymer anodes. J Appl Phys 102:073719
185. Cheng X, Noh Y-Y, Wang J, Tello M, Frisch J, Blum R-P, Vollmer A, Rabe JP, Koch N, Sirringhaus H (2009) Controlling electron and hole charge injection in ambipolar organic field-effect transistors by self-assembled monolayers. Adv Funct Mater 19:2407
186. Wolf U, Bässler H (1999) Enhanced electron injection into light-emitting diodes via interfacial tunneling. Appl Phys Lett 74:3848
187. Abkowitz M, Facci JS, Stolka M (1993) Time-resolved space charge-limited injection in a trap-free glassy polymer. Chem Phys 177:783
188. Fong HH, Papadimitratos A, Hwang J, Kahn A, Malliaras GG (2009) Hole injection in a model fluorene–triarylamine copolymer. Adv Funct Mater 19:304
189. Arkhipov VI, Heremans P, Bässler H (2003) Why is exciton dissociation so efficient at the interface between a conjugated polymer and an electron acceptor? Appl Phys Lett 82:4605
190. Albrecht U, Bässler H (1995) Yield of geminate pair dissociation in an energetically random hopping system. Chem Phys Lett 235:389
191. Emelianova EV, van der Auweraer M, Bässler H (2008) Hopping approach towards exciton dissociation in conjugated polymers. J Chem Phys 128:224709
192. Rubel O, Baranovskii SD, Stolz W, Gebhard F (2008) Exact solution for hopping dissociation of geminate electron-hole pairs in a disordered chain. Phys Rev Lett 100:196602
193. Deibel C, Strobel T, Dyakonov V (2009) Origin of the efficient polaron-pair dissociation in polymer-fullerene blends. Phys Rev Lett 103:036402
194. Bredas JL, Norton JE, Cornil J, Coropceanu V (2009) Molecular understanding of organic solar cells: the challenges. Acc Chem Res 42:1691
195. Marsh RA, Hodgkiss JM, Friend RH (2010) Direct measurement of electric field-assisted charge separation in polymer: fullerene photovoltaic diodes. Adv Mater 22:3672
196. Hodgkiss JM, Campbell AR, Marsh RA, Rao A, Albert-Seifried S, Friend RH (2010) Subnanosecond geminate charge recombination in polymer-polymer photovoltaic devices. Phys Rev Lett 104:177701
197. Schubert M, Yin CH, Castellani M, Bange S, Tam TL, Sellinger A, Horhold HH, Kietzke T, Neher D (2009) Heterojunction topology versus fill factor correlations in novel hybrid small-molecular/polymeric solar cells. J Chem Phys 130:094703

Top Curr Chem (2012) 312: 67–126
DOI: 10.1007/128_2011_224

Published online: 28 September 2011

Frontiers of Organic Conductors and Superconductors

Gunzi Saito and Yukihiro Yoshida

Abstract We review the development of conductive organic molecular assemblies including organic metals, superconductors, single component conductors, conductive films, conductors with a switching function, and new spin state (quantum spin liquid state). We emphasize the importance of the ionicity phase diagram for a variety of charge transfer systems to provide a strategy for the development of functional organic solids (Mott insulator, semiconductor, superconductor, metal, complex isomer, neutral-ionic system, alignment of chemical potentials, etc.). For organic (super) conductors, the electronic dimensionality of the solids is a key parameter and can be designed based on the self-aggregation ability of a molecule. We present characteristic structural and physical properties of organic superconductors.

Keywords Charge transfer solid · Electronic dimensionality · Functional organic solid · Ionicity diagram · Organic metal · Organic superconductor · Phase transition · Quantum spin liquid state · Switching

Contents

G. Saito (✉) and Y. Yoshida (✉)
Research Institute, Meijo University, Shiogamaguchi 1-501 Tempaku-ku, Nagoya 468-8502, Japan
e-mail: gsaito@meijo-u.ac.jp; yyoshida@meijo-u.ac.jp

1 Introduction

In this review we introduce the present status of conducting organic assemblies including single component conductors, organic metals and superconductors of the charge transfer type, exotic conductors having switching function, and new spin state (quantum spin liquid state) that neighbors a superconducting phase, with a focus on our achievements. We also briefly describe metallic and superconducting polymers. Ionic conduction, including proton conduction, is one of the essential transport phenomena in organic materials and has already been extensively reviewed [1] and therefore will not be discussed further in the present review.

2 Single Component Conductors

Since the discovery of the first metallic charge transfer (CT) solid, TTF·TCNQ (Scheme 1) by Ferraris, Cowan, et al. in 1973 [2], much attention has been devoted to organic (super)conductors by studies of several component CT solids based on TTF and TCNQ derivatives [3–18]. Besides numerous studies on multicomponent CT solids, several single-component organic conductors have been demonstrated based on (1) closed shell organic solids under high pressure [19–23] and those having peripheral chalcogen atoms [24, 25] or long alkyl chain (fastener effect) [26–29] functionalized TTF derivatives (Scheme 2), (2) organic neutral π-radicals (Scheme 3) [30–37], (3) betainic (zwitterionic) radicals of TCNQ [38–40] and TTF derivatives [41–45] (Scheme 4), and (4) transition metal complexes of phthalocyanine (Pc) [46, 47] and TTF-dithiolate [48, 49] ligands (Scheme 5). Among them, metallic behavior has been reported on three kinds of materials: closed shell organic solids under extremely high pressure [19–23], transition metal complexes of

S S S S TTF NC NC CN CN TCNQ

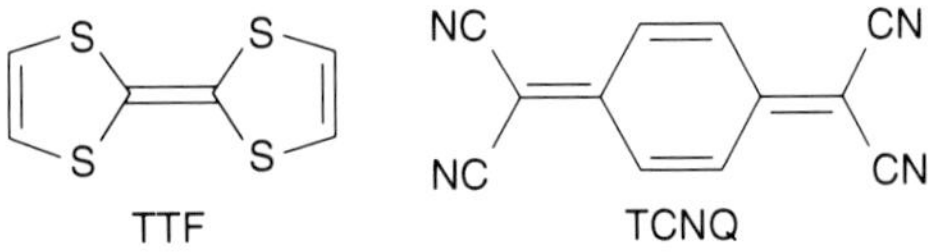

Scheme 1 Chemical structures of TTF and TCNQ

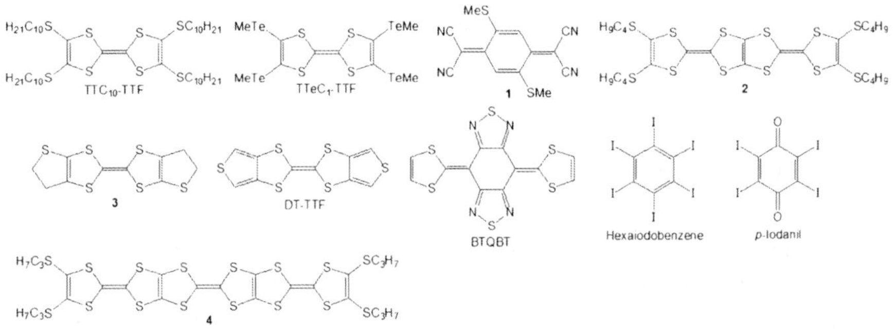

Scheme 2 Chemical structures of closed shell molecules of organic conductors

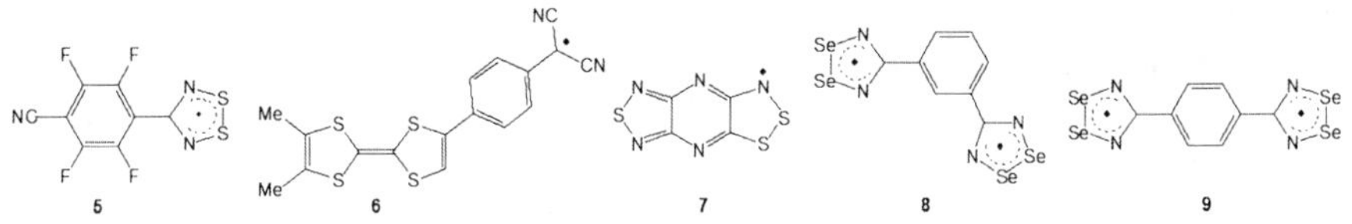

Scheme 3 Chemical structures of neutral π-radical and biradical molecules of organic conductors

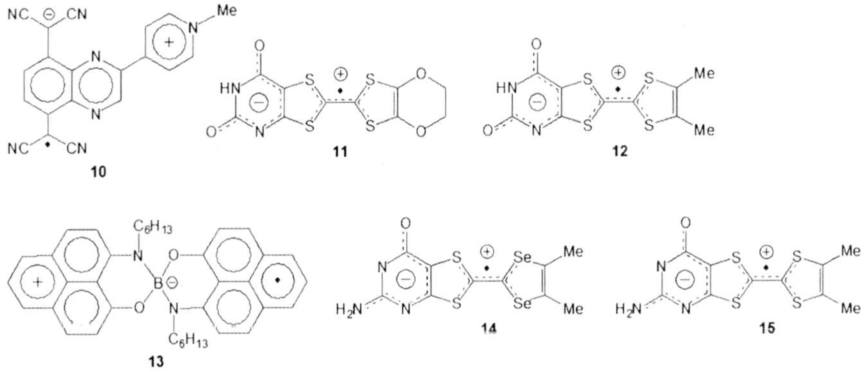

Scheme 4 Chemical structures of betainic π-radical molecules of organic conductors

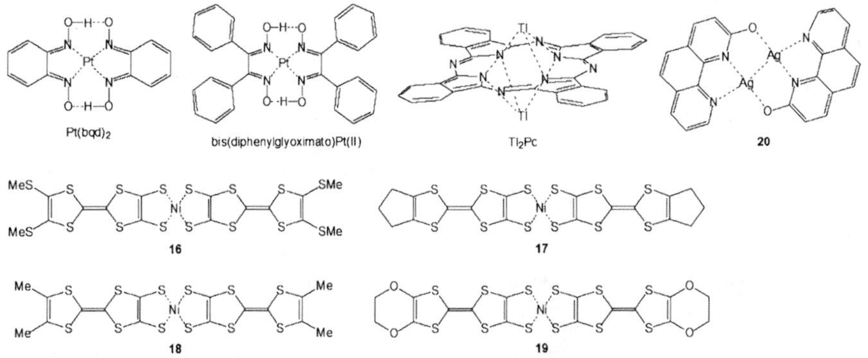

Scheme 5 Chemical structures of transition metal complexes of organic conductors

phthalocyanines [46], and transition metal complexes of TTF-dithiolate ligands [48, 49]. Superconductivity was observed in closed shell organic solids under high pressure [20, 21, 23].

2.1 Closed Shell Neutral Solids

Even though pentacene is known to be the first organic metal (semimetal) showing a decrease of resistivity down to ca. 200 K at 21.3 GPa [19], no superconductivity has been reported so far for solids composed only of aromatic hydrocarbons. Though Schön et al. reported superconductivity on polyacenes (anthracene, tetracene, and pentacene) in field effect transistor (FET) devices at low temperatures (T_c < 4 K), the paper was retracted [50]. Electric conductivity increases by the enhancement of intermolecular interactions by appropriate use of hetero-atomic contacts [24, 25, 51–56] to exert an atomic-wire effect and by peripheral addition of alkyl or alkylchalcogen groups [26–29, 53, 56, 57] to exert a fastener effect. The atomic-wire effect [51, 56] or fastener effect [26] afforded conduction paths in the solids, giving rise to high-mobility materials [24, 25, 29, 51, 55–59]. Table 1 summarizes selected conductors of closed shell molecules. However, so far at ambient pressure, the conductivity has only reached ~10^{-3} S cm^{-1} (bis(thiadiazolo)quino-TTF (BTQBT) and **4**). There are two single-component superconductors under extremely high pressure, *p*-iodanil ($\sigma_{RT} = 1 \times 10^{-12}$ S cm^{-1}

Table 1 Selected organic conductors of closed shell molecules

Molecule	σ_{RT}/S cm^{-1} at ambient pressure (AP)[a]	Characteristics[b]	Reference
TTC_{10}–TTF	2.7×10^{-6}	Fastener effect, $\mu = 9$–20 cm^2 V^{-1} s^{-1} (time of flight)	[28, 29]
$TTeC_1$–TTE	1.4×10^{-5}	Hetero-atomic contacts, $\mu = 19$–29 cm^2 V^{-1} s^{-1} (time of flight)	[51]
1	2×10^{-5}	Hetero-atomic contacts, $\mu = 6 \times 10^{-4}$ cm^2 V^{-1} s^{-1}	[52]
2	1.7×10^{-4}	Fastener effect	[53]
3	2.7×10^{-4}	Hetero-atomic contacts	[54]
DT–TTF	6×10^{-4}	Hetero-atomic contacts, $\mu = 1.4$ cm^2 V^{-1} s^{-1} (FET)	[55]
BTQBT	1×10^{-3}	Hetero-atomic contacts, $\mu = 0.2$ cm^2 V^{-1} s^{-1} (FET)	[24, 25]
4	2.5×10^{-3}	Fastener effect	[57]
p-Iodanil	1×10^{-12} at AP → 2×10 at 25 GPa, T_c = ca. 2 K at 52 GPa	Hetero-atomic contacts	[20]
Hexaiodobenzene	$T_c = 0.6 - 0.7$ K at ca. 33 GPa, T_c = ca. 2.3 K at 58 GPa	Hetero-atomic contacts	[21]

[a] σ_{RT} electric conductivity at room temperature (RT)
[b] μ mobility

Table 2 Selected organic conductors of neutral π-radical molecules

Molecule	σ_{RT}/S cm^{-1} at AP	Characteristics	Reference
5	No data, Mott insulator	Push–pull effect, weak ferromagnet ($T_c = 35.5$ K at AP, $T_c = 64.5$ K at 1.6 GPa)	[60–62]
6	5.7×10^{-6}	Push–pull effect	[63]
7	10^{-4}	Dimer	[30]
8	2×10^{-4}	Extension of π-system, α-form: zigzag chain, β-form: dimer	[64, 65]
9	$<10^{-2}$ (polymerized)	Extension of π-system, dimer	[66]

at ambient pressure, $\sigma_{RT} = 2 \times 10$ S cm^{-1} at 25 GPa, and superconductivity at $T_c =$ ca. 2 K at 52 GPa), and hexaiodobenzene ($T_c = 0.6$–0.7 K at around 33 GPa and ca. 2.3 K at 58 GPa). Both have peripheral chalcogen atoms, iodine, which may increase the electronic dimensionality of the solid under pressure.

2.2 *Neutral π-Radical Solids*

When the effective on-site Coulomb repulsive energy (U_{eff}) of the solid composed of π-radical molecules is smaller than the bandwidth (W), then the solid becomes a half-filled metal provided that the molecules stack uniformly without dimerization and can be described by a band picture. So far, no such radical molecules have been prepared. In order to decrease U_{eff} and stabilize radical molecules chemically, a push–pull effect and an extension of the π-system have been implemented, though a large U_{eff} and high reactivity (polymerization) are still crucial for the metallic transport. Table 2 summarizes selected organic conductors of neutral π-radical molecules.

2.3 *Zwitterionic (Betainic) π-Radical Solids*

A betainic structure is very effective in decreasing U_{eff} according to the LeBlanc's proposal for the TCNQ anion radical salt (eq. 1), where α is the molecular polarizability of the cation and r is the distance between TCNQ anion radical and a cation [67]:

$$U_{eff} = (1 - \alpha/r^3)U. \tag{1}$$

For a single component betaine, the cation moiety is connected or fused with the anion moiety by chemical bond and this fusion is more appropriate to decrease r. Pyrimido fused TTF betaines (**11**, **12**, **14**, and **15** in Scheme 4, $r = 4 - 5$ Å) are such examples compared with the single-bonded betaine (**10**, $r > 10$ Å). Table 3 summarizes selected organic conductors of betainic π-radical molecules. Very high σ_{RT} values for **12**, **14**, and **15** even on compacted pellet sample have been ascribed to strong intermolecular interactions through complementary hydrogen bonds (see Sect. 3.4.3). A phenalenyl-based betainic radical **13** shares one radical electron

Table 3 Selected organic conductors of betainic π-radical molecules

Molecule	σ_{RT}/S cm^{-1} at AP	Characteristics	Reference
10[a]	3.2×10^{-5}	Connected by single bond	[38]
11[a]	5×10^{-4}	Fused	[42, 44, 45]
12[a]	1.2×10^{-3}	Fused	[41, 43]
13[b]	5×10^{-2}	Fused, sharing one electron by two phenalenyl groups	[31, 34]
14[a]	5×10^{-2}	Fused	[44, 45]
15[a]	1.4×10^{-1}	Fused	[44, 45]

[a]Measured on compacted pellet
[b]Measured on single crystal

between two phenalenyl groups leading to +0.5 charge on each phenalenyl group that may give rise to high mobility for the radical electron even for a large-*U* system. This resembles the concept of mixed valence or partial charge.

2.4 Transition Metal Complex Solids

Some transition metal complexes are excellent conductors. Thin films of cytochrome-c_3, which contains four heme moieties coordinated by protein, exhibited a high conductivity with mixed valence state (Fe^{2+}/Fe^{3+}) and showed an increase in conductivity as the temperature was decreased (2×10^{-2} S cm^{-1} at 268 K) [68–70]. The temperature dependence of conductivity in the highly conductive region is the opposite of that of semiconductors and may preclude the ionic conduction as a dominant contribution. However, since the high conductivity is realized in the presence of hydrogenase and hydrogen, the system is not strictly a single but rather a multicomponent molecular solid.

Although numerous reports on highly conductive single component transition metal compounds have appeared since 2000 (Table 4), the characterization of some compounds is insufficient to claim being a metallic single molecular solid because of issues of purity, measurement conditions, experimental information, etc. Many transition metal complexes of phthalocyanines or TTF-dithiolate ligands (Scheme 5) were insoluble in conventional solvents that rendered purification very difficult and the residual impurity might act as dopant to form a CT solid. Note that most authors stated that their material is the first single-component metal or good conductor without mentioning preceding studies on pentacene, *p*-iodanil, and [bis(benzoquinone dioximato)Pt(II): $Pt(bqd)_2$] under pressure or Tl_2Pc at ambient pressure done before 2000.

Metallic behavior down to low temperatures was reported on a compacted pellet of Tl_2Pc [46] with very high conductivity of $\sigma_{RT} \sim 10^4$ S cm^{-1} (four-probe method using indium lead wires). A specific three-dimensional crystal architecture of Tl_2Pc is anticipated to form a three-dimensional semimetallic band. However, the

Table 4 Selected transition metal complex solids claimed to be metal composed of single component

Molecule	σ_{RT}/S cm^{-1} at AP	Characteristics	Reference	Year
$Pt(bqd)_2$	3.3×10^{-3} at AP [71], insulator–metal–semiconductor under 1.7 GPa		[72, 73]	1989
Tl_2Pc	$>10^4$, metal >5 K	Poor reproducibility	[46]	1994
16	10^{-1}, metal (300–275 K)	Analytically pure	[74]	1996
17	4×10^2, metal >0.6 K	dHvA oscillations	[48, 75]	2001
18	$3 - 4 \times 10^2$, metal >230 K		[76]	2001
19	8, metal >120 K	Dysonian EPR signal, no elemental analysis data	[77]	2003
20	14, metal	Measured by two-probe method, no SQUID signal	[78]	2003

reproducibility of the transport properties and even of the synthesis of the molecule is poor [79].

High conductivity for transition metal complexes of TTF-dithiolate ligands have long been known, due to the mixing of π-d orbitals resulting in a small HOMO–LUMO gap [80]. At present, **17** is the most reliable metal in this category ($\sigma_{RT} = 4 \times 10^2$ S cm^{-1}, metallic down to 0.6 K) based on its purity, temperature dependencies of resistivity and magnetic susceptibility, and de Haas-van Alphen (dHvA) oscillations [48, 75].

Compound **20** was reported to be the first highly conductive (14 S cm^{-1} on compaction pellet sample) single-component molecular metal different from the dithiolate-type; however, the conductivity was measured by a two-probe method and no SQUID response was observed for 200 K $> T >$ 5 K [78].

A metal complex $Pt(bqd)_2$ exhibited an insulator–metal–semiconductor transition under pressure [72, 73]. A continuous color change was observed in bis (diphenylglyoximato)Pt(II), Scheme 5 [81] under pressure and thus has been used as a pressure indicator. For these cases it is not clear that π-d orbital mixing is critical for the transport.

3 Organic Metals of Charge Transfer Type

3.1 Basic Concept for Organic Conductors

A metallic band structure is realized when the CT solids have a partial CT state and molecules form uniform segregated columns or layers. Figure 1 shows electrical conductivity data for 1:1 low-dimensional TTF·TCNQ system, as a function of redox potentials [82]. The two lines **a** and **b** are related to the equation expressing the relationship between I_D, E_A, and the Madelung energy $M(\delta)$ (δ = degree of CT) between partially charged component molecules (eq. 2) [83], where I_D and E_A are

A B C D E F G H I J K L
(I)
(M)
(N)
(a)
(b)
21 22
23
24
25
26
27
$E_1(A)$ / V
$E_1(D)$ / V
0.6
0.4
0.2
0
−0.2
0.2
0.3
0.4
0.5
0.6

Important Remarks:

1) Mixed valence state for low-dimensional CT solids
   $-0.02\ \mathrm{V} \leq \Delta E(\mathrm{DA}) \leq +0.34\ \mathrm{V},\ 0.5 \leq \delta < 1$
2) Complex isomerization (bistable CT solids)
   TMTSF·TCNQ (**24**), TSF·Et_2TCNQ (**25**)
   BEDT-TTF(ET)·TCNQ (**26**) → ET two-dimensional conductor
3) Region **I** ($\delta = 1$)
   Mott insulator (**21**, **22**), antiferromagnet, spin-Peierls, FET channel, switching (memory) system
4) Region **M** ($0.5 \leq \delta < 1$)
   metal (**23**, **27**), electrode (for FET), transparent metal of BO complex, switching, capacitor
5) Region **N** ($\delta < 0.5$)
   non-linear optics, FET channel, solar cell

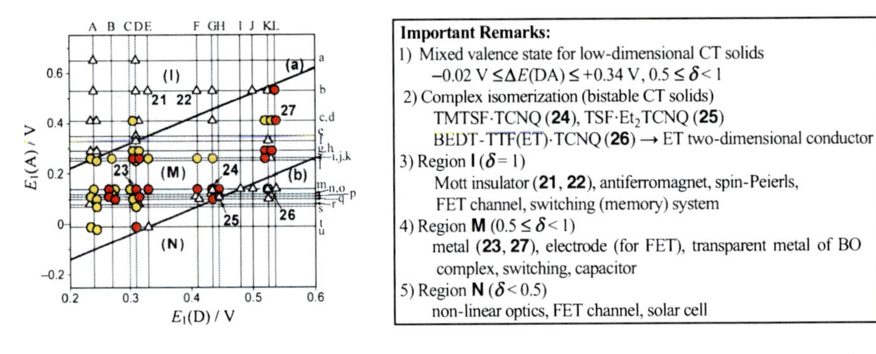

Fig. 1 Ionicity diagram for TTF·TCNQ system plotted as $E_1(\mathrm{A})$ vs $E_1(\mathrm{D})$ vs SCE after modification of the original diagram in [82]. *Open triangles* insulators or semiconductors; *yellow circles* highly conducting in compaction studies; *red circles* organic metals. Donors and acceptors are depicted in Scheme 6. Complexes **21–27** are HMTTF·F_4TCNQ, HMTSF·F_4TCNQ, TTF·TCNQ, TMTSF·TCNQ, TSF·Et_2TCNQ, ET·TCNQ, and DBTTF·Cl_2TCNQ, respectively. $E_1(\mathrm{A})$ and $E_1(\mathrm{D})$ in this figure are the peak values. Region **N**: neutral, **M**: partial CT, **I**: fully ionic. Line **a**: $\Delta E(\mathrm{DA}) = -0.02$ V, **b**: $\Delta E(\mathrm{DA}) = 0.34$ V

A: TTN
B:TMTTF
C
D: TTF
TCNQ
E: HMTTF
F: HMTSF
G: TMTSF
H: TSF
I
J: TTC1-TTF
K: BEDT-TTF(ET)
L: DBTTF
a : 2,5-(CN)2
b : F4
c : 2,5-Br2
d : 2,5-Cl2
e : 2,5-I2
f : 2,5-F2
g : Br
h : Cl
i : F
j : 2,5-Cl,Me
k : 2,5-Br,Me
l : 2,5-I,Me
m : TCNQ
n : Me
o : 2,5-(iso-Propyl)2
p : 2,5-Et2
q : 2,5-Me2
r : 2,5-OEt,SMe
s : OMe
t : 2,5-(OMe)2
u : 2,5-OMe,OEt

Scheme 6 Chemical structures of electron donor and acceptor molecules in Fig. 1

the ionization potential of an electron donor (D) and electron affinity of an electron acceptor (A), respectively (Scheme 6). The mixed valence region (**M**) is located between fully ionic (**I**) and neutral (**N**) regions. In the region **M**, the CT solids are either highly conductive (yellow circles) or metallic (red circles) when they have segregated stacks. The solids in the regions **I** and **N** are insulators (triangles), in general.

$$I_\mathrm{D} - E_\mathrm{A} = M(\delta). \tag{2}$$

So the partial CT state can be predicted and controlled by ($I_\mathrm{D} - E_\mathrm{A}$) or ΔE(DA) [$\Delta E(\mathrm{DA}) = E_1(\mathrm{D}) - E_1(\mathrm{A})$: E_1, first redox potential; $-0.02\ \mathrm{V} \leq \Delta E(\mathrm{DA}) \leq +0.34$ V for TTF·TCNQ system] for a combination of specific D and A, and the complex D·A exhibits a low lying CT band below $5 \times 10^3\ \mathrm{cm}^{-1}$.

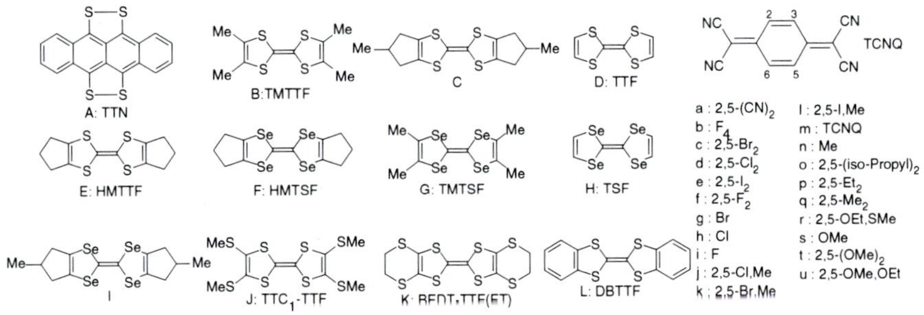

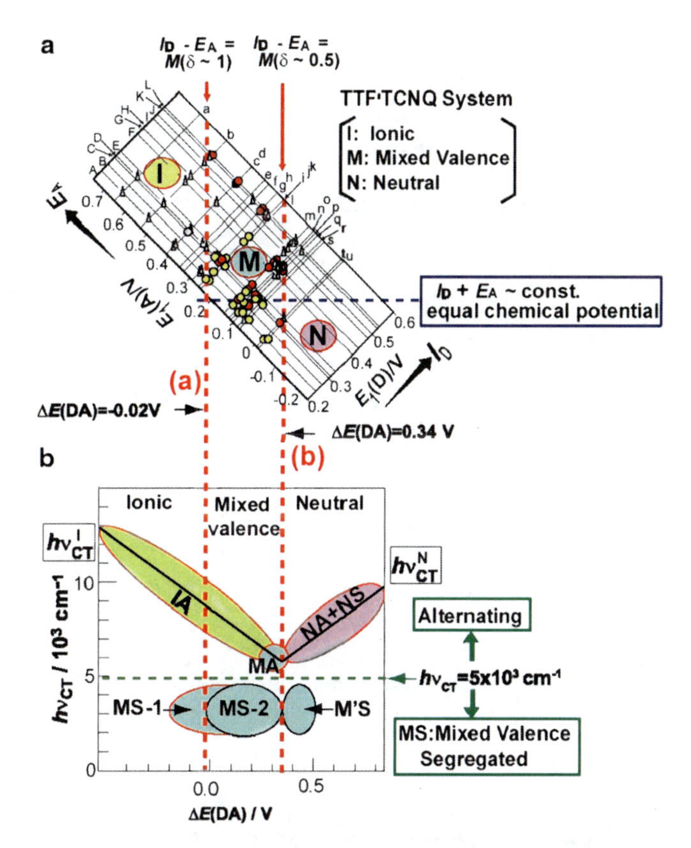

Fig. 2 (**a**) The same figure as Fig. 1 except for the scale of E_1(D or A). Organic metals on the *blue dotted line* have the same chemical potential ($I_p + E_A$ = constant). (**b**) Schematic phase diagram of ionicity, conductivity and stacking of DA CT solids. The first optical transition energy in solid ($h\nu_{CT}$) is plotted against the ΔE(DA) value. Left- and right-hand sides of the V-shaped line correspond to $h\nu_{CT}^{N} = I_D - E_A - C$ (eq. 3) and $h\nu_{CT}^{I} = -I_D + E_A + (2\alpha - 1)C$ (eq. 4) respectively, where C and αC are the Coulomb attractive energy between $D^{\bullet+}$ and $A^{\bullet-}$ and the Madelung energy, respectively. **IA** ionic alternating, **MA** mixed valence alternating, **NA** neutral alternating, **NS** neutral segregated, **MS** mixed valence segregated (**MS-1**: non 1:1 (minor component is fully ionic), **MS-2:** 1:1 low-dimensional, **M′S**: 1:1 high-dimensional). An appropriate V-shaped line for the *p*-quinone system was obtained by a parallel shift of the V-shaped line for the TCNQ system towards the lower side by 0.13–0.16 V

Figure 2 shows the relationship between Fig. 1 and another kind of ionicity phase diagram (V shaped line in Fig. 2b) proposed by Torrance for the neutral-ionic (N–I) phase transition for the alternating stacks [84]. Figure 2a was made by rotating Fig. 1 (E_1(D) and E_1(A) have the same scale here) so as to make the two borderlines **a** and **b** vertical. Then all CT solids that lie on a horizontal line in Fig. 2a have the same chemical potential ($I_D + E_A$ = constant). The organic metals in the region **M** residing on several different horizontal lines in Fig. 2a were employed as the source and drain electrodes to control the Fermi level alignment

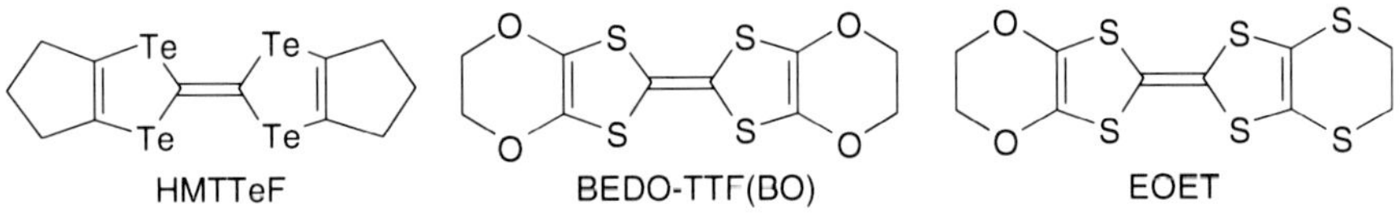

HMTTeF BEDO-TTF(BO) EOET

Scheme 7 Chemical structures of HMTTeF, BEDO-TTF(BO) and EOET

between electrodes and channel for FET, making the injection of carriers smooth and giving varied polarity in FET behavior (see Sect. 3.4.1) [85].

In the neutral region near the bottom of the V-shaped line (region **MA**), an enantiotropic phase transition system (N–I transition) is located [84]. The CT solids that have $h\nu_{\mathrm{CT}}$ bands below 5×10^3 cm^{-1} (horizontal green dotted line) belong to a different class (region **MS**) that usually includes (super)conductors and narrow-gap semiconductors having mixed valence segregated stacks or layers. The important point here is that **M′S** is for high-dimensional 1:1 CT solids, which usually extend their metallic regime toward lower δ values (higher ΔE(DA) values), like the HMTTeF [86] and BEDO-TTF (BO) [87] systems (Scheme 7), owing to their strong self-aggregation ability (see Sect. 3.3). Even with $\delta = 1/3$, some BO complexes have segregated stacks and show metallic behavior. Figures 1 and 2 illustrate a prediction on the modes of molecular stacking: alternating and segregated, and information on the chemical potentials and other functionalities as described below.

3.2 Organic Metals and Related Functional Solids

Ionicity diagrams as depicted in Figs. 1 and 2 are clues to explore functional conductors of CT type, such as molecular metals, Mott insulators, N–I systems, complex isomers, and self-aggregated two-dimensional conductors (see comments in Fig. 1 and [3]).

1. The fully ionic solids (region **I**) afforded band insulators, 1:1 Mott insulators with ground states of antiferromagnets (E·b(**21**) and F·b(**22**) in Fig. 1) or spin-Peierls systems, ferroelectrics, ferromagnets, spin-ladders, and nonlinear transport materials (switching and memory).
2. The mixed valence solids (region **M** in Fig. 1a, and **MA**, **MS-1**, **MS-2** and **M′S** in Fig. 2b) afforded (super)conductors and the following various kinds of insulators: (1) a (nearly) uniform segregated stack having spin density wave (SDW), and anion or charge ordered (=charge disproportionation) state, (2) a non-uniform segregated stack showing Peierls-type distortion, spin-Peierls distortion, and dimer-type Mott insulators including antiferromagnets, quantum spin liquid and spin-ladders, and (3) an alternating stack including N–I systems, ferroelectrics, and highly conductive semiconductors. There are hybrids from the combination of ferro-, ferri-, or paramagnetic species based on transition

metal compounds as one component and mixed valence counterparts to form magnetic CT conductors.

3. The neutral solids (regions **N** and **NA** + **NS**), in general, exhibit a CT band represented by eq. 3 in the caption to Fig. 2, regardless of the stacking modes. Since most of the CT solids in the region **N** prefer alternating stacks with a few exceptions, they are band insulators with low ionicity. Very weak CT solids having segregated stacks are potential candidates for FET channel and solar cell materials since they have good conduction paths. Hydrogen-bonding and proton-transfer between the components manifest many interesting functions: switching, ferroelectrics, etc.
4. Near the borderline **b** in Fig. 1, the bistability concerning the ionicity between the neutral and partial CT states is realized, i.e., the monotropic complex isomers G·m, H·p, and K·m (**24**, **25**, and **26** in Fig. 1, respectively). Even though K·m (ET·TCNQ) is expected to afford a neutral insulator based on its ΔE(DA) value, a highly conductive complex isomer has been prepared. This result indicates that the ET molecule has a significant self-aggregation ability to form a segregated column with increased dimensionality, which is a nature of the solids in the region **M′S** in Fig. 2b (see next section). The insulating CT solids residing near the borderline **a** or **b** have the potential to exhibit a phase transition into a highly conducting phase induced by external stimuli (electric field, photons, etc.) with smaller threshold than those placed far from the borderlines.

3.3 Molecular Design for Dimensionality: Self-Aggregation Ability

Since the metallic state in the one-dimensional electronic system is unstable, an increase in the electronic dimensionality is necessary to prevent the nesting of Fermi surfaces. Several attempts have been made through "pressure", "heavy atom substitution", "peripheral addition of alkylchalcogen groups", or "fusion of TTF skeletons" [88] (Fig. 3, Scheme 8). The latter three correspond to the enhancement of the self-aggregation ability of the molecules and hence increase the electronic dimensionality of molecular assemblies. The HMTTeF molecules afforded a stable metallic phase without any trace of superconductivity [86]. The BO molecules also afforded stable two-dimensional metals having two-dimensional Fermi surfaces (b in Fig. 3) owing to the strong self-aggregation ability (see Sect. 3.4.2) [87]. The substitution of an ethylenedioxy group with an ethylenedithio group (BO → EOET (Scheme 7) [89] → ET) destabilized the stable metallic state of BO compounds and provided unstable two-dimensional conductors (d in Fig. 3). The elimination of one ethylenedioxy group (BO → EDO) [90, 91] was found to be very effective in making a one-dimensional Fermi surface (e in Fig. 3) to afford localized (magnetic) phase or metal–insulator (MI) transition (see Sect. 4.2). Several superconductors

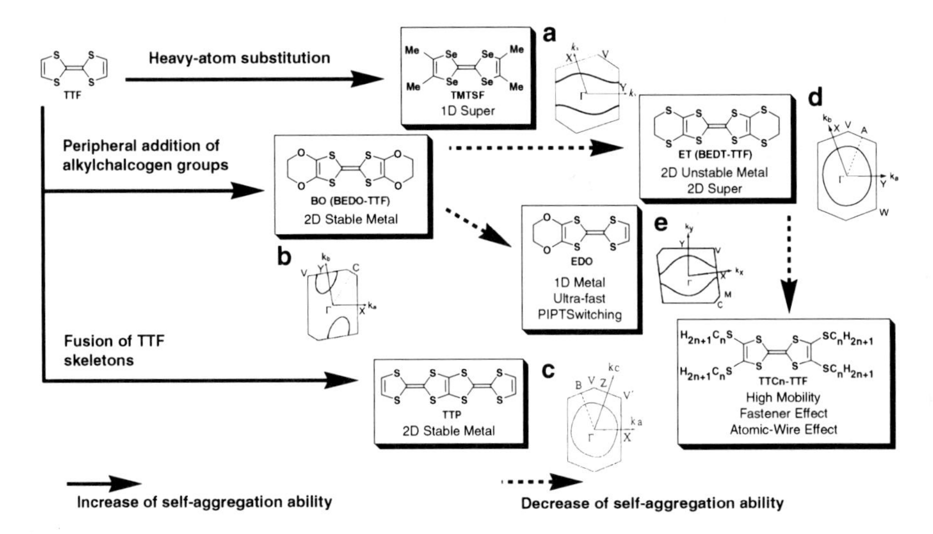

Fig. 3 Strategy for chemical modification of the TTF molecule to increase or decrease the electronic dimensionality (D) by the aid of enhance or suppress the self-aggregation ability of the donor molecules, respectively. Typical Fermi surfaces of TMTSF (**a**: $(TMTSF)_2NbF_6$), BO (**b**: $(BO)_{2.4}I_3$), TTP (**c**: $(BEDT\text{-}TTP)_2I_3$, Scheme 8) [88], ET (**d**: β-$(ET)_2I_3$), and EDO (**e**: $(EDO)_2PF_6$) CT solids are depicted. PIPT: photo-induced-phase-transition

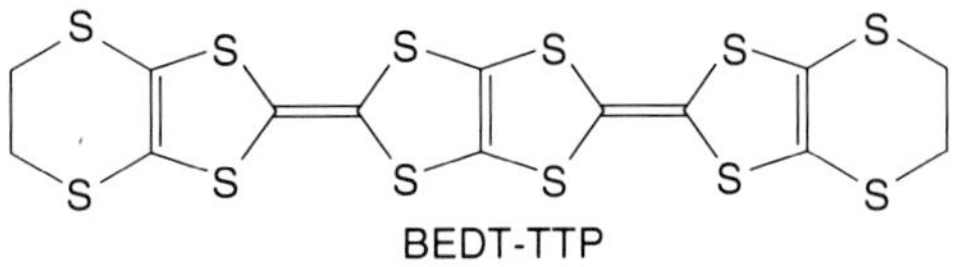

BEDT-TTP

Scheme 8 Chemical structure of BEDT-TTP

have been prepared based on TMTSF having warped one-dimensional Fermi surface (a in Fig. 3), and on two-dimensional metals of ET, BO and a variety of analogs of TTF, even though TTF itself did not afford superconductors (see Sect. 5.1).

3.4 *Variety of Conductive Charge Transfer Solids*

3.4.1 Tuning of Fermi Level of FET Electrodes

Figure 4 demonstrates the control of p-, n-, and ambipolar-type FET operations in prototypical single-crystal organic FETs by "chemically tuning" the Fermi energy in TTF·TCNQ-based organic metal electrodes [85]. Figure 4a shows a device, where the organic channel is a neutral CT solid DBTTF·TCNQ, which has an alternating stack and the valence and conduction bands are mainly derived from the

HOMO of DBTTF and the LUMO of TCNQ, respectively (Fig. 4b). Source and drain electrodes are several organic metals of the TTF·TCNQ type having different chemical potentials predicted using Fig. 4c which is the same as Fig. 2a. For the electrodes whose chemical potentials are set within the conduction band of the channel material, FET exhibited *n*-type behavior (A in Fig. 4d). When the chemical potentials of organic metals are allocated within or near the valence band of the channel, p-type behaviors were observed (E, F in Fig. 4d). When the chemical potentials of the electrodes are within the gap of the channel, FET exhibited ambipolar-type behavior (B–D in Fig. 4d). Since the channel material is the alternating CT solid, the drain current is not excellent and a Mott type insulator of DA type or almost neutral CT solid having segregated stacks is much preferable in this context.

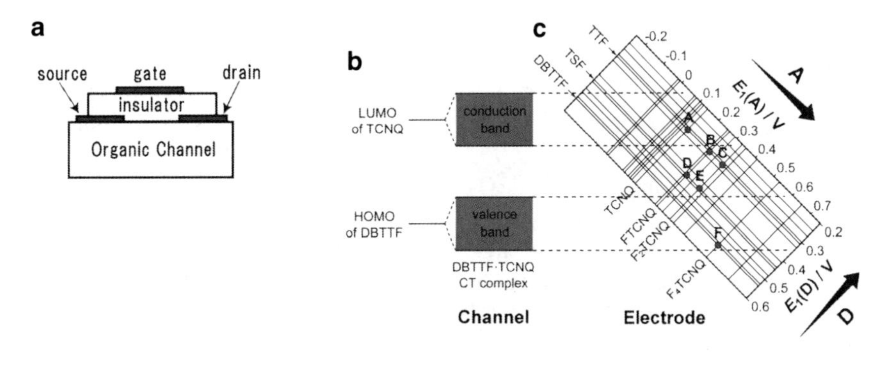

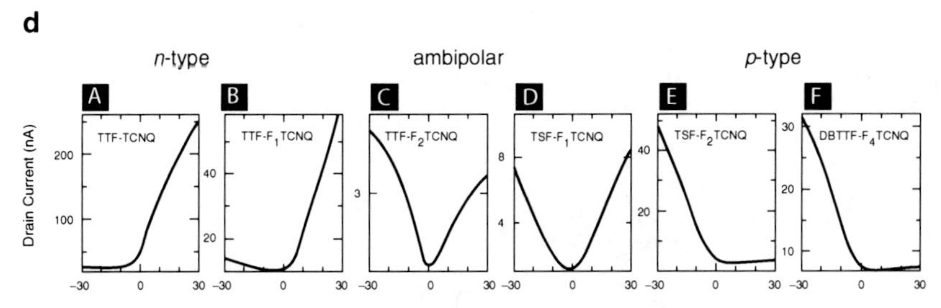

Fig. 4 Organic FET (OFET) composed of CT-complex-based organic metal electrodes. (**a**) An illustration of the device. (**b**) Interface band diagram of metal/semiconductor contact in DBTTF·TCNQ single crystal OFET with a variety of organic metal electrodes. (**c**) Modified figure of Fig. 2a. The conductive complexes **A–F** are drawn as functions of both I_D [or E_1(D)] and E_A [or E_1(A)]. (**d**) Transfer characteristics at $V_D = 5$ V of DBTTF·TCNQ single-crystal field effect transistors with the source and drain electrodes, composed of **A**: TTF·TCNQ, **B**: TTF·FTCNQ, **C**: TTF·F_2TCNQ, **D**: TSF·FTCNQ, **E**: TSF·F_2TCNQ, and **F**: DBTTF·F_4TCNQ, measured along the crystal long axes [85]

3.4.2 Two-Dimensional Stable Metals in Various Shapes

The related elements, proton (H^+), hydrogen ($H^•$), and hydride (H^-) change their physical properties (including their size) drastically by the change of the number of electrons. Hydrogen-bond and proton-transfer interactions are the key to understanding many chemical reactions, biological activities, structure of molecular assemblies and supramolecules, functionalities in the solid state, etc.

As shown in Fig. 3 the peripheral addition of alkylchalcogen groups to the TTF skeleton increases the self-aggregation ability of the molecules and hence the electronic dimensionality of molecular assemblies increases. The typical example is the BO system [87]. Figure 5a,b shows one of the common packing patterns of the BO molecules. The strong self-aggregation ability of the BO molecules arises from both the CH···O hydrogen bonds in the face-to-face direction (Fig. 5a) and robust transfer interactions in two different oblique directions owing to the strong S···S atomic contacts (Fig. 5b) since inner chalcogen atoms have much higher electronic density than outer chalcogen atoms. These afford both a wide valence range of metallic state ($0.33 \leq \delta$) and a limited number of preferable packing patterns giving rise to a stable two-dimensional metallic state.

These robust intermolecular interactions provide a wide metallic band even in strongly disordered systems such as Langmuir–Blodgett (LB) films (BO complexes of $(MeO)_2$TCNQ, C_nTCNQ (n = 10, 14), behenic acid, and stearic acid) [92–95], polycarbonate films dispersed with BO complexes (reticulate doped polymer (RDP) films with I_3 or Br salt, surface resistivity 1×10^{-3} S/□ at RT corresponding to ca. 10^2–10^3 S cm^{-1}) some of which are transparent (Br salt) [96, 97], compressed pellets with ferrimagnetic behavior [$(BO)_3[FeCr(oxalate)_3](H_2O)_{3.5}$] [98], films sensitive to the moisture in air [$(BO)_2ReO_4(H_2O)$, $(BO)_2Br(H_2O)_3$] [99–101], etc., regardless of the sort, shape, and size of acceptor or anion molecules. As a result, the BO complexes hardly exhibit any phase transition including the superconducting one (only two superconducting salts with $T_c \leq 1.5$ K; $(BO)_3Cu_2(NCS)_3$ [102] and $(BO)_2ReO_4(H_2O)$ [103]).

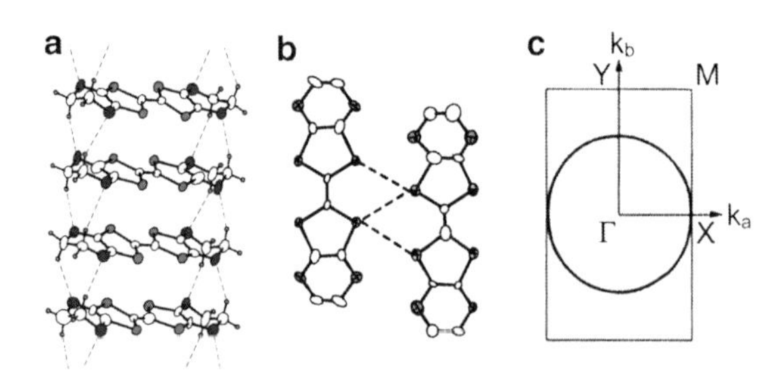

Fig. 5 Donor packing and Fermi surface for BO compounds. (**a**) Face-to-face packing (*dotted lines* indicate the CH···O hydrogen bonds), (**b**) side-by-side contacts (*dotted lines* indicate the short S···S atomic contacts), and calculated Fermi surfaces of (**c**) $(BO)_2Cl(H_2O)_3$. Calculated Fermi surface of $(BO)_{2.4}I_3$ is depicted in Fig. 3b

3.4.3 Conductors Based on Biological Materials

A variety of transport properties of biological materials, such as hemoglobin, amino acids, proteins, polypeptides, and so on, have been investigated, started with the pioneering works by Eley et al. [18, 104–113] (see also Chap. 12 in [18]). Most of them are insulating ($<10^{-11}$ S cm^{-1} at 400 K) [18, 104–113] except cytochrome-c_3 (see Sect. 2.4) [68–70]. In recent studies on biomolecular conductors, DNA is one of the most active target molecules, and numerous experiments concerning the transport properties of DNA molecule have been carried out [114–122]. In the double-stranded DNA molecules, nucleobases establish a one-dimensional π-stacking structure which was proposed to be an efficient charge conduction path within DNA.

This past decade has seen numerous controversial studies regarding electrical conduction of DNA. Some reported high conductivity [115, 116, 118] with σ_{RT} of at most 10^4 S cm^{-1} [115] or even superconducting properties [119], while others claimed that the carefully deionized DNA molecules are insulating [117, 120] in agreement with the old reports [121, 122] with σ_{RT} less than 10^{-6} S cm^{-1}. The controversy seems to have settled on a wide consensus that, apart from ionic conduction by the sodium gegenions, double-stranded DNA is an electrical insulator.

Several conductive CT solids with nucleobase skeletons have been developed in the TTF systems having uracil moieties ($\sigma_{RT} = 10^{-1}$–2 S cm^{-1}) [123–127]. Also several attempts have been undertaken to investigate the CT complexes in a variety of biochemical systems, especially using nucleobases (Scheme 9) [18, 104]. Estimation of I_P of the nucleobases, as potential components in CT complexes, indicate that they are reasonably effective π-donors particularly in the case of guanine (**G**); $I_D = 7.64$–7.85 eV vs adenine (**A**, 7.80–8.26 eV), cytosine (**C**, 8.45–8.74 eV), and thymine (**T**, 8.74–8.87 eV) [128–131].

In the complex formation of nucleobases with *p*-chloranil, only **G** gave a CT solid which is in the neutral ground state estimated from the optical spectrum [132]. The mobility of electrons was described with regard to the transport properties on the CT solid of TCNQ with **G** [133]. **C** and 1-methylcytosine gave dark blue TCNQ radical anion salts with a 2:1 stoichiometry [134]. The examination of the CT solids of **C**, which has a weak electron donating ability ($E_p^{ox} = +1.90$ V vs SCE in water) as well as medium proton donating and strong proton accepting abilities (pK_a = 4.55 and 12.2) [135] with several TCNQs (RTCNQ) revealed the following [136–140].

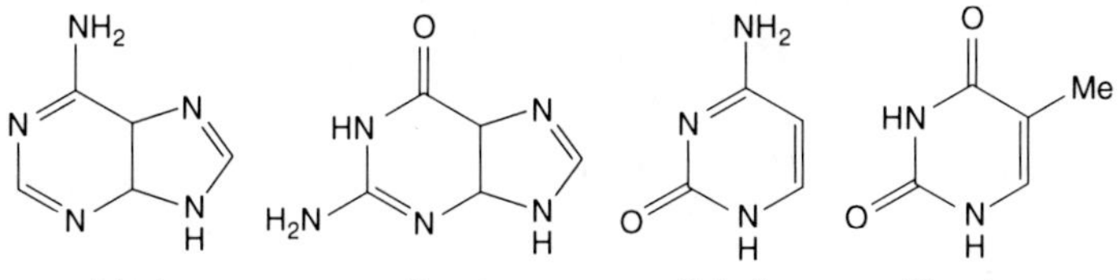

Adenine Guanine Cytosine Thymine

Scheme 9 Chemical structures of nucleobases; adenine, guanine, cytosine, and thymine

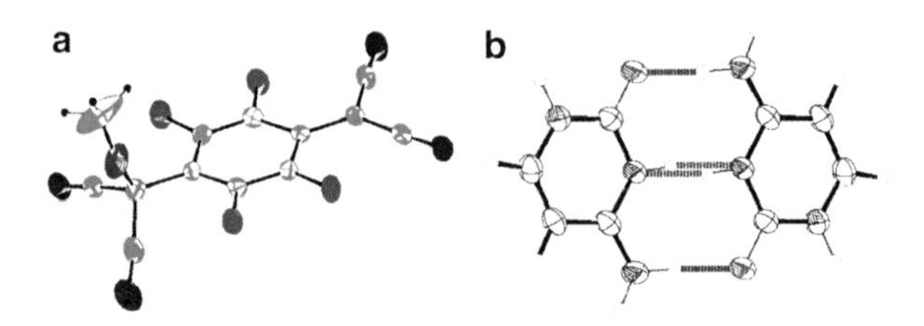

Fig. 6 Molecular structures of $F_4TCNQ\text{-}OMe^-$ (**a**) and hemiprotonated cytosine pair CHC^+ (**b**): *dotted lines* indicate the hydrogen bonds

Reaction between **C** in methanol and RTCNQ in acetonitrile yielded three kinds of ionic solids: (1) insulators composed of methoxy substituted RTCNQ anions such as ($\mathbf{CHC^+}$)[$F_4TCNQ\text{-}OMe^-$](H_2O) (Fig. 6) [136], (2) semiconducting CT solids with fully ionic RTCNQ radical anions such as ($\mathbf{CHC^+}$)($TCNQ^{\bullet -}$) [137, 138], and (3) conducting CT solids of partially ionic or mixed valent RTCNQ radical anions such as ($\mathbf{CHC^+}$)($MeTCNQ^{0.5\bullet -}$)$_2$ [138], where $\mathbf{CHC^+}$ is the hemiprotonated cytosine pair (Fig. 6b). Cation units in all products were found to be protonated cytosine species, most commonly $\mathbf{CHC^+}$, where H^+ comes from methanol. This result suggests that the intrinsic transport properties of DNA should be studied not in protic solvents but under strictly dried conditions.

Crystal structural analysis of ($\mathbf{CHC^+}$)($TCNQ^{\bullet -}$) revealed the segregated structure with a uniform stacking pattern (Fig. 7a). The interplanar separations of TCNQ and $\mathbf{CHC^+}$ columns were 3.14 and 3.32 Å, respectively. The $\mathbf{CHC^+}$ pairs formed a one-dimensional ribbon structure (Fig. 7b). The hydrogen bonds between $\mathbf{CHC^+}$ ribbon and TCNQ molecules constructed the layered structure. In addition, the self-aggregation ability of **C** strengthened the uniform arrangement of the crystal resulting in both the high conductivity ($\sigma_{RT} = 3.2 \times 10^{-2}$ S cm^{-1} on single crystal), which is one of the best among the conventional Mott type TCNQ salts and the absence of spin-Peierls type structural distortion down to 10 K. Transport property of ($\mathbf{CHC^+}$)($TCNQ^{\bullet -}$) was examined under high pressures up to about 7 GPa using a diamond anvil cell. The activation energy ε_a of 0.14 eV at ambient pressure decreased monotonically by a rate of 0.013 eV GPa^{-1} [140]. The partially ionic salt of MeTCNQ in Group 3 exhibited the highest conductivity of 2 S cm^{-1} so far observed for CT complexes based on biological molecules. This study revealed that the protonated states of **C**, especially the $\mathbf{CHC^+}$ species, are extraordinary stable and furthermore the characteristic pattern of the complementary hydrogen bonds between the cytosine molecules contribute to allow effective molecular packing and to control the electronic structure of TCNQ molecules for electronic conductors.

3.4.4 Two-Dimensional Metal Based on C_{60}

CT solids of fullerene C_{60} with a number of different inorganic cations have shown metallic or superconducting properties (for superconductivity, see Sect. 5.2.3). Among the fullerene metals, the best known families are MC_{60} anion radical salts

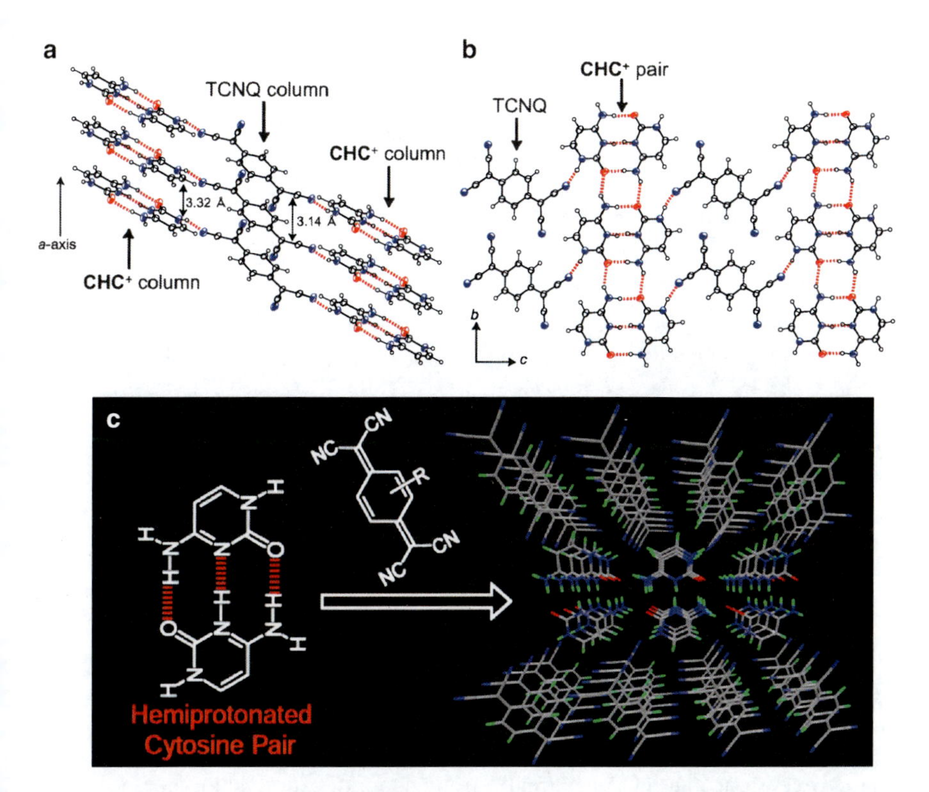

Fig. 7 Crystal structure of (**CHC**$^+$)(TCNQ) salt. (**a**) Uniform segregated stacks of **CHC**$^+$ and TCNQ. (**b**) **CHC**$^+$ ribbons by complementary hydrogen bonds and TCNQ form a layer within a *bc*-plane (hydrogen bonds: *red dotted lines*). (**c**) Formation of three-dimensional structure (//*a*) between hemiprotonated cytosine pair and RTCNQ species

(M = K, Rb, and Cs), which contain linearly polymerized $C_{60}^{\bullet-}$, and superconducting M_3C_{60} salts (M = alkali metals), obtained by doping C_{60} with alkali metals [141–143]. As metal cations expand the three-dimensional lattice of the initial C_{60} framework, M_3C_{60} salts exhibit three-dimensional metallic conductivity, whereas MC_{60} salts are either three-dimensional (when M = K) or quasi-one-dimensional metals (when M = Rb or Cs) [144]. Two-dimensional fullerene metals have not been obtained by conventional doping methods except for Na_4C_{60} where C_{60} has a two-dimensional polymeric structure [145], whilst the various possible ways of modifying M_xC_{60} CT solids have almost been exhausted. With conventional organic donor molecules, C_{60} is too weak an acceptor molecule to afford ionic solids. A very strong organic donor molecule, tetrakis(dimethylamino) ethylene (TDAE, Scheme 10), did yield a completely ionic solid – a ferromagnet with T_c = 16 K [146].

A multicomponent approach for synthesizing ionic fullerene compounds $D_I^{+}{\cdot}D_{II}{\cdot}(\text{fullerene})^{\bullet-}$ is very effective in developing various functional and structural fullerene CT solids, including σ- and π-type dimers of fullerenes and an η-type

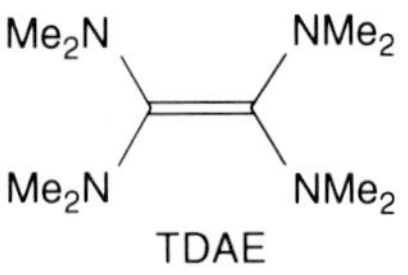

Me$_2$N NMe$_2$
Me$_2$N NMe$_2$
TDAE

Scheme 10 Chemical structure of TDAE

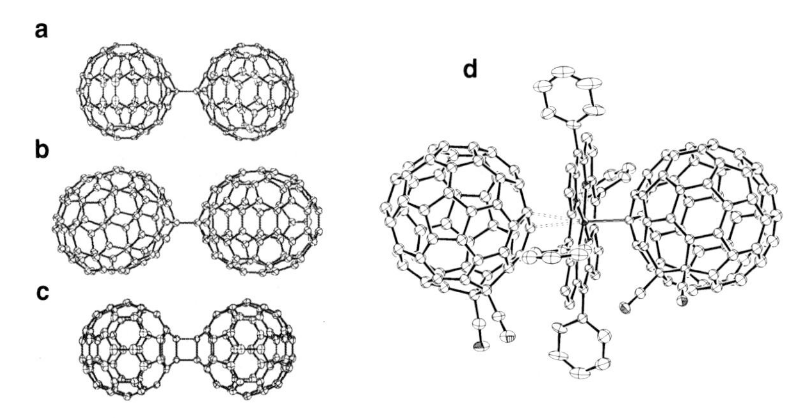

Fig. 8 (**a**) σ-type C_{60} dimer, (**b**) σ-type C_{70} dimer, (**c**) π-type C_{60} dimer, and (**d**) η^1-coordination of cobalt(II) tetraphenylporphyrin with $C_{60}(CN)_2$

complex (Fig. 8), where D_I^+ is a small, strong donor or cation that ionizes fullerene and determines its charged state, whereas D_{II} is a large, neutral molecule that defines the crystal packing of the complex [147–154]. In order to exhibit metallic properties, the C_{60} sublattice should have a close-packed structure with appropriate $C_{60}\cdots C_{60}$ distances. Otherwise, diamagnetic single bonded $(C_{60}^-)_2$ dimers were formed when the distance is short, or strong spin frustration will be created based on the trilateral triangle spin geometry of $C_{60}^{\bullet-}$ when the distance is large just like κ-(ET)$_2$Cu$_2$(CN)$_3$ in Sect. 6.1. The triptycene (TPC) molecule (Scheme 11) afforded a suitable geometrical space and spatial regulation as the D_{II} component for $C_{60}^{\bullet-}$ ions, giving rise to a close-packed fullerene two-dimensional sublattice in which the $C_{60}^{\bullet-}$ monomers preferentially form a two-dimensional honeycomb network of $C_{60}^{\bullet-}$. Namely, the TPC molecules form hexagonal layers with voids that accommodate the *N*-methyldiazabicyclooctane cation (MDABCO$^+$, D_I^+) (Scheme 11, Fig. 9a). Docking $C_{60}^{\bullet-}$ into the periodic hollow sites in the (MDABCO$^+$)·TPC network (Fig. 9b,c) leads to the two-dimensional organic metal (MDABCO$^+$)·TPC·($C_{60}^{\bullet-}$) [154]. There are two kinds of C_{60} layers, layer *A* (Fig. 9f) and layer *B*. $C_{60}^{\bullet-}$ molecules in layer *B* are rotationally disordered above 200 K and layer *B* is not metallic and exhibits spin frustration. The ordering of $C_{60}^{\bullet-}$ in the *B* layers with MDABCO$^+$ at around 200 K triggers a transition from a nonmetallic and antiferromagnetically frustrated state to a metallic state in layer *B*, whilst the ordered $C_{60}^{\bullet-}$ in layer *A* keeps its two-dimensional itinerancy over the entire temperature range. It exhibits a metallic state down to 1.9 K, which was

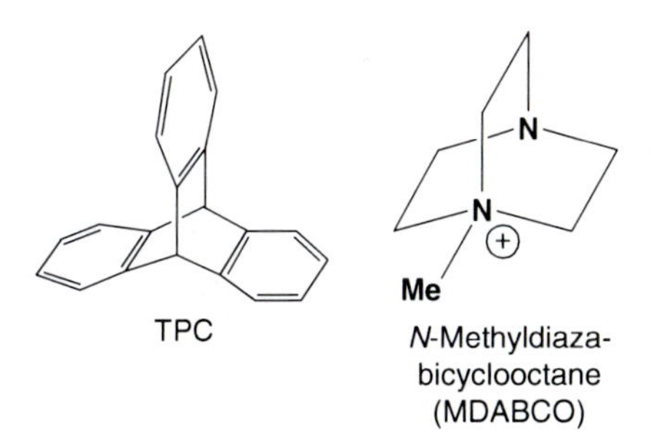

Scheme 11 Chemical structures of TPC and MDABCO

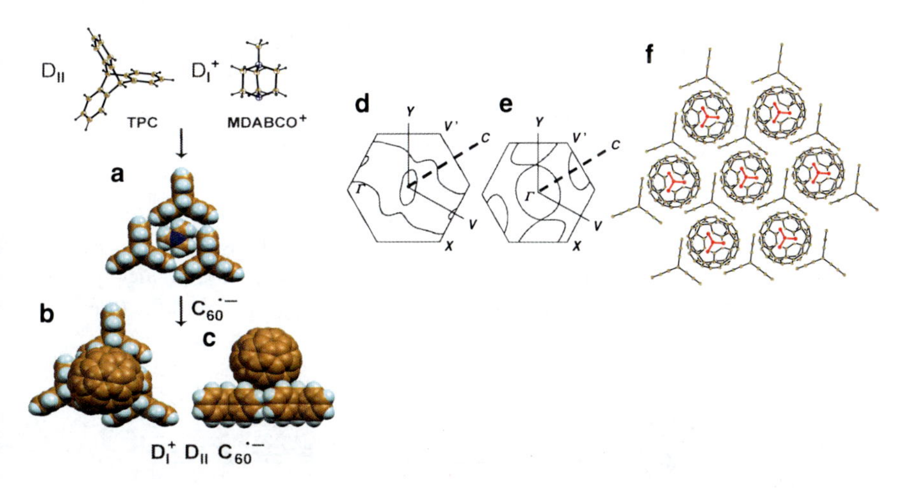

Fig. 9 Molecular structures of TPC (D_{II}) and MDABCO$^+$ (D_I^+) [154]. Crystal structure packing in (MDABCO$^+$)·TPC·($C_{60}^{\bullet -}$). (**a**) TPC molecules form a hexagonal hollow and the MDABCO$^+$ cation fits into the hollow; the $C_{60}^{\bullet -}$ molecules are docked into the hollow in TPC layer in a key–keyhole relationship [*top view* (**b**) and *side view* (**c**)] to form $D_I^+D_{II}C_{60}^{\bullet -}$. Colors: C, *dark yellow*: H, *pale blue*; and N, *dark blue*. Calculated Fermi surface at 160 K in (**d**) layer *A* and (**e**) layer *B*. (**f**) Projection of the (MDABCO$^+$)·TPC layer on the C_{60} layer *A* (*red color*: MDABCO$^+$)

consistent with the calculated Fermi surfaces (Fig. 9d,e). This metal is composed of only light elements (C, H, and N).

4 Exotic Conductors with a Switching Function

4.1 Basic Aspects

All of the organic molecules have multifunctional natures and provide plural intermolecular interactions depending on the nature of counter component, morphology (solid, films, uni-molecule, etc.), and external circumstances. For example, the CT interaction between D and A molecules in solid are broken down into two kinds of interactions: Interaction I – electron transfer from neutral D to A molecules that costs ($I_D - E_A$) and Interaction II – Madelung energy M, as described in Sect. 3.1 (Fig. 10a). One can expect that the balance between the interactions I and II can be controlled easily by external stimuli. The controllability increases as the system approaches to a boundary area and the system shows a variety of phase transitions (i.e., metal ↔ insulator, Mott insulator ↔ metal, quantum spin liquid ↔ superconductor, neutral ↔ ionic, valence tautomerization), monotropic (e.g., complexes **24–26** in Fig. 1) and enantiotropic (e.g., TTF·*p*-chloranil) complex isomerizations, and switching or memory effect depending on the potential depths and barrier height ΔE in Fig. 10b. Therefore, possible candidates for such phase transition can be selected based on the diagrams in Fig. 10c, d.

The switching or memory phenomena induced by electric field application or photo irradiation have been studied on Mott insulators, charge ordered insulators, and N–I transition systems and were found to be fast phase transitions in general. For the former two systems, the phase transitions caused a pronounced change in reflectance and conductivity from insulating to metallic features. The third system also exhibited a change in conductivity and dielectric response connected with the transports of solitons and/or domain walls, dynamic dimerization, and

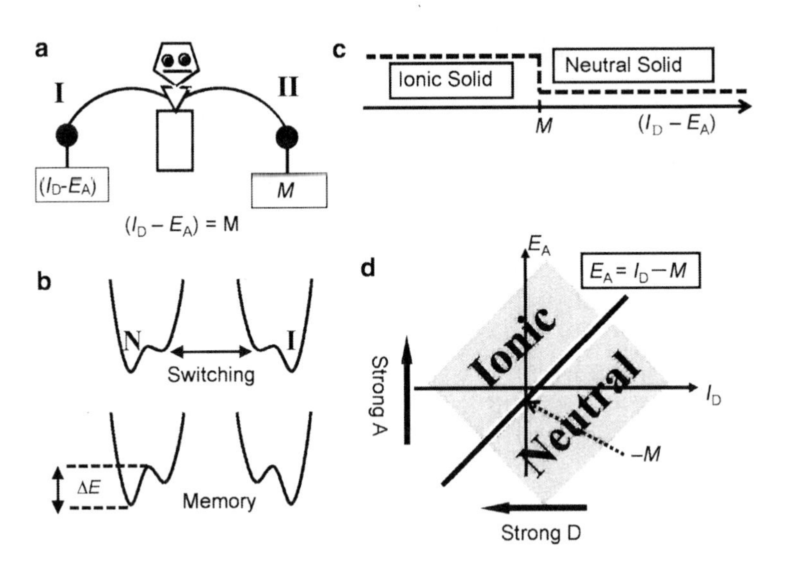

Fig. 10 (**a**) A schematic balance between ionization energy ($I_D - E_A$) and Madelung cohesive energy M for CT solids. (**b**) Model double minimum potential for the N–I system for switching (*upper*, potential barrier between stable and metastable states is small and thermally accessible) and memory (*lower*, potential barrier is rather high $\Delta E >> k_BT$). One- (**c**) and two-dimensional (**d**) diagrams for searching for the boundary zone and functional materials. (**d**) is a schematic diagram corresponding to Fig. 1

ferroelectricity. It should be emphasized that most of the π-molecules are, in Pearson nomenclature, soft acids or soft bases. Thus they have a huge electronic polarizability, and are susceptible to external stimuli when the ionicity and mutual orientation of the molecules are appropriately designed. Table 5 summarizes CT solids having switching behavior and are compared with superconductor $(TMTSF)_2ClO_4$ showing sliding SDW. So far among the switching systems by electric-field, $(NT)_3GaCl_4$ (Scheme 12) exhibited the lowest threshold (1×10^2 V cm^{-1}) [166], which is smaller by one order of magnitude than the well known Cu·TCNQ [155, 156], but much larger than that of sliding SDW. As for the response to the electric field, the ability to show thyristor action was demonstrated for the charge-ordered systems based on TTF derivatives [167].

For these transition systems, the following five parameters are important for the development of materials:

1. *Response time* which can be controlled by taking into account the origin of the transition (1) fs for pure electronic transition, (2) addition of molecular deformation leads to ps response time, (3) further addition of lattice deformation leads to a slower response time > ps.
2. *Coherence* (transition efficiency) depends on the degree of electron–phonon or electron–molecular vibration coupling.
3. *Response temperature* (operating temperature) may decrease in the following order: bond formation and cleavage → molecular deformation → lattice deformation → electronic deformation such as SDW and charge-order melting.
4. *Threshold* of the external stimuli should not be zero to have clear switching and memory.
5. *Durability* of the system.

4.2 Ultrafast Photo-Induced Phase Transition in $(EDO)_2X$

To destabilize the metallic state of the BO complexes, the elimination of one ethylenedioxy group (BO → EDO) [90, 91] was very efficient owing to the weakened self-aggregation ability (Fig. 3). The $(EDO)_2X$ salts (X = PF_6, AsF_6, and SbF_6) are three-quarters-filled band conductors with a quasi-one-dimensional Fermi surface (Fig. 3e) and exhibit a first-order MI transition (Fig. 11a, b) at rather high temperatures (240–280 K) [90, 173]. The T_{MI} was tuned by chemical modifications. Deuteration of EDO (d_2-EDO, Scheme 13) increased T_{MI} by ca. 2.5 K, while complexation with larger counter anions decreased T_{MI} in the order X = ClO_4 (>337 K) > PF_6 (278 K) > AsF_6 (ca. 268 K) > SbF_6 (ca. 240 K) [173–176].

The phase transition consists of a cooperative mechanism with charge-ordering, anion order–disorder, Peierls-like lattice distortion, which induces a doubled lattice periodicity giving rise to $2k_F$ nesting, and molecular deformation (Fig. 11c). The high temperature metallic phase is composed of flat EDO molecules with +0.5 charge, while the low temperature insulating phase is composed of both flat monocations

Table 5 Selected CT solids having switching behavior under electric field and/or photon irradiation are compared with the sliding SDW[a]

Mechanism	CT solid	Electric field			Photon irradiation				Reference
		E_{th} (V cm^{-1})	T_{oper} (K)	Reference	Photon density	Sensitivity	Response time	T_{oper} (K)	
Mott or spin-Peierls	Cu·TCNQ	4×10^3	RT	[155, 156]	1,500 W cm^{-2}			RT	[157]
Insulator↔Mixed valency or metal[b]	Cu·TNAP	8×10^3	RT	[155]	–	–	–	–	
	K·TCNQ	$>10^3$	<230	[158]		20 dimers	ps[c]	<394	[159]
	Rb·TCNQ	–	–			<10	1.5 ps	RT	Okamoto H private communication
	Ag(DMDCNQI)$_2$	–	–				3–5 days	RT	[160]
Mott + Peierls Insulator ↔ Metal	Cu(d_6-DMDCNQI)$_2$	–	–		~10^8 W cm^{-2}	100e	20 ps	<78	[161]
CO insulator↔Metal	(EDO)$_2$PF$_6$	–	–		2×10^{18} cm^{-3}	500	1.5–2 ps	270	[162, 163]
	α-(ET)$_2$I$_3$	–	–				120 ns	4	[164]
	θ-(ET)$_2$CsZn(SCN)$_4$	3×10^2	<20	[165]	–	–	–	–	
	(NT)$_3$GaCl$_4$	1×10^2	RT	[166]					
	θ-(ET)$_2$CsCo(SCN)$_4$	1.4×10^2	4.2	[167]					
Neutral↔Ionic	TTF·QCl$_4$	3×10^3	190	[168, 169]	1.8×10^8 cm^{-3}	280–2,800	20–100 ps	77	[170, 171]
	TTeC$_1$–TTF·TCNQ						ps[c]	<300	[159]
Sliding SDW	(TMTSF)$_2$ClO$_4$	$<5 \times 10^{-4}$	1.5	[172]	–	–	–	–	

CO: charge ordered; E_{th} threshold electric field; T_{oper} operating temperature

[a]No experiments

[b]Mixed valency or metallic behavior has not been observed by thermal variation except Ag(DMDCNQI)$_2$

[c]According to the time-dependence of photo-induced reflectivity change [159], the response times for TTeC$_1$–TTF·TCNQ and K·TCNQ are a few times and a few hundred times faster, respectively, than that for TTF·*p*-chloranil

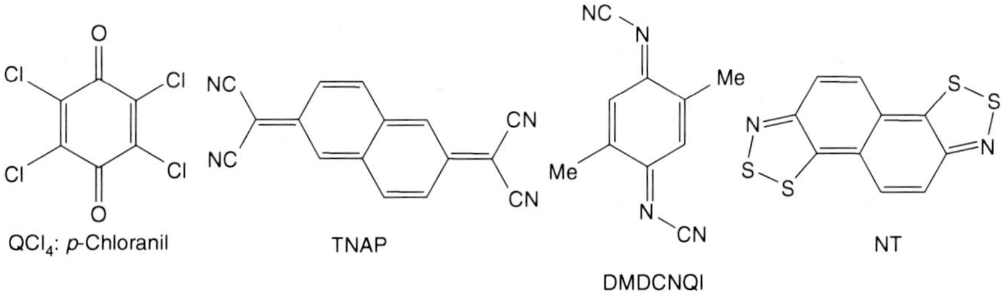

Scheme 12 Chemical structures of *p*-chloranil, TNAP, DMDCNQI, and NT

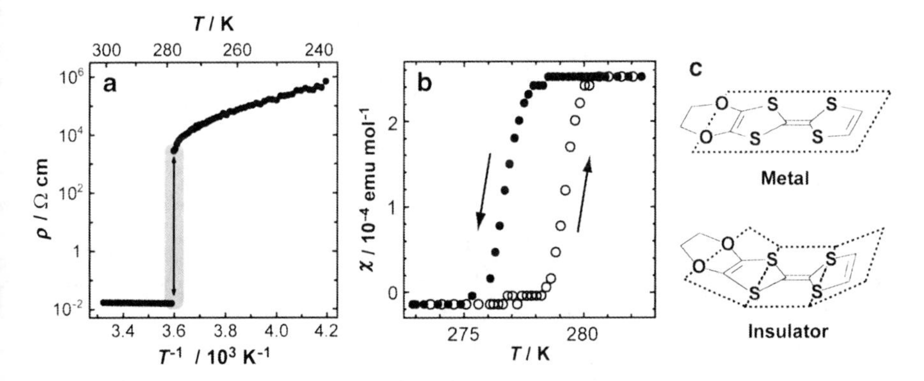

Fig. 11 Temperature dependence of (**a**) resistivity and (**b**) magnetic susceptibility of $(EDO)_2PF_6$. *Gray shadow* in (**a**) and *arrows* in (**b**) indicate the MI transition. (**c**) Molecular structures of neutral and monocationic EDO molecules [90]. Calculated Fermi surface of $(EDO)_2PF_6$ is depicted in Fig. 3e

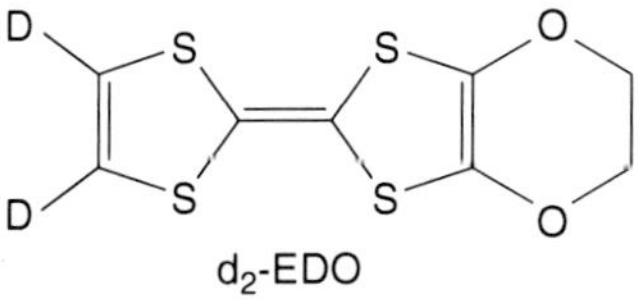

Scheme 13 Chemical structure of d_2-EDO

and bent neutral EDO molecules with charge-ordered stripes (+1, +1, 0, 0) [177, 178]. This stripe is different from the so far known (0, +1, 0, +1) stripe for θ-$(ET)_2MM'(SCN)_4$ [179], indicating that the neighbor-site Coulomb repulsion energy is not dominant compared to the transfer energy within the $(EDO^{1+})_2$ dimer.

Laser irradiation onto the insulating $(EDO)_2PF_6$ crystal induces a phase transition to the highly conductive state within a few picoseconds [162, 180]. The crystal surface was excited by laser irradiation with a pulse width of 0.12 ps. The excitation photon energy (1.55 eV) was nearly resonant to the CT band at 11.1×10^3 cm^{-1} (1.37 eV), directly reflecting the excitation of the charge ordered state.

The reflectance change $\Delta R/R$ from insulating to conductive states exhibits negative and positive maxima at the probe photon energy of 1.38 and 1.72 eV, respectively (Fig. 12). The life-time of the photo-induced conductive phase strongly depends on the excitation intensity. In case of 2×10^{18} cm^{-3} excitation condition, the reflectance change occurred within only about 1.5 ps. Therefore, it is said that the melting of the charge ordered state accompanied by the insulator-to-conductor phase conversion occurs within 1.5 ps just after excitation with threshold-like behavior (threshold photon density is 10^{18} cm^{-3}). The excitation intensity corresponds to a single excitation photon for ca. 500 molecules. Within the resolution time (1 μs), the electric conductivity was largely enhanced (more than five orders of magnitude) just after photo-excitation.

So far, among the switching systems by photo irradiation, TTF·*p*-chloranil, which is known to have strong electron–lattice coupling, has the highest sensitivity (280–2,800 molecules per photon) and fast response time (20–100 ps), though the operating temperature is low (77 K). The EDO system has high sensitivity (500 molecules per photon), fast response time (1.5–2 ps), and moderately high operating temperature (270 K for X = PF_6). To realize a molecular phase-switching device controllable by light irradiation with 1 ps response time (i.e., THz region), it is essential to develop a material that shows highly sensitive and ultra-fast PIPT phenomena near RT with high repeatability and durability. Such an ultra-fast transition has been observed in a purely electronic origin. Although the ultra-fast transition (within a few hundred fs) accompanied by the molecular conformational change has been observed for systems such as retinal in rhodopsin [181], this is a unimolecular nano-system. As for the meso-size scale switch, an electron–lattice coupled coherent system is necessary for an ultra-fast transition.

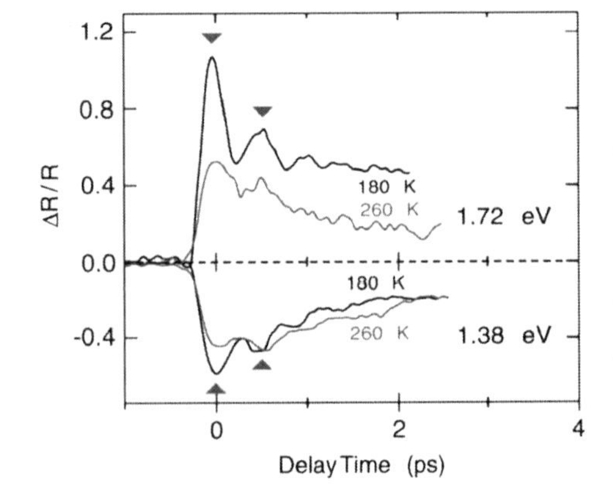

Fig. 12 Probe photon energy dependence of the time profile for the reflectance change $\Delta R/R$ observed at 180 K and 260 K. The pump photon energy (*E*//*b*) was 1.55 eV and the probe photon energy (*E*//*b*) was 1.72 and 1.38 eV for the *upper* and *lower panels*, respectively. The oscillations in $\Delta R/R$ relate molecular deformation modes [162]

5 Organic Superconductors of Charge Transfer Type

There are several comprehensive textbooks and review articles devoted to organic superconductors [3, 4, 182–186] which describe design and preparation, crystal and band structures, chemical, transport, magnetic, optical, and thermal properties, and theory.

5.1 Superconductors Based on Donor Molecules

5.1.1 TMTSF Superconductors

The most successful molecule by "heavy atom substitution" (Fig. 3) is TMTSF. Since the discovery of the first organic superconductor $(TMTSF)_2PF_6$ ($T_c = 1.1$ K at 0.65 GPa) by Jérome and Bechgaard [187, 188], eight quasi-one-dimensional superconductors $(TMTSF)_2X$ (X = PF_6, AsF_6, SbF_6, NbF_6, TaF_6, ClO_4, ReO_4, and FSO_3) with $T_c < 3$ K have been prepared [187–189]. Among them, the superconducting NbF_6 salt can only be prepared by using ionic liquid (1-ethyl-3-methylimidazolium)NbF_6 [189] and others by using more common tetrabutylammonium salts as the electrolyte. Salts with octahedral anions exhibited an MI transition at 11–17 K at ambient pressure due to SDW; superconductivity appeared with an on-set T_c of ca. 1 K at 0.6–1.2 GPa. Salts with (pseudo)tetrahedral anions exhibited an order–disorder transition for the anions which induced an MI transition for X = ReO_4 and FSO_3 at 177 K and 88 K, respectively, and superconductivity appeared at 1.2 K and 3 K, respectively, under 0.5–1 GPa. The isomorphous $(TMTTF)_2X$ salts displayed superconductivity under high pressure of 2.6–9 GPa with T_c less than 3 K for X = PF_6, SbF_6, BF_4, and Br [190–193].

$(TMTSF)_2ClO_4$ is the only ambient pressure superconductor among them; it did not show the Hebel–Slichter coherence peak, indicating that its superconductivity is of the non-*s*-wave type [194]. A generalized phase diagram, including $(TMTTF)_2X$ and $(TMTSF)_2X$, indicates that the superconducting phase neighbors the magnetic SDW phase (Fig. 13) [195].

5.1.2 ET Two-Dimensional Conductors and Superconductors

κ-Type Superconductors

TTF derivatives with peripheral addition of alkylchalcogen groups were found to be effective in increasing dimensionality of CT solids and suppressing the Peierls-type MI transition. The first ET two-dimensional organic metal down to low

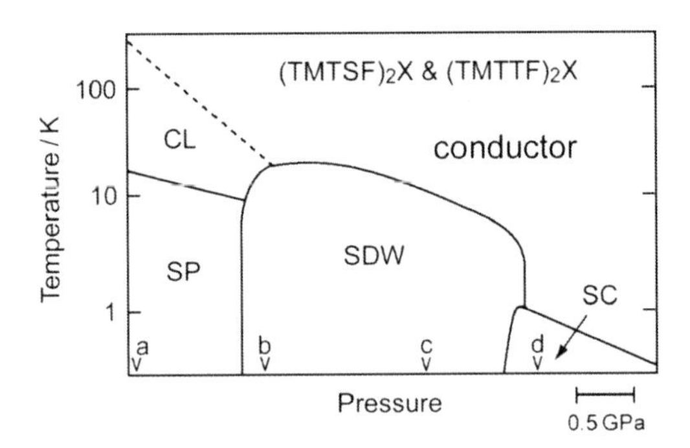

Fig. 13 Generalized phase diagram for the $(TMTSF)_2X$ and $(TMTTF)_2X$ by Jérome [195]. CL, SP, SDW, and SC refer to charge-localized (which corresponds to charge-ordered state), spin-Peierls, spin density wave, and superconducting states, respectively. (**a**) $(TMTTF)_2PF_6$, (**b**) $(TMTTF)_2Br$, (**c**) $(TMTSF)_2PF_6$, (**d**) $(TMTSF)_2ClO_4$

temperatures was $(ET)_2ClO_4$(1,1,2-trichloroethane) [196]. Since then, hundreds of ET solids have been prepared. Different kinds of ET···ET (π–π, S···S) and ET···anion (hydrogen bonds) intermolecular interactions, large conformational freedom of ethylene groups, flexible molecular framework, fairly narrow bandwidth (W), and strong electron correlations (U_{eff}) gave a rich variety of complexes with different crystal and electronic structures ranging from insulators to superconductors. About 60 ET superconductors are known so far. Currently β'-$(h_8\text{-ET})_2ICl_2$ (on-set T_c = 14.2 K at 8.2 GPa [197], mid-point T_c of 13.4 K is estimated), and κ-$(d_8\text{-ET})_2Cu[N(CN)_2]Cl$ (T_c = 13.1 K at 0.03 GPa) [198] show the highest T_c under pressure, while both are Mott insulators at ambient pressure. At ambient pressure, κ-$(d_8\text{-ET})_2Cu(CN)[N(CN)_2]$ shows the highest T_c of 12.3 K [199] followed by κ-$(h_8\text{-ET})_2Cu[N(CN)_2]Br$ (T_c = 11.8 K) [200].

The four κ-type superconductors κ-$(ET)_2CuL_1L_2$ (L_1, L_2 = Cl, Br, NCS, and N$(CN)_2$), which were discovered by us and the Argonne group [200–203], share some common structural and physical properties. Table 6 summarizes the two kinds of ligand in a salt, T_c of H- and D-salts (salt using h_8-ET and d_8-ET, respectively; Scheme 14), ratio of transfer interactions t'/t for triangle geometry of ET dimers (see Sect. 6.1), U/W, and year of discovery. Figure 14 shows the crystal structure of the prototype κ-$(ET)_2Cu(NCS)_2$, anion structures, calculated Fermi surface, and micrograph of single crystals. Although these ET salts have similar structural aspects, their transport properties differ (Fig. 15). κ-$(ET)_2Cu(CN)[N(CN)_2]$ (**29**) showed a monotonic decrease of resistivity with upper curvature down to T_c. κ-$(ET)_2Cu[N(CN)_2]Br$ (**31**) exhibited a similar behavior to that of κ-$(ET)_2Cu(NCS)_2$ (**30**) except that a metallic regime near RT was observed in **30**. κ-$(ET)_2Cu[N(CN)_2]Cl$ (**32**) showed a semiconductor (ε_g = 24 meV)–semiconductor (ε_g = 104 meV) transition at ca. 42 K due to an antiferromagnetic (AF) fluctuation resulting in a weak ferromagnet below 27 K [208, 209]. Under a weak pressure, it showed a similar temperature dependence to that of κ-$(ET)_2Cu[N(CN)_2]Br$.

Table 6 Four typical κ-type superconductors with T_c above 10 K (**29** – **32**) and a Mott insulator (**34**)

Number in Fig. 15 and Anion	Ligand		T_c/K		t'/t	U/W	Year	Reference
	L_1	L_2	H-salt	D-salt				
(**30**) $Cu(NCS)_2$	SCN	NCS	10.4	11.2 [3]	0.81–0.86	0.94	1988	[201]
(**31**) $Cu[N(CN)_2]Br$	$N(CN)_2$	Br	11.8	11.2 [204]	0.67	0.92	1990	[200]
(**32**) $Cu[N(CN)_2]Cl$	$N(CN)_2$	Cl	12.8 (0.03GPa)	13.1 [198]	0.75	0.90	1990	[202]
(**29**) $Cu(CN)[N(CN)_2]$	CN	$N(CN)_2$	11.2	12.3 [3, 199]	0.66–0.71	0.87	1991	[203]
(**34**) $Cu_2(CN)_3$	CN	CN (or NC)	6.8–7.3		1.06	0.9	1991	[203, 205–207]

Ligand L_1 forms infinite chain by the coordination to Cu(I). Ligand L_2 coordinates to Cu(I) as pendant
The t and t' values were calculated by the extended Hückel method

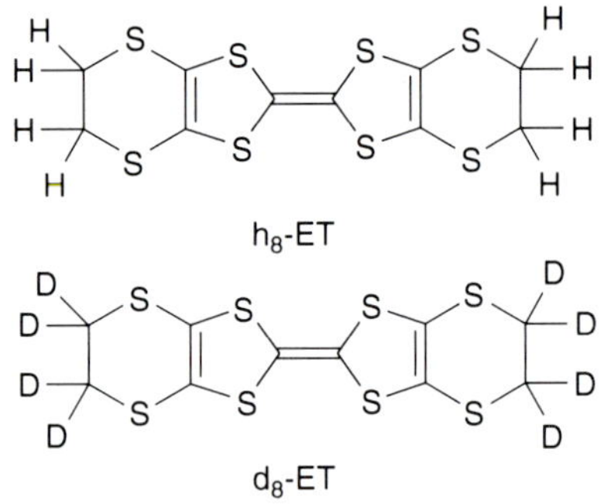

h_8-ET

d_8-ET

Scheme 14 Chemical structures of h_8-ET and d_8-ET

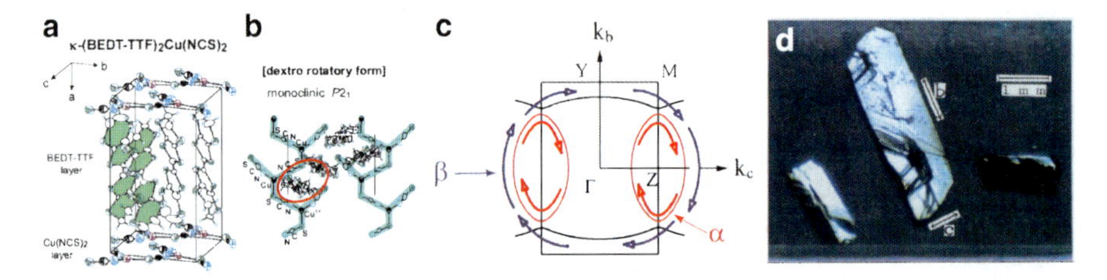

Fig. 14 κ-$(ET)_2Cu(NCS)_2$. (**a**) Crystal structure: two-dimensional conducting ET layer is sandwiched by the insulating anion layers along the *a*-axis (*bc*-plane is two-dimensional conducting plane). Two kinds of layers are Josephson coupled. (**b**) Anion structure: Cu···SCN···Cu···SCN··· forms zigzag infinite chain along the *b*-axis and other ligand SCN (Ligand L_2 in Table 6) coordinates to Cu(I) by N atom to make a space (indicated by *red ellipsoid*) to which an ET dimer fits. Picture is the dextrorotatory form. (**c**) Reflecting the crystal symmetry, the calculated Fermi surfaces of the $P2_1$ salts (κ-$(ET)_2Cu(NCS)_2$, κ-$(ET)_2Cu(CN)[N(CN)_2]$) showed the certain energy gap between a one-dimensional electron like Fermi surface and a two-dimensional cylindrical hole like one (α orbit), while such a gap is absent in the *Pnma* salts (κ-$(ET)_2Cu[N(CN)_2]Br$, κ-$(ET)_2Cu[N(CN)_2]Cl$). For κ-$(ET)_2Cu(NCS)_2$, electrons move along the closed ellipsoid (α-orbit) then at higher magnetic field (>20 T) electron hops from the ellipsoid to open Fermi surface to show circular trajectory (β-orbit, Magnetic breakdown). (**d**) Single crystals by electrooxidation

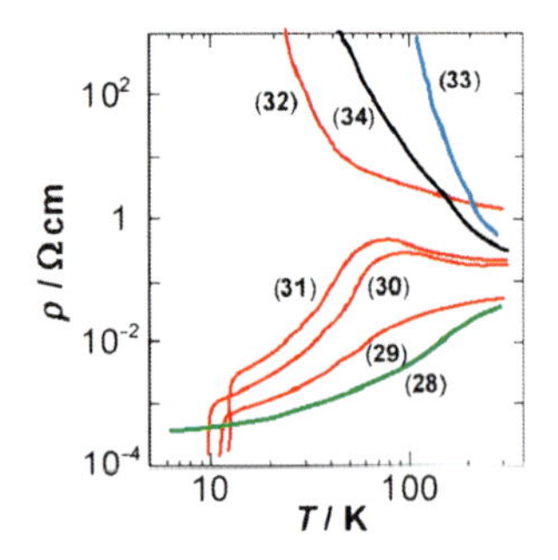

Fig. 15 Temperature dependences of resistivity of 10 K class superconductors κ-$(ET)_2Cu(CN)[N(CN)_2]$ (**29**), κ-$(ET)_2Cu(NCS)_2$ (**30**), κ-$(ET)_2Cu[N(CN)_2]Br$ (**31**), and κ-$(ET)_2Cu[N(CN)_2]Cl$ (**32**) are compared with those of a good metal with low T_c β-$(ET)_2AuI_2$ (**28**), strongly electron correlated insulator θ-$(ET)_2Cu_2(CN)[N(CN)_2]_2$ (**33**), and a Mott insulator κ-$(ET)_2Cu_2(CN)_3$ (**34**) at ambient pressure

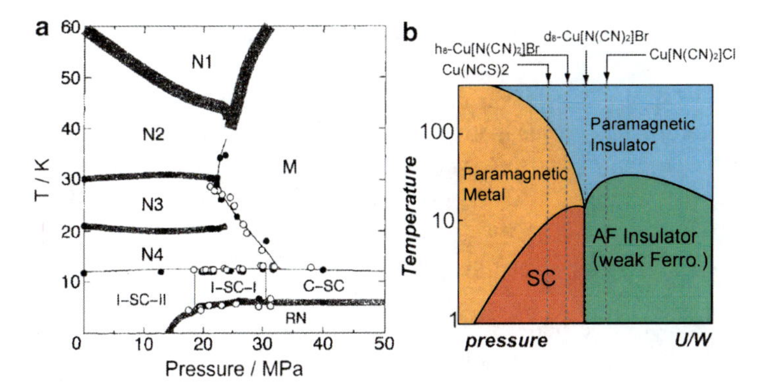

Fig. 16 (**a**) Phase diagram of κ-(h$_8$-ET)$_2$Cu[N(CN)$_2$]Cl determined from conductivity and magnetic measurements [213, 217, 218]. N1–N4: nonmetallic phase, M: metallic phase, RN: reentrant nonmetallic phase, I-SC-I, II: incomplete superconducting phase, S-SC: complete superconducting phase. N2 shows the low-dimensional AF fluctuation. N3 shows growth of three-dimensional AF ordered phase. N4: weak ferromagnetic phase. (**b**) Proposed phase diagram [211, 212]

κ-(h$_8$-ET)$_2$Cu[N(CN)$_2$]Cl showed a complicated T–P phase diagram as elucidated by Ishiguro and Ito et al. (Fig. 16a) [213–222]. Thoroughgoing studies under He gas pressure showed firm evidence of the coexistence of superconducting (**I-SC-II** phase: **I-SC** = incomplete superconducting) and AF phases [210, 216, 223, 224], where the radical electrons of ET molecules played both roles of localized and itinerant ones. Under a pressure of ca. 20–30 MPa another incomplete superconducting phase (**I-SC-I**) appeared and the complete superconducting (**C-SC**) phase resides adjacent to this phase at higher pressures. Below these superconducting phases, a reentrant nonmetallic (**RN**) phase was observed. Similar T–P phase diagrams were obtained for κ-(d$_8$-ET)$_2$X (X = Cu[N(CN)$_2$]Cl [218] and X = Cu[N(CN)$_2$]Br [219–221]) with a parallel shift of pressure. They occur at the higher and lower pressure sides of the κ-(h$_8$-ET)$_2$Cu[N(CN)$_2$]Cl for the Br and Cl salts, respectively. In contrast to the H salt, κ-(d$_8$-ET)$_2$Cu[N(CN)$_2$]Cl did not exhibit a coexistence of the superconducting and AF phases, and hence afforded AF resonance [222]. Increasing the distance between the ET dimers in Fig. 14a,b causes the transfer interactions between ET dimers to decrease; this may correspond to the decrease of band-width and to the increase of density of states at Fermi level $D(\varepsilon_F)$, and consequently T_c is expected to increase. According to this line of thought, higher T_c is expected for the salt having a larger anion spacing. Such a κ-type salt may be found near the border between poor metals and Mott insulators.

The Fermi surfaces of these salts have been studied by measuring the quantum oscillations [183] such as SdH (Shubnikov–de Haas) and dHvA and geometrical oscillations (AMRO, angle-dependent magnetoresistance oscillation) ([4], Appendix, pp 445–448). The Fermi surface of κ-(ET)$_2$Cu(NCS)$_2$ (Fig. 14c) calculated based on the crystal structure is in good agreement with those observed data [225].

Superconducting Characteristics of κ-Type Superconductors

1. H_{C2}. κ-$(ET)_2Cu(NCS)_2$ gave higher upper critical magnetic field H_{c2} values in the two-dimensional plane than the Pauli limited magnetic field H_{Pauli} [226, 227].
2. *Symmetry of superconducting state*. No Hebel–Slichter coherence peak was observed in either κ-$(ET)_2Cu(NCS)_2$ or κ-$(ET)_2Cu[N(CN)_2]Br$ in ^1H NMR measurements, ruling out a BCS s-wave state. The symmetry of the superconducting state of κ-$(ET)_2Cu(NCS)_2$ had been controversially described as normal BCS-type or non-BCS type; however, scanning tunneling spectroscopy showed d-wave symmetry with line nodes along the direction near $\pi/4$ from κ_a- and κ_c-axes [228, 229], and thermal conductivity measurements were consistent with this result [230]. κ-$(ET)_2Cu$ $[N(CN)_2]Br$ showed the same symmetry [231].
3. *Inverse isotope effect*. The inverse isotope effect has so far been observed for κ-$(ET)_2Cu(NCS)_2$ [3–18, 232] (Fig. 53 in [3]), κ-$(ET)_2Cu(CN)[N(CN)_2]$ [3, 199], and κ-$(ET)_2Cu[N(CN)_2]Cl$ [198], and the normal isotope effect for κ-$(ET)_2Cu[N(CN)_2]Br$ [204, 233].
4. *Phase diagram*. A proposed T–P phase diagram for κ-$(ET)_2CuL_1L_2$ by Kanoda (Fig. 16b), where only the parameter U/W is taken into account, includes the salts κ-$(ET)_2Cu(NCS)_2$, κ-$(ET)_2Cu[N(CN)_2]Br$, and κ-$(ET)_2Cu[N(CN)_2]Cl$ [211, 212]. However, the metallic behaviors of κ-$(ET)_2Cu(NCS)_2$ **30** above 270 K and of κ-$(ET)_2Cu(CN)[N(CN)_2]$ **29**, the whole nature of κ-$(ET)_2Cu_2(CN)_3$ **34**, and the low-temperature reentrant behavior of κ-$(ET)_2Cu[N(CN)_2]Br$ and κ-$(ET)_2Cu[N(CN)_2]Cl$ cannot be allocated in this diagram. This diagram is a simplified one, compared with the experimentally observed phase diagram [213, 222], but is convenient and useful to explain the general trends for these salts. The phase diagram and "geometrical isotope effect" [234] point out that T_c decreases with increasing pressure if only the parameter U/W or $D(\varepsilon_F)$ is taken into account. This tendency has been observed under hydrostatic pressure but not under uni-axial pressure (see next Section and Sects. 5.1.3 and 6.2).

Other ET Superconductors

One of the most intriguing ET superconductors is the salt with I_3 anion, which afforded α, α_t, β_L, β_H, δ, ε, γ, θ, and κ-type salts with different crystal and electronic structures. Among them, α, α_t, β_L, β_H, γ, θ, and κ-type salts are superconductors with $T_c = 7.2$, ~8, 1.5, 8.1, 2.5, 3.6, and 3.6 K, respectively [235–249]. The α-$(ET)_2I_3$ exhibited nearly temperature independent resistivity down to 135 K [235], at which charge-ordered MI transition occurred [236]. It has been claimed that α-$(ET)_2I_3$ has a zero-gap state with a Dirac cone type energy dispersion, hence with zero-effective mass and infinite mobility in the metallic state like graphene [237, 238]. Under hydrostatic pressure it became two-dimensional metal down to low temperatures (at 2 GPa). Very interestingly, however, it became superconductor under the uniaxial pressure along the a-axis (0.2 GPa, $T_c = 7.2$ K on-set), though along the b-axis it remained metal down to low temperature

(0.3–0.5 GPa) [239]. α-$(ET)_2I_3$ was able to be converted to mosaic polycrystal with $T_c \sim 8$ K by tempering at 70–100 °C for more than 3 days. The $T_c \sim 8$ K phase thus obtained exhibited a similar NMR pattern to that of the β_H-salt; however, it was isolated at ambient pressure so was designated as α_t-salt [240]. RDP films composed of ET in polycarbonate (2 wt%) were treated with CH_2Cl_2/I_2 vapors and then annealed at 137 or 155 °C to convert the α-$(ET)_2I_3$ to the superconducting α_t-$(ET)_2I_3$. The film is metallic and exhibits a broad superconducting transition below 7 K [241].

The β_L-salt was the first ambient pressure superconductor in the ET family with $T_c = 1.5$ K, reported by Yagubskii et al. [242]. The β_L-salt is characterized by having a superlattice appearing at 175 K with incommensurate modulations of ET and I_3 to each other [243]. Then the orientationally disordered ethylene groups near I_3 are ordered so as to make a new periodicity according to the incommensurate superlattice periodicity. The β_L-salt was converted to the high T_c phase, β_H-salt $T_c = 8.1$ K, by pressurizing (hydrostatic pressure) above 0.04 GPa by the suppression of the superlattice and then by depressurizing while keeping the sample below 125 K [244, 245]. The β_H-salt returned to β_L-salt when the salt was kept above 125 K at ambient pressure. The T_c of β_H-salt decreased with hydrostatic pressure monotonically; however, under the uniaxial stress the further T_c increase taking a maximum at a piston pressure of 0.3–0.4 GPa is observed for both directions parallel and perpendicular to the donor stack [246].

5.1.3 Superconductors of Other Donor Molecules

Besides ET, BO (two low T_c superconductors; see Sect. 3.4.2), TMTSF, and TMTTF superconductors, there are other superconductors (Scheme 15, numbers in bracket are the total members of each superconductor and the highest T_c) of CT salts based on symmetric (BETS [250], BEDSe-TTF [251], and BDA-TTP [252–255]) and asymmetric donors (ESET-TTF [256], *S,S*-DMBEDT-TTF [257], *meso*-DMBEDT-TTF [258, 259], DMET [260], DMEDO-TSeF [261, 262], DODHT [263], TMET-STF [264], DMET-TSeF [265], DIETS [266], EDT-TTF [267], MDT-TTF [268, 269], MDT-ST [270, 271], MDT-TS [272], MDT-TSF [273–276], MDSe-TSF [277], and DTEDT [278]). The reported T_cs of most superconductors recently prepared are the on-set T_c values that are approximately 0.5–1 K higher than the mid-point T_c values. T_cs of them are less than 10 K.

DMEDO-TSeF afforded eight superconductors. Six of them are κ-(DMEDO-TSeF)$[Au(CN)_2]$(solvent) and their T_cs (1.7–5.3 K) are tuned by the use of cyclic ethers as solvent of crystallization [262].

β-$(BDA\text{-}TTP)_2X$ ($X = SbF_6$, AsF_6) exhibited a slight T_c increase at the initial stage of uniaxial strain parallel to the donor stack and interlayer direction while T_c decreased perpendicular to the donor stack [253]. θ-$(DIETS)_2[Au(CN)_4]$ exhibited superconductivity under uniaxial strain parallel to the c-axis ($T_c = 8.6$ K at 1 GPa), though under hydrostatic pressure a sharp MI transition remained even at 1.8 GPa [266].

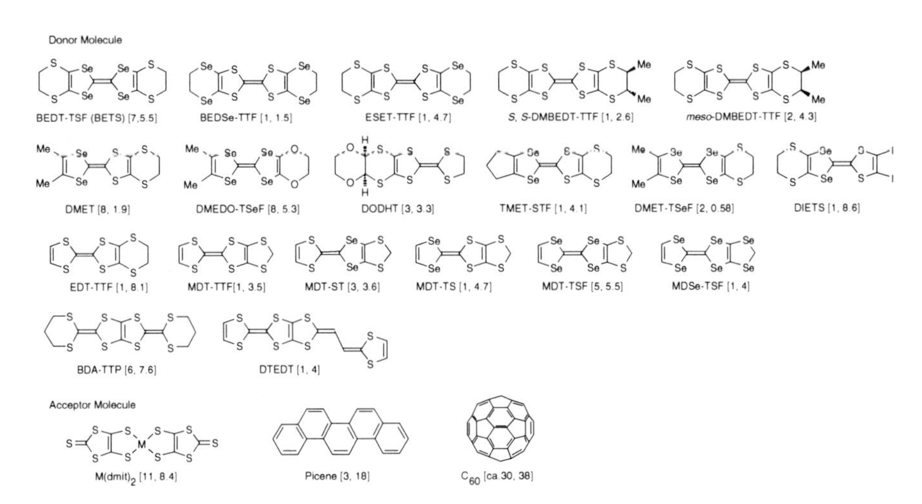

Scheme 15 Component molecules for molecular superconductors except TMTSF, TMTTF, ET and BO systems. *Numbers* in *bracket* are the total members of each superconductor and the highest T_c (in K)

MDT-ST, MDT-TS, and MDT-TSF superconductors have noninteger ratio of donor and anion molecules such as (MDT-TS)$(AuI_2)_{0.441}$ making the Fermi level different from the conventional 3/4 filled band for TMTSF and ET 2:1 salts [270–276]. The Fermi surface topology of (MDT-TSF)X (X = $(AuI_2)_{0.436}$, $(I_3)_{0.422}$), and (MDT-ST)$(I_3)_{0.417}$ has been studied by SdH and AMRO [271, 274–276].

κ-(MDT-TTF)$_2$AuI$_2$ ($T_c = 3.5$ K) exhibited a Hebel–Slichter coherent peak just below T_c, indicating a BCS-type gap with *s*-symmetry [269], while *d*-wave like superconductivity has been suggested for β-(BDA-TTP)$_2$SbF$_6$ [254, 255].

The most intriguing phenomenon is the reentrant superconductor–insulator–superconductor transition under a magnetic field for (BETS)$_2$FeCl$_4$. Kobayashi et al. have developed BETS salts formed with tetrahedral anions MX_4 (M: Fe and Ga, X: Cl and Br) [279–283]. Especially salts with λ-[284, 285] or κ-type packings [279–281] have been studied in terms of the competition of magnetic ordering and superconductivity. The initial notable finding is for λ-(BETS)$_2$GaCl$_4$ with a superconducting transition at 8 K (mid-point 5.5 K) and for λ-(BETS)$_2$FeCl$_4$ with a coupled AF and MI transitions at 8.3 K. For the FeCl$_4$ salt, a relaxor ferroelectric behavior in the metallic state below 70 K [286] and a firm nonlinear electrical transport associated with the negative resistance effect in the magnetic ordered state have been observed [287]. Moreover, it has been found by Uji et al. that the FeCl$_4$ salt shows a field-induced superconducting transition under a magnetic field of 18–41 T applied exactly parallel to the conducting layers [285]. Interestingly, the λ-(BETS)$_2$Fe$_x$Ga$_{1-x}$X$_4$ passes through a superconducting to insulating transition on cooling [288]. The κ-(BETS)$_2$FeX$_4$ (X = Cl, Br) are AF superconductors, where the transition temperatures were $T_N = 2.5$ K and $T_c = 1.1$ K for the Br salt, and $T_N = 0.45$ K and $T_c = 0.17$ K for the Cl salt [289]. Similar phenomena, namely AF, ferromagnetic, or field induced superconductivity, have been observed in several inorganic solids such as the Chevrel phase [290] and heavy-fermion systems [291]. Recently it has been reported that λ-$(BETS)_2GaCl_4$ exhibited superconductivity in the minute size of four pairs of $(BETS)_2GaCl_4$ based on the scanning tunneling microscopy study [292].

5.2 *Superconductors Based on Acceptor Molecules*

There are three kinds of superconductors based on acceptor molecules: $M(dmit)_2$ (Scheme 15, M = Ni, Pd) [293], picene [294], and C_{60} [141–143]. C_{60} system has the highest T_c for molecular superconductors ($T_c = 38$ K) followed by the picene one ($T_c = 18$ K); however, those two systems are very unstable chemically (caused partly by very weak electron accepting ability) and decompose immediately at ambient condition.

5.2.1 dmit System

Eleven superconductors were prepared based on $M(dmit)_2$ (three for M = Ni, eight for M = Pd) and their T_cs are less than 8.4 K. Only one showed superconductivity at ambient pressure (EDT-TTF[$Ni(dmit)_2$], $T_c = 1.3$ K) [295]. Superconducting LB films of dimethylbis(tetradecyl)ammonium[$M(dmit)_2$] ($T_c < 3.9$ K) have been reported [296].

5.2.2 Picene System

Very recently, Kubozono and his coworkers reported new organic superconductors: alkali-metal doped picene compounds [294]. Although their shielding fractions are relatively small (<15%), the bulk superconducting phase was observed below 7.0 K for $K_{2.9}$picene, 18 K for $K_{3.3}$picene, and 6.9 K for $Rb_{3.1}$picene, in which the LUMO + 1 orbital for picene would be in a half-filled electronic state. For $K_{3.3}$picene, T_c is significantly higher than that of K-doped graphite ($T_c \sim 5.5$ K) [297] and comparable to that of K_3C_{60} ($T_c = 18$ K) [141]. The Pauli-like paramagnetic susceptibility is higher than that of $K_{2.9}$picene with lower T_c, suggesting BCS-type superconductivity. At present, although the crystal structures of the doped compounds are unclear, the refined lattice parameters are indicative of the deformation of the herringbone structure of pristine picene and the intercalation of alkali dopants within the two-dimensional picene layers [294].

5.2.3 C_{60} System

The icosahedral C_{60} molecule with I_h symmetry has triply degenerate LUMO and LUMO + 1 orbitals with t_{1u} and t_{1g} symmetries, respectively, and C_{60} can accept up to 12 electrons.

Immediately after the isolation of macroscopic quantities of C_{60} solid [298], highly conducting [299] and superconducting [141] behaviors were verified for the K-doped compounds prepared by a vapor–solid reaction (Haddon, Hebard, et al.). Crystallographic study based on the powder X-ray diffraction profile revealed that the composition of the superconducting phase is K_3C_{60} and the diffraction pattern can be indexed to be a face-centered cubic (fcc) structure with a three-dimensional electronic pathway [300]. The lattice parameter ($a = 14.24$ Å) is apparently expanded relative to the undoped cubic C_{60} ($a = 14.17$ Å). The superconductivity has been observed for many A_3C_{60} (A: alkali metal), e.g., Rb_3C_{60} ($T_c = 29$ K [301]), Rb_2CsC_{60} ($T_c = 31$ K [302]), and $RbCs_2C_{60}$ ($T_c = 33$ K; the highest T_c among the ambient pressure C_{60} superconductors, reported by Tanigaki et al. [302]), and their structures are analogous to that of K_3C_{60} with varying lattice constants. The T_c varies monotonically with the lattice constant, independently of the type of the alkali dopant [302, 303]. This behavior can be interpreted in terms of BCS theory, in agreement with the observation of Hebel–Slichter coherence peaks for NMR [304] and μSR [305] and the normal isotope effect; namely T_c decreases by the isotopic substitution $^{12}C \rightarrow {}^{13}C$ [306].

Keeping the C_{60} valence invariant (−3), the intercalation of NH_3 molecules (e.g., $(NH_3)K_3C_{60}$) results in a lattice distortion from cubic to orthorhombic, accompanied by the appearance of AF ordering instead of superconductivity [307]. Changing the valence in cubic system also has a pronounced effect on T_c. For example, T_c in $Rb_{3-x}Cs_xC_{60}$ prepared in liquid ammonia gradually increases as the mixing ratio approaches $x = 2$ [308]. Further increasing the nominal ratio of Cs leads to a sizable decrease of T_c, despite the fact that the lattice keeps the fcc structure for $x < 2.65$. Such a band-filling control has been realized for $Na_2Cs_xC_{60}$ ($0 \leq x \leq 1$) [309] and Li_xCsC_{60} ($2 \leq x \leq 6$) [310], and shows that T_c decreases sharply as the valence state on C_{60} deviates from −3.

According to the relationship between the lattice volume and T_c as described, cubic Cs_3C_{60} would be an ultimate candidate for a higher T_c superconductor, but the conventional vapor–solid reaction affords only the thermodynamically stable CsC_{60} and Cs_4C_{60} phases. In 1995, noncubic Cs_3C_{60} was obtained by a solution process in liquid ammonia, and the superconductivity was observed below 40 K under an applied hydrostatic pressure of 1.4 GPa [311].

In 2008, the A15 or body-centered cubic (bcc) Cs_3C_{60} phase, which shows bulk superconductivity under applied hydrostatic pressure, was obtained, together with a small amount of by-products of body-centered orthorhombic (bco) and fcc phases, by a solution process in liquid methylamine (Prassides, Rosseinsky, et al.) [312]. Interestingly, the lattice contraction with respect to pressure results in an increase in T_c up to around 0.8 GPa, above which T_c gradually decreases. The highest T_c is

38 K, which exceeds the value of $RbCs_2C_{60}$ (33 K). The trend in the initial pressure range is not explicable within the simple BCS theory. Under an ambient pressure, on the other hand, the A15 Cs_3C_{60} shows an AF ordering below 46 K, verified by means of ^{133}Cs NMR and μSR [313]. Very recently, it has been found that the fcc phase also shows an AF ordering at 2.2 K under an ambient pressure, and a superconducting transition at 35 K under an applied hydrostatic pressure of about 0.7 GPa [314]. Note that T_c of both phases follows the universal relationship for A_3C_{60} superconductors in the vicinity of the Mott boundary.

Some noncubic superconductors have been obtained for $Yb_{2.75}C_{60}$ ($T_c = 6$ K) [315], $Sm_{2.75}C_{60}$ ($T_c = 8$ K) [316], Ba_4C_{60} ($T_c = 6.7$ K) [317, 318], and Sr_4C_{60} ($T_c = 4.4$ K) [318]. Eu_6C_{60} with bcc packing undergoes a ferromagnetic transition at 12 K, arising from Eu^{2+} cations with $S = 7/2$ spin [319], and shows a giant negative magnetoresistance arising from a significant π-f coupling between the conduction electrons on C_{60} and localized 4*f*-electrons on Eu [320]. Ce_xC_{60} shows a coexistence of superconductivity and ferromagnetism below 13.5 K, although its crystal structure and composition are currently unclear [321].

C_{60} doped with K ($T_c < 8.1$ K) [322] and Rb ($T_c < 23$ K) [323] exhibit superconductivity on LB films, which was detected by the AC complex magnetic susceptibility or low magnetic field microwave absorption measurements. However, both the structural disorder inherent to the LB films and the low-dimensional nature of the thin-layer structure severely prohibit the observation of superconductivity by resistivity measurements.

As mentioned, the (super)conductors based on C_{60} and picene are chemically very unstable, and immediately decompose on exposure to air. Crystal engineering to protect such (super)conductors against air and moisture is essential for further investigation.

5.3 *Metallic Doped Polymers and Unidentified Organic Superconductors*

5.3.1 Polymer Superconductors and Metallic Polymer

Little's 1964 proposal for high T_c superconductivity was based on a polymer system having both a conduction path and highly polarizable pendants, which mediate the formation of Cooper pairs in the conduction path by electron–exciton coupling [324]. There are at least two inorganic polymer superconductors, poly(sulfur nitride) $(SN)_x$ [325–327] and black phosphorus [328], with crystalline forms. The covalent bond of the golden crystal of $(SN)_x$ has an ionic character by a partial electron transfer (0.4 *e*) from S to N. Weak interchain interactions between SN polymer chains give it a quasi-one-dimensional nature. Along the polymer chain, the σ_{RT} value is 1–4 × 10^3 S cm^{-1}, rising by a factor of ca. 10^2 at 4.2 K and superconductivity appeared at $T_c = 0.26$ K and T_c increased under pressure ($T_c \leq 3$ K). Black phosphorus has a two-dimensional layer structure with $\sigma_{RT} \sim 1$ S cm^{-1} and exhibited superconductivity under pressure ($T_c \sim 6$ K, 16 GPa). When the sample was pressurized after cooling the sample at ambient pressure to 4.2 K, T_c increased considerably ($T_c = 10.7$ K, 29 GPa). These are metallic and superconducting polymers without doping.

After Little's proposal, many researchers have pursued such an exciting system in vain. Even metallic behavior was rarely seen in doped organic polymers, gels, and actuators. As mentioned in Sect. 3.4.4, MC_{60} with linearly polymerized $C_{60}^{\bullet-}$ exhibited one-dimensional (M = Rb, Cs) or three-dimensional (M = K) metallic behavior [144]. Recently a doped polyaniline was reported to exhibit a metallic temperature dependence for a crystalline polymer; chemical oxidation of monomers grew crystallite polyaniline [329]; early doping studies on polypyrrole (PF_6) and poly(3,4-ethylene-dioxythiophene)X (X = PF_6, BF_4, and CF_3SO_3) prepared by electrooxidation at low temperatures also showed a metallic temperature dependence below 10–20 K (Scheme 16) [330, 331].

So far no organic polymers have been confirmed to show superconductivity, in spite of several unconfirmed polymer superconductors mentioned below. It should be remembered that the sample should be well oriented, otherwise the disorder inherent to organic polymers will destroy the superconductivity.

5.3.2 Unidentified Superconducting Organics

There have been several experimental reports describing very high T_cs or very fascinating materials with low T_c though these data are not reproducible.

The most puzzling one was the Na salt of cholanate with $T_c = 277.0$ K reported in 1976, though cholanic acid does not have π-electrons (Scheme 17) [332]. The superconductivity was detected by conductivity and magnetic measurements. The authors mentioned that the salt showed an insulator to superconducting transition and the transition was fractional. Very recently, a biological compound (double-stranded DNA) was reported to exhibit proximity-induced superconductivity below 1 K [119]. The deionized DNA molecules are insulating and superconductivity cannot be expected.

In 1978, aniline black was reported to show possible superconductivity by the irreversible drop of resistivity by 10^6 in the I–V measurements at RT around 250 V (T_c = 295.5 K) [333]. In 1989, a resistance drop by nine orders of magnitude and a strong

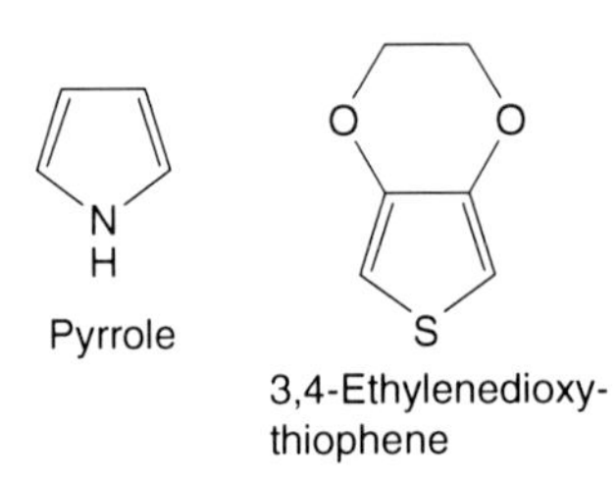

Scheme 16 Chemical structures of pyrrole and 3,4-ethylene-dioxythiophene

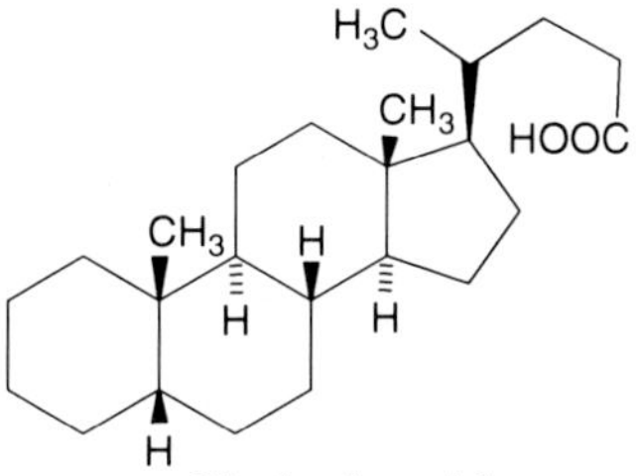

Cholanic acid

Scheme 17 Chemical structure of cholanic acid

diamagnetism which was destroyed by magnetic field were reported on polypropylene oxidized for 3 years (T_c = 293 K) [334]. Oligo- and polyphthalocyanines have been reported to exhibit T_c of 83 and 92 K, detected by LFMA (low-field magnetic absorption) [335], which is very sensitive but sometimes gives false signals.

6 Spin Disordered State (Quantum Spin Liquid State) Neighboring Superconductivity

Figures 13 and 16b, and also the phase diagrams for electron-correlated C_{60} [336], cuprate oxide, iron pnictide, and heavy fermion systems [337], indicate that a magnetic ordered state (SDW, AF) can be located in a phase diagram next to the superconducting state. However, for the system having very strong electron frustrations a new exotic magnetic state (quantum spin liquid state without any magnetic order) appeared. That spin liquid state has only been predicted theoretically [338]. Furthermore, the superconducting state neighbors directly to the spin liquid state for the Mott insulator κ-$(ET)_2Cu_2(CN)_3$ **34** (in Table 6), which has a larger anion space than those for four 10 K class superconductors **29–32** in Table 6. Its Fermi surface has a similar shape to that of κ-$(ET)_2Cu(NCS)_2$ [339–341].

6.1 New Spin State Originated from Strong Spin Frustrations: Quantum Spin Liquid State

Figure 17a shows the packing motif of κ-$(ET)_2Cu_2(CN)_3$ where an ET dimer, which is encircled by an ellipsoid, is a unit with $S = 1/2$ spin. The geometry of the spin lattice (Fig. 17b) is a triangular lattice with two kinds of transfer integrals, $t = (|t_p| + |t_q|)/2$ and $t' = t_{b2}/2$ [343, 344], suggesting that the system has a strong spin frustration. All κ-type salts have such triangular spin lattice, though the magnitude of the frustration depends on the shape of the triangle and the ratio t'/t is a good parameter to estimate the frustration. The t'/t values for κ-$(ET)_2X$ in Table 6 calculated by extended Hückel method indicate that the spins in κ-$(ET)_2Cu_2(CN)_3$ ($t'/t = 1.06$) are severely frustrated compared with the other κ-type salts ($t'/t = 0.7$–0.9). Even though the previous density-function theory (DFT) calculations [345, 346] gave a little higher t'/t and the recent DFT calculations using a generalized-gradient-approximation gave the smaller t'/t (~0.8) [347, 348] than unity, κ-$(ET)_2Cu_2(CN)_3$ exhibited the unprecedented features caused by strong spin frustration.

Figure 17c,d compares the line shapes of ^{1}H NMR absorption of κ-$(ET)_2$ $Cu_2(CN)_3$ and κ-$(ET)_2Cu[N(CN)_2]Cl$, respectively. κ-$(ET)_2Cu[N(CN)_2]Cl$ exhibited a drastic change below 27 K owing to the formation of three-dimensional AF ordering, while, the absorption band of κ-$(ET)_2Cu_2(CN)_3$ remained almost invariant down to 32 mK, indicating a nonspin-ordered state: the quantum spin liquid state [342, 349–357].

The three Mott insulators, κ-$(ET)_2Cu[N(CN)_2]Cl$, deuterated κ-$(ET)_2Cu[N(CN)_2]Br$, and κ-$(ET)_2Cu_2(CN)_3$, have nearly the same U_{eff}/W (~0.9); however, the electronic ground states of them are different. The spins in the former two salts condensed into the AF state because of the less frustrated spin geometry in κ-$(ET)_2Cu[N(CN)_2]Cl$ ($t'/t \sim 0.75$) and D-salt of κ-$(ET)_2Cu[N(CN)_2]Br$ ($t'/t = 0.68$ for the H-salt). Since the spin frustration is quite significant in κ-$(ET)_2Cu_2(CN)_3$ because of the equilateral triangle spin geometry, the formation of the AF and superconducting states is suppressed at ambient pressure and the unprecedented spin liquid state appears instead.

Controversial discussions ensued concerning the magnitude of the gap of the spin liquid state. Specific heat measurements suggested a gapless nature [358], while thermal conductivity measurements suggested a small gap [359]. Furthermore, there is an abnormality in lattice near 5–6 K which was detected by ^{13}C NMR [357] and thermal expansion [360] measurements, indicating that the lattice is not frozen even at 5–6 K.

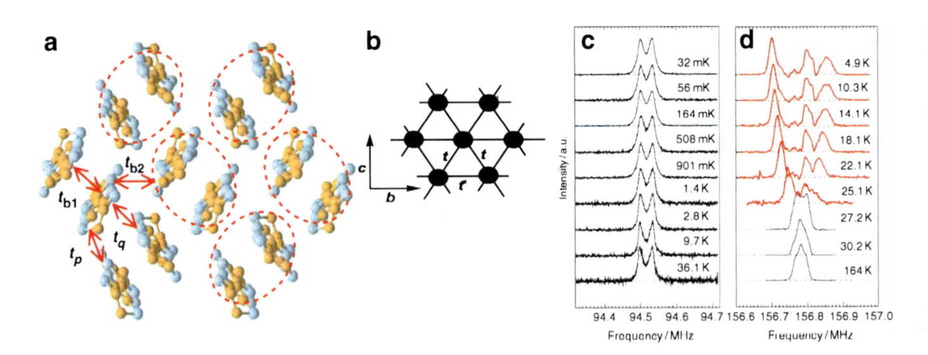

Fig. 17 (**a**) Donor packing pattern of κ-$(ET)_2Cu_2(CN)_3$ along the *a*-axis (transfer integrals; $t_{b1} = 22$ meV, $t_{b2} = 12$ meV, $t_p = 8$ meV, and $t_q = 3$ meV) and (**b**) triangular spin lattice ($t'/t = 1.06$; $t' = t_{b2}$, $t = (|t_p| + |t_q|)/2$) composed of the ET dimer which is encircled by an *ellipsoid* in (**a**) and represented by *closed circle* in (**b**). Line shape of ^{1}H NMR of (**c**) κ-$(ET)_2Cu_2(CN)_3$ [342] and (**d**) κ-$(ET)_2Cu[N(CN)_2]Cl$ [209]

6.2 Emergence of Superconducting State Next to Spin Liquid State

The uni-axial strain method can apply strain only along one direction. For κ-$(ET)_2Cu_2(CN)_3$, uni-axial strain changed the temperature dependence of resistivity from that depicted in Fig. 15 to be similar to those of **30** and **31**, namely semiconductor–metal–superconducting behavior. A superconducting state readily appeared nearly above 0.1 GPa in both directions along the *b*- (Fig. 18 right figure: t'/t increases in this direction) and *c*- (Fig. 18 left figure: t'/t decreases in this direction) axes [361], without passing through the spin-ordered state. The appearance of the superconducting state is ascribed to the release of the strong spin frustrations since the t'/t deviates from unity in both directions. Within the *bc*-plane, the superconducting state appeared above 0.1 GPa [$T_c = 3.8$ K (//*b*), 5.8 K (//*c*)] and T_c increased up to 6.8 K (//*b*, 0.5 GPa) and 7.2 K (//*c*, 0.3 GPa). Along the *a**-axis, the superconducting state appeared above 0.3 GPa.

A plot of T_c vs T_{IM} (Fig. 19), which is a Mott insulator–metal transition temperature, indicates that the pressure dependence of T_c behaves similarly in both directions. However, the uni-axial results are considerably different from those resulting under hydrostatic pressure, which extinguish the superconducting phase above 0.3 GPa. κ-$(ET)_2Cu_2(CN)_3$ was converted to a metal and superconductor by applying hydrostatic pressure through a Mott insulator–metal transition at 13–14 K with a resistivity drop by 10^5 [354, 362]. The critical pressure and T_c under hydrostatic pressure differ within the literature [203, 205–207, 354, 362], reflecting anisotropic nature of T_c and high sensitivity to the inclusion of Cu^{2+} and $N(CN)_2$ anion as described in the next section. The uni-axial method afforded (1) a much higher T_c value, (2) an increase of T_c at the initial pressure region, (3) an anisotropic pressure dependence, and (4) superconducting phase remaining at higher pressure compared with that of hydrostatic results. There have been many hydrostatic pressure studies on systems having very anisotropic electronic structures. According to the results in Fig. 19 where the hydrostatic pressure results do not agree with

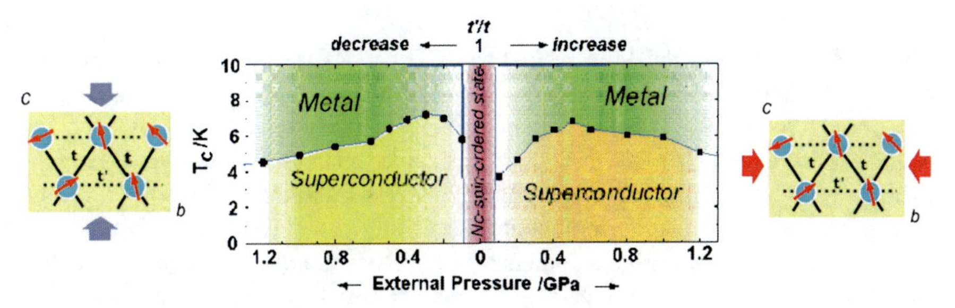

Fig. 18 Temperature–uniaxial pressure phase diagram in the low temperature region of κ-$(ET)_2Cu_2(CN)_3$ [361]. The strain along the *c*-axis corresponds to decrease t'/t (*left side*), while the stress along the *b*-axis increases t'/t (*right side*)

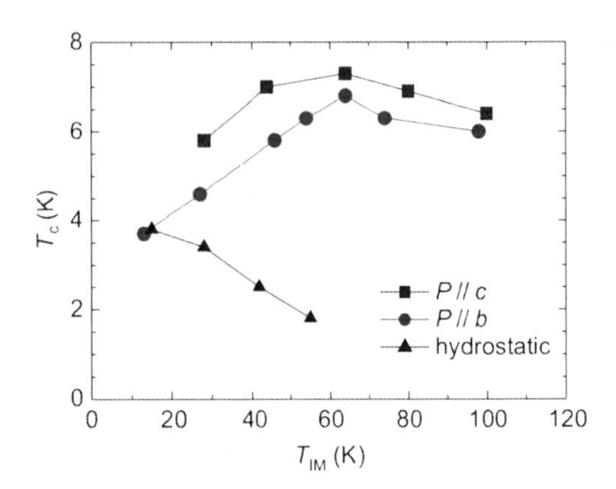

Fig. 19 Pressure dependence of on-set T_c of κ-$(ET)_2Cu_2(CN)_3$ by the uni-axial strain and hydrostatic pressure methods

any of those along the principal axes or their averaged ones, it is very difficult to understand logically the hydrostatic pressure results.

The emergence of the superconducting state is interpreted by both the increase of U_{eff}/W and the deviation of t'/t from unity. The increase of T_c in the initial pressure regime is ascribed to the reduction of the spin frustration. The following decrease of T_c in whole measured directions is explained by the decrease of $D(\varepsilon_F)$ owing to the increase of W. The appearance of superconducting state immediately after the release of the spin frustration in the spin liquid state is an indication of the importance of the magnetic mediation for superconductivity. The uni-axial strain experiments, which included other κ-type superconductors, clearly revealed that T_c increased as the U/W approaches unity and as the t'/t departs from unity (Fig. 20) [363].

Following κ-$(ET)_2Cu_2(CN)_3$, five materials [364], including $EtMe_3Sb[Pd(dmit)_2]_2$ as an organic solid [365], have been found to have quantum spin liquid states, however, superconductivity has been confirmed only for κ-$(ET)_2Cu_2(CN)_3$.

6.3 *Control of* U/W *and Band Filling:* κ′-$(ET)_2Cu_2(CN)_3$

The anion structure of the Mott insulator κ-$(ET)_2Cu_2(CN)_3$ **34** in Fig. 21 revealed the disorder in the position of C and N atoms of the C≡N groups (L_2 part; Table 6, Fig. 21a) [205–207, 370], due to the existence of an inversion center. However, ^{13}C NMR experiments observed very sharp resonance lines due to the homogeneous local field in the metallic state [371], which suggests that the C/N disorder, if any, does not work as the disorder potential in the conduction layer.

Owing to the very similar geometrical shape, size, and equal charge between $Cu(CN)_2$[N≡C–Cu–C≡N for X–Cu–Y, with obtuse bond angle of 120.1°] and $N(CN)_2$[N≡C–N–C≡N, bond angle 116.7°], they were nearly freely replaceable with each other in the anion layer, resulting in comparable lattice parameters among κ-$(ET)_2Cu(CN)_3$, κ-$(ET)_2Cu(CN)[N(CN)]_2$, and their alloy, κ′-salt (Fig. 21a–c).

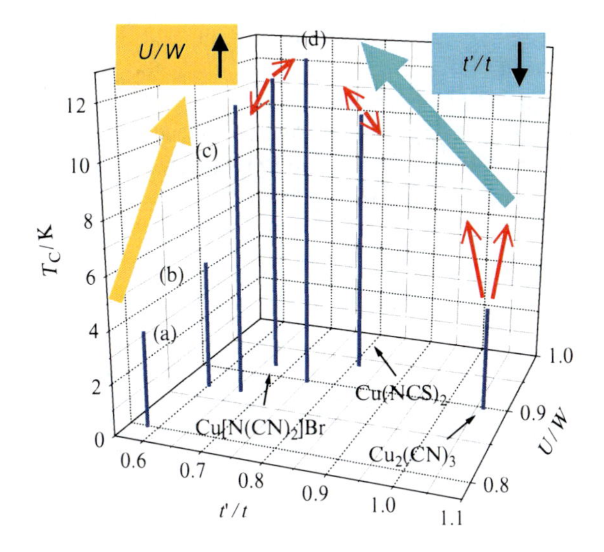

Fig. 20 T_c of κ-$(ET)_2X$ salts are plotted as function of t'/t and U/W [363]. X = I_3 (**a**), $Ag(CN)_2{\cdot}H_2O$ (**b**), Cu(CN)$[N(CN)_2]$ (**c**), and Cu$[N(CN)_2]$Cl (**d**). *Blue* and *yellow arrows* indicate the direction of t'/t decreases and U/W increases, respectively. *Red arrows* correspond to the change of T_c by applying uni-axial strain

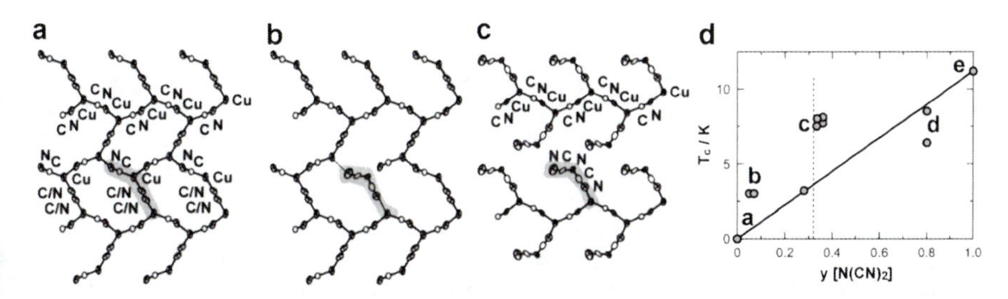

Fig. 21 The anion structures of (**a**) κ-$(ET)_2Cu_2(CN)_3$ and (**c**) κ-$(ET)_2Cu(CN)[N(CN)_2]$. (**b**) A schematic figure of κ-$(ET)_2(Cu^{+}{}_{2-x-y}Cu^{2+}{}_{x})\{(CN)_{3-2y}[N(CN)_2]_y\}$ with $y \sim 0.1$. (**d**) Relation between the content of $N(CN)_2$, y and T_c in several crystals of κ' salt: κ-$(ET)_2(Cu^{1+})_{2-x-y}(Cu^{2+})_x(CN)_{3-2y}[N(CN)_2]_y$. As for points **a**–**e**, see text. *Dashed line* indicates the samples of $y \sim 0.3$ [362, 366–369]

It was found that the exact chemical formula of κ'-$(ET)_2Cu(CN)_3$ was κ-$(ET)_2(Cu^{1+}{}_{2-x-y}Cu^{2+}{}_{x})\{(CN)_{3-2y}[N(CN)_2]_y\}$ and its transport natures were governed by the amount of Cu^{2+} (x) and ligand $[NC{-}N{-}CN]^-$ (y) [368]. At $x = 0$ and $y = 0$, the salt is a Mott insulator κ-$(ET)_2Cu_2(CN)_3$ (point **a** in Fig. 21d), while the other extreme side ($x = 0$, $y = 1$) is κ-$(ET)_2Cu(CN)[N(CN)_2]$ with $T_c = 11.2$ K at ambient pressure (point **e**). By changing both x (80–1,200 ppm) and y [preferential values of y are 0.05 (point **b**), 0.3–0.4 (**c**), 0.8 (**d**)], the T_c was tuned from 3 to 11 K. At $y = 0.3$–0.4, the T_c ranged from 3 to 10 K and the crystals with different T_c had different x values, indicating that the charge of ET was modified from +0.5 to +0.5 $(1 - x)$, that corresponds to the change of chemical potential, i.e., band-filling. T_c increased with increasing x (= the content of Cu^{2+}) up to 400 ppm, and then T_c decreased.

These experimental facts indicate that this system can be an excellent model of band-filling control, and will be a good candidate for making a superlattice composed of Mott insulator/superconductor hetero-junctions. It should be emphasized that their lattice parameters are nearly kept constant through such an anion modification, which is the most essential feature for achieving the successful tuning of T_c in an organic superconductor.

References

1. Wasserscheid P, Welton T (2003) Ionic liquids in synthesis. Wiley, Weinheim
2. Ferraris JP, Cowan DO, Walatka V, Perlstein JH (1973) Electron transfer in a new highly conducting donor-acceptor complex. J Am Chem Soc 95:948–949
3. Saito G, Yoshida Y (2007) Development of conductive organic molecular assemblies: organic metals, superconductors, and exotic functional materials. Bull Chem Soc Jpn 80:1–137
4. Ishiguro T, Yamaji K, Saito G (1998) Organic superconductors, 2nd edn. Springer, Berlin
5. Pouget JP (1993) Physical features of low dimensional organic super-conductors. Mol Cryst Liq Cryst 230:101–131
6. Cowan DO, Fortkort JA, Metzger RM (1991) Design constraints for organic metals and superconductors. In: Metzger RM, Day P, Papavassiliou GC (eds) Lower-dimensional systems and molecular electronics, NATO ASI Ser. B248. Plenum, New York
7. Delhaes P (1991) Review of the radical ion salts. In: Metzger RM, Day P, Papavassiliou GC (eds) Lower-dimensional systems and molecular electronics, NATO ASI Ser. B248. Plenum, New York
8. Cowan DO (1989) New opportunities in solid state chemistry: systematic trends and the design and study of new organic conductors. In: Yoshida Z, Shiba T, Ohshiro Y (eds) Proceedings of the 4th International Kyoto Conference on new aspects of organic chemistry. Kodansha, Tokyo
9. Conwell E (1988) Semiconductors and semimetals, vol 27. Academic, New York
10. Sandman DJ, Ceasar GP (1986) Donors for organic metals: molecular aspects. Israel J Chem 27:293–299
11. Wudl F (1982) Is there a relationship between aromaticity and conductivity? Pure Appl Chem 54:1051–1058
12. Torrance JB (1979) The difference between metallic and insulating salts of tetracyanoquinodimethone (TCNQ): how to design an organic metal. Acc Chem Res 12:79–86
13. Wheland RC (1976) Correlation of electrical conductivity in charge-transfer complexes with redox potentials, steric factors, and heavy atom effects. J Am Chem Soc 98:3926–3930
14. Wudl F (1976) The chemistry of anisotropic organic materials. In: Keller HJ (ed) Chemsitry and physics of one-dimensional metals. Plenum, New York
15. Bloch AN (1974) Design and study of one-dimensional organic conductors I: the role of structural disorder. In: Masuda K, Silver M (eds) Energy and charge transfer in organic semiconductors. Plenum, New York
16. Shchegolev IF (1972) Electric and magnetic properties of linear conducting chains. Phys Status Solidi A 12:9–45
17. Herbstein FH (1971) Crystalline π-molecular compounds (chemistry, spectroscopy and crystallography). In: Dunitz JD, Ibers JA (eds) Perspectives in structural chemistry, vol 4. Wiley, New York
18. Foster R (1969) Organic charge-transfer complexes. Academic, New York
19. Aust RB, Bentley WH, Drickamer HG (1964) Behavior of fused-ring aromatic hydrocarbons at very high pressure. J Chem Phys 41:1856–1864

20. Yokota T, Takeshita N, Shimizu K, Amaya K, Onodera A, Shirotani I, Endo S (1996) Pressure-induced superconductivity of iodanil. Czech J Phys 46:817–818
21. Iwasaki E, Shimizu K, Amaya K, Nakayama A, Aoki K, Carlon RP (2001) Metallization and superconductivity in hexaiodobenzene under high pressure. Synth Met 120:1003–1004
22. Shirotani I, Onodera A, Kamura Y, Inokuchi H, Kawai N (1976) Electrical conductivity of hexaiodobenzene and tetrahalo-p-benzoquinone under very high pressures. J Solid State Chem 18:235–239
23. Amaya K, Shimizu K, Takeshita N, Eremets MI, Kobayashi TC, Endo S (1998) Observation of pressure-induced superconductivity in the megabar region. J Phys Condens Matter 10:11179–11190
24. Yamashita Y, Tanaka S, Imaeda K, Inokuchi H, Sano M (1992) Preparation and properties of bis[1,2,5]thiadiazolo-p-quinobis(1,3-dithiole) (BTQBT) and its derivatives. novel organic semiconductors. J Org Chem 57:5517–5522
25. Imaeda K, Yamashita Y, Li Y, Mori T, Inokuchi H, Sano M (1992) Hall-effect observation in the new organic semiconductor bis(1,2,5-thiadiazolo)-p-quinobis(1,3-dithiole)(BTQBT). J Mater Chem 2:115–118
26. Inokuchi H, Saito G, Seki K, Wu P, Tang TB, Mori T, Imaeda K, Enoki T, Higuchi Y, Inaka K, Yasuoka N (1986) A novel type of organic semiconductors. molecular fastener. Chem Lett 15:1263–1266
27. Wu N, Saito G, Imaeda K, Shi Z, Mori T, Enoki T, Inokuchi H (1986) Uncapped alkylthio substituted tetrathiafulvalenes (TTC_n-TTF) and their charge transfer complexes. Chem Lett 15:441–444
28. Imaeda K, Enoki T, Shi Z, Wu P, Okada N, Yamochi H, Saito G, Inokuchi H (1987) Electrical conductivities of tetrakis(alkylthio)tetrathiafulvalene (TTC_n-TTF) and tetrakis(alkyltelluro) tetrathiafulvalenes ($TTeC_n$-TTF). Bull Chem Soc Jpn 60:3143–3147
29. Li Y, Nakano C, Imaeda K, Inokuchi H, Maruyama Y, Iwasawa N, Saito G (1990) Charge-carrier drift mobilities and phase transition in tetrakis(octylthio)tetrathiafulvalene, TTC_8-TTF. Bull Chem Soc Jpn 63:1857–1859
30. Barclay TM, Cordes AW, Haddon RC, Itkis ME, Oakley RT, Reed RW, Zhang H (1999) Preparation and characterization of a neutral π-radical molecular conductor. J Am Chem Soc 121:969–976
31. Chi X, Itkis ME, Patrick BO, Barclay TM, Reed RW, Oakley RT, Cordes AW, Haddon RC (1999) The first phenalenyl-based neutral radical molecular conductor. J Am Chem Soc 121:10395–10402
32. Brusso JL, Cvrkalj K, Leitch AA, Oakley RT, Reed RW, Robertson CM (2006) Resonance stabilized bisdiselenazolyls as neutral radical conductors. J Am Chem Soc 128:15080–15081
33. Leitch AA, Reed RW, Robertson CM, Britten JF, Yu X, Secco RA, Oakley RT (2007) An alternating π-stacked bisdithiazolyl radical conductor. J Am Chem Soc 129:7903–7914
34. Itkis ME, Chi X, Cordes AW, Haddon RC (2002) Magneto-opto-electronic bistability in a phenalenyl-based neutral radical. Science 296:1443–1445
35. Pal SK, Itkis ME, Tham FS, Reed RW, Oakley RT, Haddon RC (2005) Resonating valence-bond ground state in a phenalenyl-based neutral radical conductor. Science 309:281–284
36. Pal SK, Itkis ME, Tham FS, Reed RW, Oakley RT, Donnadieu B, Haddon RC (2007) Phenalenyl-based neutral radical molecular conductors: substituent effects on solid-state structures and properties. J Am Chem Soc 129:7163–7174
37. Kubo T, Shimizu A, Sakamoto M, Uruichi M, Yakushi K, Nakano M, Shiomi D, Sato K, Takui T, Morita Y, Nakasuji K (2005) Synthesis, intermolecular interaction, and semiconductive behavior of a delocalized singlet biradical hydrocarbon. Angew Chem Int Ed 44:6564–6568
38. Tsubata Y, Suzuki T, Miyashi T, Yamashita Y (1992) Single-component organic conductors based on neutral radicals containing the pyrazino-TCNQ skeleton. J Org Chem 57:6749–6755
39. Suzuki T, Miyanari S, Tsubata Y, Fukushima T, Miyashi T, Yamashita Y, Imaeda K, Ishida T, Nogami T (2001) Single-component organic semiconductors based on novel radicals that

exhibit electrochemical amphotericity: preparation, crystal structures, and solid-state properties of *N,N'*-dicyanopyrazinonaphthoquinodiiminides substituted with an *N*-alkylpyridinium unit. J Org Chem 66:216–224
40. Suzuki T, Miyanari S, Kawai H, Fujiwara K, Fukushima T, Miyashi T, Yamashita Y (2004) Pyrazino-tetracyanonaphthoquinodimethanes: sterically deformed electron acceptors affording zwitterionic radicals. Tetrahedron 60:1997–2003
41. Neilands O, Tilika V, Sudmale I, Grigorjeva I, Edzina A, Fonavs E, Muzikante I (1997) Dioxo- and aminooxopyrimidotetrathiafulvalenes: π-electron donors for design of conducting materials containing intramolecular hydrogen bonds of nucleic acid base pair type. Adv Mater Opt Electron 7:39–43
42. Balodis K, Khasanov S, Chong C, Maesato M, Yamochi H, Saito G, Neilands O (2003) Single component betainic conductor: pyrimido fused TTF derivatives having ethylenedioxy group. Synth Met 133–134:353–355
43. Neilands O (2001) Dioxo-and aminooxopyrimido-fused tetrathiafulvalenes-base compounds for novel organic semiconductors and for design of sensors for recognition of nucleic acid components. Mol Cryst Liq Cryst 355:331–349
44. Saito G, Balodis K, Yoshida Y, Maesato M, Yamochi H, Khasanov S, Murata T (2007) Mesomeric fused betainic radicals as organic conductors. Multifunctional conducting molecular materials. RSC, Cambridge
45. Murata T, Balodis K, Saito G (2008) Single-component organic conductors based on neutral betainic radicals of *N*-methyl substituted dioxo- and aminooxo-pyrimido-fused TTFs. Synth Met 158:497–505
46. Janczak J, Kubiak R, Zaleski A, Olejniczak J (1994) Metallic conductivity and phase transition in Tl_2Pc. Chem Phys Lett 225:72–75
47. Morimoto K, Inabe T (1995) Conducting neutral radical crystals of axially substituted phthalocyanine: crystal structures, dimensionality and electrical conductivity. J Mater Chem 5: 1749–1752
48. Tanaka H, Okano Y, Kobayashi H, Suzuki W, Kobayashi A (2001) A three-dimensional synthetic metallic crystal composed of single-component molecules. Science 291:285–287
49. Kobayashi A, Fujiwara E, Kobayashi H (2004) Single-component molecular metals with extended-TTF dithiolate ligands. Chem Rev 104:5243–5264
50. Schön JH, Kloc Ch, Batlogg B (2000) Superconductivity in molecular crystals induced by charge injection. Nature 406:702–704
51. Inokuchi H, Imaeda K, Enoki T, Mori T, Maruyama Y, Saito G, Okada N, Yamochi H, Seki K, Higuchi Y, Yasuoka N (1987) Tetrakis(methyltelluro)tetrathiafulvalene ($TTeC_1TTF$), a high-mobility organic semiconductor. Nature 329:39–40
52. Zambounis JS, Mizuguchi J, Rihs G, Chauvet O, Zuppiroli L (1994) Optical and electrical properties of evaporated 2,5-bis-methylthio-7,7′,8,8′-tetracyanoquinodimethane. J Appl Phys 76:1824–1829
53. Kimura S, Kurai H, Mori T (2002) 2,5-Bis(1,3-dithiol-2-ylidene)-1,3,4,6-tetrathiapentalene (TTP) derivatives having four long alkylthio chains. Tetrahedron 58:1119–1124
54. Rovira MC, Novoa JJ, Tarres J, Rovira C, Veciana J, Yang S, Cowan DO, Canadell E (1995) Bis(ethylenethio)tetrathiafulvalene (BET-TTF), an organic donor with high electrical conductivity. Adv Mater 7:1023–1027
55. Mas-Torrent M, Durkut M, Hadley P, Ribas X, Rovira C (2004) High mobility of dithiophene-tetrathiafulvalene single-crystal organic field effect transistors. J Am Chem Soc 126:984–985
56. Saito G, Yoshida Y, Murofushi H, Iwasawa N, Hiramatsu T, Otsuka A, Yamochi H, Isa K, Mineo-Ota E, Konno M, Mori T, Imaeda K, Inokuchi H (2010) Preparation, structures, and physical properties of tetrakis(alkylthio)-tetraselenafulvalene (TTC_n-TSeF, n = 1-15). Bull Chem Soc Jpn 83:335–344

57. Ashizawa M, Kimura S, Mori T, Misaki Y, Tanaka K (2004) Tris-fused tetrathiafulvalenes (TTF): highly conducting single-component organics and metallic charge-transfer salt. Synth Met 141:307–313
58. Ebata H, Izawa T, Miyazaki E, Takimiya K, Ikeda M, Kuwabara H, Yui T (2007) Highly soluble [1]benzothieno[3,2-b]benzothiophene (BTBT) derivatives for high-performance, solution-processed organic field-effect transistors. J Am Chem Soc 129:15732–15733
59. Kang MJ, Doi I, Mori H, Miyazaki E, Takimiya K, Ikeda M, Kuwabara H (2011) Alkylated dinaphtho[2,3-b:2′,3′-f]thieno[3,2-b]thiophenes (C_n-DNTTs): organic semiconductors for high-performance thin-film transistors. Adv Mater 23:1222–1225
60. Banister AJ, Bricklebank N, Lavender I, Rawson JM, Gregory CI, Tanner BK, Clegg W, Elsegood MRJ, Palacio F (1996) Spontaneous magnetization in a sulfur-nitrogen radical at 36 K. Angew Chem Int Ed Engl 35:2533–2535
61. Palacio F, Antorrena G, Castro M, Burriel R, Rawson J, Smith JNB, Bricklebank N, Novoa J, Ritter C (1997) High-temperature magnetic ordering in a new organic magnet. Phys Rev Lett 79:2336–2339
62. Mito M, Kawae T, Takeda K, Takagi S, Matsushita Y, Deguchi H, Rawson JM, Palacio F (2001) Pressure-induced enhancement of the transition temperature of a genuine organic weak-ferromagnet up to 65 K. Polyhedron 20:1509–1512
63. Suzuki T, Yamada M, Ohkita M, Tsuji T (2001) Generation and amphoteric redox properties of novel neutral radicals with the TTF-TCNQ hybrid structure. Heterocycles 54:387–394
64. Andrews MP, Cordes AW, Douglass DC, Fleming RM, Glarum SH, Haddon RC, Marsh P, Oakley RT, Palstra TTM, Schneemeyer LF, Trucks GW, Tycko R, Waszczak JV, Young KM, Zimmerman NM (1991) One-dimensional stacking of bifunctional dithia- and diselenadiazolyl radicals: preparation and structural and electronic properties of 1,3-[$(E_2N_2C)C_6H_4(CN_2E_2)$] (E = sulfur, selenium). J Am Chem Soc 113:3559–3568
65. Cordes AW, Haddon RC, Hicks RG, Oakley RT, Palstra TTM, Schneemeyer LF, Waszczak JV (1992) Polymorphism of 1,3-phenylene bis(diselenadiazolyl). Solid-state structural and electronic properties of β-1,3-[$(Se_2N_2C)C_6H_4(CN_2Se_2)$]. J Am Chem Soc 114:1729–1732
66. Cordes AW, Haddon RC, Oakley RT, Schneemeyer LF, Waszczak JV, Young KM, Zimmerman NM (1991) Molecular semiconductors from bifunctional dithia- and diselenadiazolyl radicals. Preparation and solid-state structural and electronic properties of 1,4-[$(E_2N_2C)C_6H_4(CN_2E_2)$] (E = sulfur, selenium). J Am Chem Soc 113:582–588
67. LeBlanc OH Jr (1965) On the electrical conductivities of tetracyanoquinodimethan anion-radical salts. J Chem Phys 42:4307–4308
68. Nakahara Y, Kimura K, Inokuchi H, Yagi T (1979) Electrical conductivity of solid state proteins: simple proteins and cytochrome c_3 as anhydrous film. Chem Lett 8:877–880
69. Kimura K, Nakahara Y, Yagi T, Inokuchi H (1979) Electrical conduction of hemoprotein in the solid phase: anhydrous cytochrome c_3 film. J Chem Phys 70:3317–3323
70. Kimura K, Nakajima S, Niki K, Inokuchi H (1985) Determination of formal potentials of multihemoprotein, cytochrome c_3 by ^1H nuclear magnetic resonance. Bull Chem Soc Jpn 58:1010–1012
71. Megnamisi-Belombe M (1977) Evidence for intrinsic electrical conduction in the linear metal-chain semiconductor bis(1,2-benzoquinonedioximato)platinum(II), $Pt(bqd)_2$. J Solid State Chem 22:151–156
72. Shirotani I, Konno M, Taniguchi Y (1989) Electrical and optical properties and crystal structure of bis(dimethylglyoximato)Pt(II) at high pressures. Synth Met 29:123–128
73. Shirotani I, Kawamura A, Suzuki K, Utsumi W, Yagi T (1991) Insulator-to-metal-to-semiconductor transitions of one-dimensional bis(1,2-dione dioximato)Pt(II) complexes at high pressures. Bull Chem Soc Jpn 64:1607–1612
74. Le Narvor N, Robertson N, Weyland T, Kilburn JD, Underhill AE, Webster M, Svenstrup N, Becher J (1996) Synthesis, structure and properties of nickel complexes of 4,5-tetrathiafulvalene dithiolates: high conductivity in neutral dithiolate complexes. Chem Commun 1363–1364

75. Tanaka H, Tokumoto M, Ishibashi S, Graf D, Choi ES, Brooks JM, Yasuzuka S, Okano Y, Kobayashi H, Kobayashi A (2004) Observation of three-dimensional Fermi surfaces in a single-component molecular metal, $[Ni(tmdt)_2]$. J Am Chem Soc 126:10518–10519
76. Kobayashi A, Tanaka H, Kobayashi H (2001) Molecular design and development of single-component molecular metals. J Mater Chem 11:2078–2088
77. Fujiwara E, Kobayashi A, Kobayashi H (2003) Structures and physical properties of nickel complexes with TTF-type ligands. Synth Met 135–136:535–536
78. Zheng SL, Zhang JP, Wong WT, Chen XM (2003) A novel, highly electrical conducting, single-component molecular material: $[Ag_2(ophen)_2]$ (Hophen = 1H-[1,10]phenanthrolin-2-one). J Am Chem Soc 125:6882–6883
79. Suga T, Isoda S, Kobayashi T (1999) Characterization and electrical conductivity of dithallium phthalocyanine (Tl_2Pc). J Porphyrins Phthalocyanines 3:397–405
80. Cassoux P, Valade L, Kobayashi H, Kobayashi A, Clark RA, Underhill AE (1991) Molecular metals and superconductors derived from metal complexes of 1,3-dithiol-2-thione-4,5-dithiolate (dmit). Coord Chem Rev 110:115–160
81. Shirotani I, Inagaki Y, Utsumi W, Yagi T (1991) Pressure-sensitive absorption spectra of thin films of bis(diphenylglyoximato)platinum(II), $Pt(dpg)_2$: potential application as an indicator of pressure. J Mater Chem 1:1041–1043
82. Saito G, Ferraris JP (1980) Requirements for an "organic metal". Bull Chem Soc Jpn 53:2141–2145
83. McConnell HM, Hoffman BM, Metzger RM (1965) Charge transfer in molecular crystals. Proc Natl Acad Sci USA 53:46–50
84. Torrance JB, Vazquez JE, Mayerle JJ, Lee VY (1981) Discovery of a neutral-to-ionic phase transition in organic materials. Phys Rev Lett 46:253–257
85. Takahashi Y, Hasegawa T, Abe Y, Tokura T, Saito G (2006) Organic metal electrodes for controlled p- and n-type carrier injections in organic field-effect transistors. Appl Phys Lett 88:073504/1-4
86. Pac SS, Saito G (2002) Peculiarity of hexamethylenetetratellurafulvalene (HMTTeF) charge transfer complexes of donor-acceptor (D-A) type. J Solid State Chem 168:486–496
87. Horiuchi S, Yamochi H, Saito G, Sakaguchi K, Kusunoki M (1996) Nature and origin of stable metallic state in organic charge-transfer complexes of bis(ethylenedioxy)tetrathiafulvalene. J Am Chem Soc 118:8604–8622
88. Mori T, Misaki Y, Yamabe T, Mori H, Tanaka S (1995) Structure and properties of an organic metal $(BEDT\text{-}TTP)_2I_3$. Chem Lett 24:549–550
89. Saito G, Sasaki H, Aoki T, Yoshida Y, Otsuka A, Yamochi H, Drozdova OO, Yakushi K, Kitagawa H, Mitani T (2002) Complex formation of ethylenedioxyethylenedithiotetrathiafulvalene (EDOEDT-TTF: EOET) and its self-assembling ability. J Mater Chem 12: 1640–1649
90. Ota A, Yamochi H, Saito G (2002) A novel metal-insulator phase transition observed in $(EDO\text{-}TTF)_2PF_6$. J Mater Chem 12:2600–2602
91. Ota A, Yamochi H, Saito G (2002) Preparation and physical properties of conductive EDO-TTF complexes. Mol Cryst Liq Cryst 376:177–182
92. Nakamura T, Yunome G, Azumi R, Tanaka M, Tachibana H, Matsumoto M, Horiuchi S, Yamochi H, Saito G (1994) Structure and electrical properties of the metallic Langmuir-Blodgett film without secondary treatments. J Phys Chem 98:1882–1887
93. Ogasawara K, Ishiguro K, Horiuchi S, Yamochi H, Saito G (1996) A new metallic Langmuir-Blodgett film formed with BO_2-$(MeO)_2TCNQ$, where BO is bisethylenedioxytetrathiafulvalene and $(MeO)_2TCNQ$ is dimethoxytetracyanoquinodimethane. Jpn J Appl Phys 35: L571–L573
94. Izumi M, Yartsev VM, Ohnuki H, Vignau L, Delhaes P (2001) Conducting Langmuir-Blodgett films: state of the art and recent progress. Recent Res Dev Phys Chem 5:37–75

95. Ishizaki Y, Izumi M, Ohnuki H, Kalita-Lipinska K, Imakubo T, Kobayashi K (2001) Formation of two-dimensional weak localization in conducting Langmuir-Blodgett films. Phys Rev B63:134201/1-5
96. Jeszka JK, Tracz A, Sroczynska A, Kryszewski M, Yamochi H, Horiuchi S, Saito G, Ulanski J (1999) Metallic polymer composites with bis(ethylenedioxy)-tetrathiafulvalene salts. Preparation-properties relationship. Synth Met 106:75–83
97. Horiuchi S, Yamochi H, Saito G, Jeszka JK, Tracz A, Sroczynska A, Ulanski J (1997) Highly-oriented BEDO-TTF molecules in metallic polymer composites. Mol Cryst Liq Cryst 296:365–382
98. Yamochi H, Kawasaki T, Nagata Y, Maesato M, Saito G (2002) BEDO-TTF complexes with magnetic counter ions. Mol Cryst Liq Cryst 376:113–120
99. Khasanov S, Zorba LV, Shibaeva RP, Kushch ND, Yagubskii EB, Rousseau R, Canadell E, Barrans Y, Gaultier J, Chasseau D (1999) Structural aspects of the phase transitions in $(BEDO\text{-}TTF)_2ReO_4{\cdot}H_2O$. Synth Met 103:1853–1856
100. Tracz A (2002) Ion exchange in $(BO)_{2.4}I_3$ microcrystals: a method for obtaining colorless, transparent, metallically conductive polymer films. J Appl Polym Sci 86:1465–1472
101. Haneda T, Tracz A, Saito G, Yamochi H (2011) Continuous and discontinuous water release/intake of $(BEDO\text{-}TTF)_2Br(H_2O)_3$ micro-crystals embedded in polymer film. J Mater Chem 21:1621–1626
102. Beno MA, Wang HH, Kini AM, Carlson KD, Geiser U, Kwok WK, Thompson JE, Williams JM, Ren J, Whangbo MH (1990) The first ambient pressure organic superconductor containing oxygen in the donor molecule, $\beta_m\text{-}(BEDO\text{-}TTF)_3Cu_2(NCS)_3$. $T_c = 1.06$ K. Inorg Chem 29:1599–1601
103. Kahlich S, Schweitzer D, Heinen I, Lan SE, Nuber B, Keller HJ, Winzer K, Helberg HW (1991) $(BEDO\text{-}TTF)_2ReO_4{\cdot}(H_2O)$: a new organic superconductor. Solid State Commun 80:191–195
104. Slifkin MA (1971) Charge transfer interactions of biomolecules. Academic, Boston
105. Pullman B, Pullman A (1963) Quantum biochemistry. Interscience, New York, London
106. Kosower EM (1962) Molecular biochemistry. McGraw-Hill, New York
107. Eley DD (1989) Studies of organic semiconductors for 40 years–I the mobile π-electron–40 years on. Mol Cryst Liq Cryst 171:1–21 and references therein
108. Möglich F, Schön M (1938) Energy migration in crystals and molecular complexes. Naturwiss 26:199
109. Szent-Györgyi A (1941) Towards a new biochemistry? Science 93:609–611
110. Eley DD (1940) The conversion of parahydrogen by porphyrin compounds, including hemoglobin. Trans Faraday Soc 35:500–505
111. Eley DD, Parfitt GD, Perry MJ, Taysum DH (1953) The semiconductivity of organic substances. Part 1. Trans Faraday Soc 49:79–86
112. Cardew MH, Eley DD (1959) The semiconductivity of organic substances. Part 3. Haemoglobin and some amino acids. Discuss Faraday Soc 27:115–128
113. Eley DD, Spivey DI (1960) The semiconductivity of organic substances. Part 6. A range of proteins. Trans Faraday Soc 56:1432–1442
114. Giese B (2000) Long-distance charge transport in DNA: the hopping mechanism. Acc Chem Res 33:631–636 and references therein
115. Fink HW, Schönenberger C (1999) Electrical conduction through DNA molecules. Nature 398:407–410
116. Porath D, Bezryadin A, Vries S, Dekker C (2000) Direct measurement of electrical transport through DNA molecules. Nature 403:635–638
117. de Pablo PJ, Moreno-Herrero F, Colchero J, Gómez-Herrero J, Herrero P, Baró AM, Ordejón P, Soler JM, Artacho E (2000) Absence of dc-conductivity in λ-DNA. Phys Rev Lett 85:4992–4995
118. Tran P, Alavi B, Grüner G (2000) Charge transport along the λ-DNA double helix. Phys Rev Lett 85:1564–1567

119. Kasumov AY, Kociak M, Guéron S, Reulet B, Volkov VT, Klinov DV, Bouchiat H (2001) Proximity-induced superconductivity in DNA. Science 291:280–282
120. Zhang Y, Austin RH, Kraeft J, Cox EC, Ong NP (2002) Insulating behavior of λ-DNA on the micron scale. Phys Rev Lett 89:198102/1-4
121. Ladik J (1960) Investigation of the electronic structure of desoxiribonucleic acid. Acta Phys Acad Sci Hung 11:239–258
122. Burnel ME, Eley DD, Subramanyan V (1969) Semiconduction in nucleic acid and its components. Ann NY Acad Sci 158:191–209
123. Morita Y, Maki S, Ohmoto M, Kitagawa H, Okubo T, Mitani T, Nakasuji K (2002) Hydrogen-bonded charge-transfer complexes of TTF containing a uracil moiety: crystal structures and electronic properties of the hydrogen cyananilate and TCNQ complexes. Org Lett 4:2185–2188
124. Ohmoto M, Maki S, Morita Y, Kubo T, Kitagawa H, Okubo T, Mitani T, Nakasuji K (2003) Complementary double hydrogen-bonded CT complexes of TTF having uracil moiety. Synth Met 133–134:337–339
125. Miyazaki E, Morita Y, Umemoto Y, Fukui K, Nakasuji K (2005) Transformation of double hydrogen-bonding motifs of TTF-uracil system by redox change. Chem Lett 34:1326–1327
126. Morita Y, Miyazaki E, Umemoto Y, Fukui K, Nakasuji K (2006) Two-dimensional networks of ethylenedithiotetrathiafulvalene derivatives with the hydrogen-bonded functionality of uracil, and channel structure of its tetracyanoquinodimethane complex. J Org Chem 71:5631–5637
127. Morita Y, Maki S, Ohmoto M, Kitagawa H, Okubo T, Mitani T, Nakasuji K (2003) New TTF having adenine moiety as a nucleic acid base. Synth Met 135–136:541–542
128. Orlov YM, Smirnov AN, Varshavsky YM (1976) Transanulare umlagerungen von 5-cyclodecinylderivaten bei solvolysereaktionen. Tetrahedron Lett 18:4377–4378
129. Dougherty D, Younathan ES, Voll R, Abdulnur S, McGlynn SP (1978) Photoelectron spectroscopy of some biological molecules. J Electron Spectrosc Relat Phenom 13:379–393
130. Lias SG, Bartmess JE, Liebman JF, Holms LJ, Levin RD, Mallard WG (1988) Gas-phase ion and neutral thermochemistry. J Phys Chem Suppl 1 17:1–861
131. Wetmore SD, Boyd RJ, Eriksson LA (2000) Electron affinities and ionization potentials of nucleotide bases. Chem Phys Lett 322:129–135 and references therein
132. Slifkin MA, Kushelevsky AP (1971) The interaction of purines and pyrimidines with chloranil. Spectrochim Acta A 27:1999–2003
133. Bazhina IN, Verzilov VS, Grechishkin VS, Grechishkina RV, Gusarov VM (1973) EPR and Hall effect in organic charge transfer complexes. Zh Strukt Khim 14:930–932
134. Sheina GG, Radchenko ED, Blagoi YP, Verkin BI (1978) Charge-transfer complexes of nucleic acid bases. Dokl Akad Nauk SSSR 240:463–466
135. Dawson RMC, Elliot WH, Jones KM (1986) Data for biochemical research, 3rd edn. Clarendon, Oxford
136. Murata T, Saito G (2006) Properties of reaction products between cytosine and F_4TCNQ in MeOH: two hemiprotonated cytosine salts with F_4TCNQ radical anion and methoxy adduct anion. Chem Lett 35:1342–1343
137. Murata T, Nishimura K, Saito G (2007) Organic conductor based on nucleobase: structural and electronic properties of a charge transfer solid composed of TCNQ anion radical and hemiprotonated cytosine. Mol Cryst Liq Cryst 466:101–112
138. Murata T, Saito G, Nishimura K, Enomoto Y, Honda G, Shimizu Y, Matsui S, Sakata M, Drozdova OO, Yakushi K (2008) Complex formation between a nucleobase and tetracyanoquinodimethane derivatives: crystal structures and transport properties of charge transfer solids of cytosine. Bull Chem Soc Jpn 81:331–344
139. Murata T, Enomoto Y, Saito G (2008) Exploration of charge-transfer complexes of a nucleobase: crystal structure and properties of cytosine-Et_2TCNQ salt. Solid State Sci 10: 1364–1368

140. Sakata M, Maesato M, Miyazaki T, Nishimura K, Murata T, Yamochi H, Saito G (2008) High-pressure transport study of a charge-transfer salt based on cytosine and TCNQ using a diamond anvil cell. J Phys Conf Ser 132:012011/1-4
141. Hebard AF, Rosseinsky MJ, Haddon RC, Murphy DW, Glarum SH, Palstra TTM, Ramirez AP, Kortan AR (1991) Superconductivity at 18 K in potassium-doped C_{60}. Nature 350:600–601
142. Rosseinsky MJ (1995) Fullerene intercalation chemistry. J Mater Chem 5:1497–1513
143. Bommeli F, Degiorgi L, Wachter D, Legeza Ö, Jánossy A, Oszlanyi G, Chauvet O, Forro L (1995) Metallic conductivity and metal-insulator transition in $(AC_{60})_n$ (A = K, Rb, and Cs) linear polymer fullerides. Phys Rev B51:14794–14797
144. Prassides K (2000) Polymer and dimer phases in doped fullerenes. In: Andreoni W (ed) The physics of fullerene-based and fullerene-related materials. Kluwer Academic, Boston
145. Oszlányi G, Baumgartner G, Faigel G, Forró L (1997) Na_4C_{60}: an alkali intercalated two-dimensional polymer. Phys Rev Lett 78:4438–4441
146. Allemand PM, Khemani KC, Koch A, Wudl F, Holczer K, Donovan S, Grüner G, Thompson JD (1991) Organic molecular soft ferromagnetism in a fullerene C_{60}. Science 253:301–303
147. Konarev DV, Khasanov SS, Otsuka A, Yoshida Y, Saito G (2002) Synthesis and crystal structure of ionic multicomponent complex: $\{[Cr^I(PhH)_2]^{\bullet+}\}_2[Co^{II}TTP(C_{60}(CN)_2)]^-$ $[C_{60}(CN)_2]^{\bullet-}\cdot 3(o\text{-}C_6H_4Cl_2)$ containing $C_{60}(CN)_2^{\bullet-}$ radical anion and σ-bonded diamagnetic $Co^{II}TTP(C_{60}(CN)_2)^-$ anion. J Am Chem Soc 124:7648–7649
148. Konarev DV, Khasanov SS, Otsuka A, Saito G (2002) The reversible formation of a single-bonded $(C_{60}^-)_2$ dimer in ionic charge transfer complex: $Cp^*_2Cr\cdot C_{60}(C_6H_4Cl_2)_2$. The molecular structure of $(C_{60}^-)_2$. J Am Chem Soc 124:8520–8521
149. Konarev DV, Khasanov SS, Saito G, Otsuka A, Yoshida Y, Lyubovskaya RN (2003) Formation of single-bonded $(C_{60}^-)_2$ and $(C_{70}^-)_2$ dimers in crystalline ionic complexes of fullerenes. J Am Chem Soc 125:10074–10083
150. Konarev DV, Khasanov SS, Saito G, Lyubovskaya RN (2004) The formation of σ-bonded $(Fullerene^-)_2$ dimers and $(Co^{II}TPP\cdot fullerene^-)$ anions in ionic complexes of C_{60}, C_{70}, and $C_{60}(CN)_2$. Recent Res Dev Chem 2:105–140
151. Konarev DV, Kovalevsky AY, Otsuka A, Saito G, Lyubovskaya RN (2005) Neutral and ionic complexes of C_{60} with metal dibenzyldithiocarbamates. Reversible dimerization of $C_{60}^{\bullet-}$ in ionic multicomponent complex $[Cr^I(C_6H_6)_2^{\bullet+}]\cdot(C_{60}^{\bullet-})\cdot 0.5[Pd(dbdtc)_2]$. Inorg Chem 44:9547–9553
152. Konarev DV, Khasanov SS, Otsuka A, Saito G, Lyubovskaya RN (2006) Negatively charged π-$(C_{60}^-)_2$ dimer with biradical state at room temperature. J Am Chem Soc 128:9292–9293
153. Konarev DV, Khasanov SS, Saito G, Otsuka A, Lyubovskaya RN (2007) Ionic and neutral C_{60} complexes with coordination assemblies of metal tetraphenylporphyrins, $M^{II}TPP_2\cdot DMP$ (M = Mn, Zn). coexistence of $(C_{60}^-)_2$ dimers bonded by one and two single bonds in the same compound. Inorg Chem 46:7601–7609
154. Konarev DV, Khasanov SS, Otsuka A, Maesato M, Saito G, Lyubovskaya RN (2010) A two-dimensional organic metal based on fullerene. Angew Chem Int Ed 49:4829–4832
155. Potember RS, Poehler TO, Cowan DO, Bloch AN (1980) Electrical switching and memory phenomena in semiconducting organic charge-transfer complexes. In: Alcacer L (ed) The physics and chemistry of low dimensional solids. D. Reidel, Dordrecht
156. Heintz RA, Zhao H, Ouyang X, Grandinetti G, Cowen J, Dunbar KR (1999) New insight into the nature of Cu(TCNQ): solution routes to two distinct polymorphs and their relationship to crystalline films that display bistable switching behavior. Inorg Chem 38:144–156
157. Benson RC, Hoffman RC, Potember RS, Bourkoff E, Poehler TO (1983) Spectral dependence of reversible optically induced transitions in organometallic compounds. Appl Phys Lett 42:855–857 and references therein
158. Kumai R, Okimoto Y, Tokura Y (1999) Current-induced insulator-metal transition and pattern formation in an organic charge-transfer complex. Science 284:1645–1647

159. Koshihara S, Tokura Y, Iwasa Y, Koda T, Saito G, Mitani T (1991) Domain-wall excitation in organic charge transfer compounds investigated by photo-reflectance spectroscopy. Synth Met 41–43:2351–2354
160. Naito T, Inabe T, Niimi H, Asakura K (2004) Light-induced transformation of molecular materials into devices. Adv Mater 16:1786–1790
161. Karutz FO, von Schutz JU, Wachtel H, Wolf HC (1998) Optically reversed Peierls transition in crystals of Cu(dicyanoquinonediimine)$_2$. Phys Rev Lett 81:140–143
162. Chollet M, Guerin L, Uchida N, Fukaya S, Shimoda H, Ishikawa T, Matsuda K, Hasegawa T, Ota A, Yamochi H, Saito G, Tazaki R, Adachi S, Koshihara S (2005) Gigantic photoresponse in 1/4-filled-band organic salt (EDO-TTF)$_2$PF$_6$. Science 307:86–89
163. Uchida N, Koshihara S, Ishikawa T, Ota A, Fukuya S, Chollet CM, Yamochi H, Saito G (2004) Ultrafast photo-response in (EDO)$_2$PF$_6$. J Phys IV 114:143–145
164. Tajima N, Fujisawa J, Naka N, Ishihara T, Kato R, Nishio Y, Kajita K (2005) Photo-induced insulator-metal transition in an organic conductor α-(BEDT-TTF)$_2$I$_3$. J Phys Soc Jpn 74: 511–514
165. Inagaki K, Terasaki I, Mori H, Mori T (2004) Large dielectric constant and giant nonlinear conduction in the organic conductor θ-(BEDT-TTF)$_2$CsZn(SCN)$_4$. J Phys Soc Jpn 73: 3364–3369
166. Okamoto K, Tanaka T, Fujita W, Awaga K, Inabe T (2006) Low-field negative-resistance effect in a charge-ordered state of thiazyl-radical crystals. Angew Chem Int Ed 45:4516–4518
167. Mori T, Terasaki I, Mori H (2007) New aspects of nonlinear conductivity in organic charge-transfer salts. J Mater Chem 17:4343–4347
168. Iwasa Y, Koda T, Koshihara S, Tokura Y, Iwasawa N, Saito G (1989) Intrinsic negative-resistance effect in mixed-stack charge-transfer crystals. Phys Rev B39:10441–10444
169. Iwasa Y, Koda T, Tokura Y, Koshihara S, Iwasawa N, Saito G (1989) Switching effect in organic charge transfer complex crystals. Appl Phys Lett 55:2111–2113
170. Koshihara S, Tokura Y, Mitani T, Saito G, Koda T (1990) Photoinduced valence instability in the organic molecular compound tetrathiafulvalene-*p*-chloranil (TTF-CA). Phys Rev B42: 6853–6856
171. Collet E, Lemée-Cailleau MH, Buron-Le Cointe M, Cailleau H, Wulff M, Luty T, Koshihara S, Meyer M, Toupet L, Rabiller P, Techert S (2003) Laser-induced ferroelectric structural order in an organic charge-transfer crystal. Science 300:612–615
172. Osada T, Miura N, Ogura I, Saito G (1987) Non-ohmic transport in the field-induced SDW state in (TMTSF)$_2$ClO$_4$. Phys Rev Lett 58:1563–1566
173. Ota A, Yamochi H, Saito G (2003) A novel metal-insulator transition in (EDO-TTF)$_2$X (X = PF$_6$, AsF$_6$). Synth Met 133–134:463–465
174. Nakano Y, Balodis K, Yamochi H, Saito G, Uruichi M, Yakushi K (2008) Isotope effect on metal-insulator transition of (EDO-TTF)$_2$XF$_6$ (X = P, As) with multi-instability of metallic state. Solid State Sci 10:1780–1785
175. Nakano Y, Yamochi H, Saito G, Uruichi M, Yakushi K (2009) Anion size and isotope effects in (EDO-TTF)$_2$XF$_6$. J Phys Conf Ser 148:012007/1-4
176. Lorenc M, Moisan N, Servol M, Cailleau H, Koshihara S, Maesato M, Shao X, Nakano Y, Yamochi H, Saito G, Collet E (2009) Multi-phonon dynamics of the ultra-fast photoinduced transition of (EDO-TTF)$_2$SbF$_6$. J Phys Conf Ser 148:012001/1-4
177. Drozdova O, Yakushi K, Yamamoto K, Ota A, Yamochi H, Saito G, Tashiro H, Tanner DB (2004) Optical characterization of 2k$_F$ bond-charge-density wave in quasi-one-dimensional 3/4-filled (EDO-TTF)$_2$X (X = PF$_6$ and AsF$_6$). Phys Rev B70:075107-1/8
178. Aoyagi S, Kato K, Ota A, Yamochi H, Saito G, Suematsu H, Sakata M, Takata M (2004) Direct observation of bonding and charge ordering in (EDO-TTF)$_2$PF$_6$. Angew Chem Int Ed 43:3670–3673
179. Mori H, Tanaka S, Mori T, Maruyama Y (1995) Crystal structures and electrical resistivities of three-component organic conductors: (BEDT-TTF)$_2$MM′(SCN)$_4$ [M = K, Rb, Cs; M′ = Co, Zn, Cd]. Bull Chem Soc Jpn 68:1136–1144

180. Onda K, Ogihara S, Yonemitsu K, Maeshima N, Ishikawa T, Okimoto Y, Shao X, Nakano Y, Yamochi H, Saito G, Koshihara S (2008) Photoinduced change in the charge order pattern in the quarter-filled organic conductor $(EDO\text{-}TTF)_2PF_6$ with a strong electron-phonon interaction. Phys Rev Lett 101:067403/1-4
181. Kobayashi T, Saito T, Ohtani H (2001) Real-time spectroscopy of transition states in bacteriorhodopsin during retinal isomerization. Nature 414:531–534
182. Williams JM, Ferraro JR, Thorn RJ, Carlson KD, Geiser U, Wang HH, Kini AM, Whangbo MH (1992) Organic superconductors (including fullerenes). Prentice Hall, Englewood Cliffs, NJ
183. Wosnitza J (1996) Fermi surfaces of low-dimensional organic metals and superconductors. Springer, Berlin
184. Singleton J (2000) Studies of quasi-two-dimensional organic conductors based on BEDT-TTF using high magnetic fields. Rep Prog Phys 63:1111–1207
185. Molecular conductors (2004) Chem Rev 104: 11
186. Mori H (2006) Materials viewpoint of organic superconductors. J Phys Soc Jpn 75:051003/1-15
187. Jerome D, Schulz HJ (1982) Organic conductors and superconductors. Adv Phys 31:299–490
188. Bechgaard K (1982) $TMTSF_2X$ salts. Preparation, structure and effect of the anions. Mol Cryst Liq Cryst 79:1–13
189. Sakata M, Yoshida Y, Maesato M, Saito G, Matsumoto K, Hagiwara R (2006) Preparation of superconducting $(TMTSF)_2NbF_6$ by electrooxidation of TMTSF using ionic liquid as electrolyte. Mol Cryst Liq Cryst 452:103–112
190. Balicas L, Behnia K, Kang W, Canadell E, Auban-Senzier P, Jerome D, Ribault M, Fabre JM (1994) Superconductivity and magnetic field induced spin density waves in the $(TMTTF)_2X$ family. J Phys I 4:1539–1549
191. Adachi T, Ojima E, Kato K, Kobayashi H, Miyazaki T, Tokumoto M, Kobayashi A (2000) Superconducting transition of $(TMTTF)_2PF_6$ above 50 kbar [TMTTF = tetramethyltetrathiafulvalene]. J Am Chem Soc 122:3238–3239
192. Auban-Senzier P, Pasquier C, Jerome D, Carcel C, Fabre JM (2003) From Mott insulator to superconductivity in $(TMTTF)_2BF_4$: high pressure transport measurements. Synth Met 133–134:11–14
193. Itoi M, Araki C, Hedo M, Uwatoko Y, Nakamura T (2008) Anomalously wide superconducting phase of one-dimensional organic conductor $(TMTTF)_2SbF_6$. J Phys Soc Jpn 77:023701/1-4
194. Takigawa M, Yasuoka H, Saito G (1987) Proton spin relaxation in the superconducting state of $(TMTSF)_2ClO_4$. J Phys Soc Jpn 56:873–876
195. Jerome D (1991) The physics of organic superconductors. Science 252:1509–1514. A more detailed diagram was then proposed; Dumm M, Loidl A, Fravel BW, Starkey KP, Montgomery LK, Dressel M (2000) Electron spin resonance studies on the organic linear-chain compounds $(TMTCF)_2X$ (C = S, Se; X = PF_6, AsF_6, ClO_4, Br). Phys Rev B61: 511–521
196. Saito G, Enoki T, Toriumi K, Inokuchi H (1982) Two-dimensionality and suppression of metal-semiconductor transition in a new organic metal with alkylthio substituted TTF and perchlorate. Solid State Commun 42:557–560
197. Taniguchi H, Miyashita M, Uchiyama K, Satoh K, Mori N, Okamoto H, Miyagawa K, Kanoda K, Hedo M, Uwatoko Y (2003) Superconductivity at 14.2 K in layered organics under extreme pressure. J Phys Soc Jpn 72:468–471. They reported T_c (on-set) = 14.2 K and T_c (mid- point) = 13.4 K at 8.2 GPa
198. Schirber JE, Overmyer DL, Carlson KD, Williams JM, Kini AM, Wang HH, Charlier HA, Love BJ, Watkins DM, Yaconi GA (1991) Pressure-temperature phase diagram, inverse isotope effect, and superconductivity in excess of 13 K in κ-$(BEDT\text{-}TTF)_2Cu[N(CN)_2]Cl$, where BEDT-TTF is bis(ethylenedithio)tetrathiafulvalene. Phys Rev B44:4666–4669

199. Saito G, Yamochi H, Nakamura T, Komatsu T, Inoue T, Ito H, Ishiguro T, Kusunoki M, Sakaguchi K, Mori T (1993) Structural and physical properties of two new ambient pressure κ-type BEDT-TTF superconductors and their related salts. Synth Met 55–57:2883–2890
200. Kini M, Geiser U, Wang HH, Carlson KD, Williams JM, Kwok WK, Vandervoort KG, Thompson JE, Stupka DL, Jung D, Wangbo MH (1990) A new ambient-pressure organic superconductor, κ-$(ET)_2Cu[N(CN)_2]Br$, with the highest transition temperature yet observed (inductive onset $T_c = 11.6$ K, resistive onset = 12.5 K). Inorg Chem 29:2555–2557
201. Urayama H, Yamochi H, Saito G, Nozawa K, Sugano T, Kinoshita M, Sato S, Oshima K, Kawamoto A, Tanaka J (1988) A new ambient pressure organic superconductor based on BEDT-TTF with T_c higher than 10 K ($T_c = 10.4$ K). Chem Lett 17:55–58
202. Williams JM, Kini AM, Wang HH, Carlson KD, Geiser U, Montgomery LK, Pyrka GJ, Watkins DM, Kommers JM, Boryschuk SJ, Crouch AV, Kwok WK, Schirber JE, Overmyer DL, Jung D, Whangbo MH (1990) From semiconductor-semiconductor transition (42 K) to the highest-T_c organic superconductor, κ-$(ET)_2Cu[N(CN)_2]Cl$ ($T_c = 12.5$ K). Inorg Chem 29:3272–3274
203. Komatsu T, Nakamura T, Matsukawa N, Yamochi H, Saito G (1991) New ambient pressure organic superconductors based on BEDT-TTF, Cu, $N(CN)_2$ and CN with $T_c = 10.7$ and 3.8 K. Solid State Commun 80:843–847
204. Komatsu T, Matsukawa N, Nakamura T, Yamochi H, Saito G (1992) Isotope effect on physical properties of BEDT-TTF based organic superconductors. Phosphorus Sulfur Silicon Relat Elem 67:295–300
205. Geiser U, Wang HH, Carlson KD, Williams JM, Charlier HA Jr, Heindl JE, Yaconi GA, Love BH, Lathrop MW, Schirber JE, Overmyer DL, Ren J, Whangbo MH (1991) Superconductivity at 2.8 K and 1.5 kbar in κ-$(BEDT\text{-}TTF)_2Cu_2(CN)_3$: the first organic superconductor containing a polymeric copper cyanide anion. Inorg Chem 30:2586–2588
206. Bu X, Frost-Jensen A, Allendoerfer R, Coppens P, Lederle B, Naughton M (1991) Structure and properties of a new κ-phase organic metal: $(BEDT\text{-}TTF)_2Cu_2(CN)_3$. Solid State Commun 79:1053–1057
207. Papavassiliou GCD, Lagouvardos J, Kakoussis VC, Terzis A, Hountas A, Hilti B, Mayer C, Zambounis JS, Pfeiffer J, Whangbo MH, Ren J, Kang DB (1992) Conducting and superconducting salts based on some symmetrical and unsymmetrical donors. Mater Res Soc Symp Proc 247:535–540
208. Welp U, Fleshler S, Kwok WK, Crabtree GW, Carlson KD, Wang HH, Geiser U, Williams JM, Hitsman VM (1992) Weak ferromagnetism in κ-$(ET)_2Cu[N(CN)_2]Cl$, where (ET) is bis (ethylenedithio)tetrathiafulvalene. Phys Rev Lett 69:840–843. This paper mistakenly described that the AF transition occurred at 45 K and weak ferromagnetic transition at 23 K. The magnetic susceptibility measurements could not identify the transition either AF or SDW
209. Miyagawa K, Kawamoto K, Nakazawa Y, Kanoda K (1995) Antiferromagnetic ordering and spin structure in the organic conductor, κ-$(BEDT\text{-}TTF)_2Cu[N(CN)_2]Cl$. Phys Rev Lett 75: 1174–1177
210. Posselt H, Muller H, Andres K, Saito G (1994) Reentrant Meissner effect in the organic conductor κ-$(BEDT\text{-}TTF)_2Cu[N(CN)_2]Cl$ under pressure. Phys Rev B49:15849–15852
211. Kanoda K (1997) Recent progress in NMR studies on organic conductors. Hyperfine Interact 104:235–249
212. Kanoda K (2006) Metal-insulator transition in κ-$(ET)_2X$ and $(DCNQI)_2M$: two contrasting manifestation of electron correlation. J Phys Soc Jpn 75:051007/1-16
213. Ishiguro T, Ito H, Yamauchi Y, Ohmichi E, Kubota M, Yamochi H, Saito G, Kartsovnik MV, Tanatar MA, Sushko YV, Logvenov GY (1997) Electronic phase diagrams and Fermi surfaces of κ-$(ET)_2X$, the high T_c organic superconductors. Synth Met 85:1471–1478
214. Sushko YV, Ito H, Ishiguro T, Horiuchi S, Saito G (1993) Magnetic-field-induced transition to resistive phase in superconducting κ-$(BEDT\text{-}TTF)_2Cu[N(CN)_2]Cl$. J Phys Soc Jpn 62: 3372–3375

215. Sushko YV, Ito H, Ishiguro T, Horiuchi S, Saito G (1993) Reentrant superconductivity in κ-(BEDT-TTF)$_2$Cu[N(CN)$_2$]Cl and its pressure phase diagram. Solid State Commun 87: 997–1000
216. Shshko YV, Murata K, Ito H, Ishiguro T, Saito G (1995) κ-(BEDT-TTF)$_2$Cu[N(CN)$_2$]Cl: magnet and superconductor. High pressure and high magnetic field experiments. Synth Met 70:907–910
217. Ito H, Ishiguro T, Kubota M, Saito G (1996) Metal-nonmetal transition and superconductivity localization in the two-dimensional conductor κ-(BEDT-TTF)$_2$Cu[N(CN)$_2$]Cl under pressure. J Phys Soc Jpn 65:2987–2993
218. Ito H, Kubota M, Ishiguro T, Saito G (1997) Metal-nonmetal transition of hydrogenated and deuterated κ-(BEDT-TTF)$_2$Cu[N(CN)$_2$]X under pressure. Synth Met 85:1517–1518
219. Ito H, Kondo T, Sasaki H, Saito G, Ishiguro T (1999) Antiferromagnetic spin resonance and magnetic phase diagram of deuterated κ-(BEDT-TTF)$_2$Cu[N(CN)$_2$]Br. Synth Met 103: 1818–1819
220. Ito H, Watanabe M, Nogami Y, Ishiguro T, Komatsu T, Saito G, Hosoito N (1991) Magnetic determination of Ginzburg-Landau coherence length for organic superconductor κ-(BEDT-TTF)$_2$X (X = Cu(NCS)$_2$, Cu[N(CN)$_2$]Br): effect of isotope substitution. J Phys Soc Jpn 60: 3230–3233
221. Ito H, Ishiguro T, Kondo T, Saito G (2000) Reentrant superconductivity of the deuterated salt of κ-(BEDT-TTF)$_2$Cu[N(CN)$_2$]Br under pressure. J Phys Soc Jpn 69:290–291
222. Kubota M, Saito G, Ito H, Ishiguro T, Kojima N (1996) Magnetism of the organic superconductor κ-(BEDT-TTF)$_2$Cu[N(CN)$_2$]Cl. Mol Cryst Liq Cryst 284:367–377
223. Lefebvre S, Wzietek P, Brown S, Bourbonnais C, Jérome D, Mézière C, Fourmigué M, Batial P (2000) Mott transition, antiferromagnetism, and unconventional superconductivity in layered organic superconductors. Phys Rev Lett 85:5420–5423
224. Posselt H, Andres K, Saito G (1995) Meissner effect in κ-ET$_2$Cu[N(CN)$_2$]Cl under pressure; observation of complete reentrant. Physica B 204:159–161
225. Oshima K, Mori T, Inokuchi H, Urayama H, Yamochi H, Saito G (1988) Shubnikov-de Haas effect and the Fermi surface in an ambient-pressure organic superconductor (bis (ethylenedithiolo)tetrathiafulvalene)$_2$Cu(NCS)$_2$. Phys Rev B 38:938–941
226. Oshima K, Urayama H, Yamochi H, Saito G (1988) Peculiar critical field behaviour in the recently discovered ambient pressure organic superconductor (BEDT-TTF)$_2$Cu(NCS)$_2$ (T_c = 10.4 K). J Phys Soc Jpn 57:730–733
227. Nam MS, Symington JA, Singleton J, Blundell SJ, Ardavan A, Perenboom JAAJ, Kurmoo M, Day P (1999) Angle dependence of the upper critical field in the layered organic superconductor κ-(BEDT-TTF)$_2$Cu(NCS)$_2$ (BEDT-TTF = bis(ethylene-dithio)tetrathiafulvalene). J Phys Condens Matter 11:L477–L484
228. Ichimura K, Arai T, Nomura K, Takasaki S, Yamada J, Nakatsuji S, Anzai H (1997) STM spectroscopy of (BEDT-TTF)$_2$Cu(NCS)$_2$. Physica C 282–287:1895–1896
229. Arai T, Ichimura K, Nomura K (2001) Tunneling spectroscopy on the organic superconductor κ-(BEDT-TTF)$_2$Cu(NCS)$_2$ using STM. Phys Rev B 63:104518/1-5
230. Izawa K, Yamaguchi H, Sasaki T, Matsuda Y (2002) Superconducting gap structure of κ-(BEDT-TTF)$_2$Cu(NCS)$_2$ probed by thermal conductivity tensor. Phys Rev Lett 88: 27002/1-4
231. Ichimura K, Takami M, Nomura K (2008) Direct observation of d-wave superconducting gap in κ-(BEDT-TTF)$_2$Cu[N(CN)$_2$]Br with scanning tunneling microscopy. J Phys Soc Jpn 77:114707/1-6
232. Saito G, Urayama H, Yamochi H, Oshima K (1988) Chemical and physical properties of a new ambient pressure organic superconductor with T_c higher than 10 K. Synth Met 27: A331–A340
233. Kini AM, Dudek JD, Carlson KD, Geiser U, Klemm RA, Williams JM, Lykke KR, Schlueter JA, Wang HH, Wurz P, Ferraro JR, Yaconi GA, Stout P (1993) Do the intramolecular C=C

stretching vibrational modes in ET mediate electron-pairing in κ-$(ET)_2X$ superconductors? Physica C 204:399–405
234. Toyota N (1996) Magnetic field effects on electronic states in BEDT-TTF salts. In: Fisk Z (ed) Physical phenomena at high magnetic fields II. World Scientific, Singapore
235. Bender K, Dietz K, Endres H, Helberg HV, Hennig I, Keller HJ, Schafer HV, Schweitzer D (1984) $(BEDT\text{-}TTF)_2J_3^-$: a two-dimensional organic metal. Mol Cryst Liq Cryst107:45–53
236. Takano Y, Hiraki K, Yamamoto HM, Nakamura T, Takahashi T (2001) Charge disproportionation in the organic conductor, α-$(BEDT\text{-}TTF)_2I_3$. J Phys Chem Solids 62:393–395
237. Tajima N, Sugawara N, Tamura M, Nishio Y, Kajita K (2006) Electronic phases in an organic conductor α-$(BEDT\text{-}TTF)_2I_3$: ultra narrow gap semiconductor, superconductor, metal, and charge-ordered insulator. J Phys Soc Jpn 75:051010/1-10
238. Katayama S, Kobayashi A, Suzumura Y (2006) Pressure-induced zero-gap semiconducting state in organic conductor α-$(BEDT\text{-}TTF)_2I_3$ salt. J Phys Soc Jpn 75:054705/1-6
239. Tajima N, Ebina-Tajima A, Tamura M, Nishio Y, Kajita K (2002) Effects of uniaxial strain on transport properties of organic conductor α-$(BEDT\text{-}TTF)_2I_3$ and discovery of superconductivity. J Phys Soc Jpn 71:1832–1835
240. Schweitzer D, Bele P, Brunner H, Gogu E, Haeberlen U, Hennig I, Klutz I, Sweitlik R, Keller HJ (1987) A stable superconducting state at 8 K and ambient pressure in α_t-$(BEDT\text{-}TTF)_2I_3$. Z Phys B 67:489–495
241. Laukhina EE, Merzhanov VA, Pesotskii SI, Khomenko AG, Yagubskii EB, Ulanski J, Kryszewski M, Jeszka JK (1995) Superconductivity in reticulate doped polycarbonate films, containing $(BEDT\text{-}TTF)_2I_3$. Synth Met 70:797–800
242. Yagubskii EB, Shchegolev IF, Laukhin VN, Kononovich PA, Karstovnik MV, Zvarykina AV, Buravov LI (1984) Normal-pressure, superconductivity in an organic metal $(BEDT\text{-}TTF)_2I_3$ [bis (ethylene dithiolo) tetrathiof ulvalene triiodide]. JETP Lett 39:12–15
243. Emge TJ, Leung PCW, Beno MA, Schultz AJ, Wang HH, Sowa LM, Williams JM (1984) Neutron and X-ray diffraction evidence for a structural phase transition in the sulfur-based ambient-pressure organic superconductor bis(ethylenedithio)tetrathiafulvalene triiodide. Phys Rev B30:6780–6782
244. Laukhin VN, Kostyuchenko EE, Sushko YV, Shchegolev IF, Yagubskii EB (1985) Effect of pressure on the superconductivity of β-$(BEDT\text{-}TTF)_2I_3$. JETP Lett 41:81–84
245. Murata K, Tokumoto M, Anzai H, Bando H, Saito G, Kajimura K, Ishiguro T (1985) Superconductivity with the onset at 8 K in the organic conductor β-$(BEDT\text{-}TTF)_2I_3$ under pressure. J Phys Soc Jpn 54:1236–1239
246. Ito H, Ishihara T, Niwa M, Suzuki T, Onari S, Tanaka Y, Yamada J, Yamochi H, Saito G (2010) Superconductivity of β-type salts under uniaxial compression. Physica B 405:S262–S264
247. Shibaeva RP, Kaminskii VF, Yagubskii EB (1985) Crystal structures of organic metals and superconductors of (BEDT-TTF)-I system. Mol Cryst Liq Cryst 119:361–373
248. Kobayashi H, Kato R, Kobayashi A, Nishio Y, Kajita K, Sasaki W (1986) A new molecular superconductor, $(BEDT\text{-}TTF)_2(I_3)_{1-x}(AuI_2)_x$ $(x < 0.02)$. Chem Lett 15:789–792
249. Kobayashi A, Kato R, Kobayashi H, Moriyama S, Nishio Y, Kajita K, Sasaki W (1987) Crystal and electronic structures of a new molecular superconductor, κ-$(BEDT\text{-}TTF)_2I_3$. Chem Lett 16:459–462
250. Kobayashi H, Kobayashi A, Cassoux P (2000) BETS as a source of molecular magnetic superconductors (BETS = bis(ethylenedithio)tetraselenafulvalene). Chem Soc Rev 29: 325–333
251. Sakata J, Sato H, Miyazaki A, Enoki T, Okano Y, Kato R (1998) Superconductivity in new organic conductor κ-$(BEDSe\text{-}TTF)_2CuN(CN)_2Br$. Solid State Commun 108:377–381
252. Okano Y, Iso M, Kashimura Y, Yamaura J, Kato R (1999) A new synthesis of Se-containing TTF derivatives. Synth Met 102:1703–1704
253. Zambounis JS, Mayer CW, Hauenstein K, Hilti B, Hofherr W, Pfeiffer J, Buerkle M, Rihs G (1992) Crystal structure and electrical properties of κ-$((S,S)\text{-}DMBEDT\text{-}TTF)_2ClO_4$. Adv Mater 4:33–35

254. Kimura S, Maejima T, Suzuki H, Chiba R, Mori H, Kawamoto T, Mori T, Moriyama H, Nishio Y, Kajita K (2004) A new organic superconductor β-(*meso*-DMBEDT-TTF)$_2$PF$_6$. Chem Commun 2454–2455
255. Kimura S, Suzuki H, Maejima T, Mori H, Yamaura J, Kakiuchi T, Sawa H, Moriyama H (2006) Checkerboard-type charge-ordered state of a pressure-induced superconductor, β-(*meso*-DMBEDT-TTF)$_2$PF$_6$. J Am Chem Soc 128:1456–1457
256. Kikuchi K, Murata K, Honda Y, Namiki T, Saito K, Ishiguro T, Kobayashi K, Ikemoto I (1987) On ambient-pressure superconductivity in organic conductors: electrical properties of (DMET)$_2$I$_3$, (DMET)$_2$I$_2$Br and (DMET)$_2$IBr$_2$. J Phys Soc Jpn 56:3436–3439
257. Shirahata T, Kibune M, Imakubo T (2006) New ambient pressure organic superconductors κ_H- and κ_L-(DMEDO-TSeF)$_2$[Au(CN)$_4$](THF). Chem Commun 1592–1594
258. Shirahata T, Kibune M, Yoshino H, Imakubo T (2007) Ambient-pressure organic superconductors κ-(DMEDO-TSeF)$_2$[Au(CN)$_4$](solv.): T_c tuning by modification of the solvent of crystallization. Chem Eur J 13:7619–7630
259. Nishikawa H, Morimoto T, Kodama T, Ikemoto I, Kikuchi K, Yamada J, Yoshino H, Murata K (2002) New organic superconductors consisting of an unprecedented π-electron donor. J Am Chem Soc 124:730–731
260. Kato R, Yamamoto K, Okano Y, Tajima H, Sawa H (1997) A new ambient-pressure organic superconductor (TMET-STF)$_2$BF$_4$ [TMET-STF = trimethylene(ethylenedithio) diselenadithiafulvalene]. Chem Commun 947–948
261. Kato R, Aonuma S, Okano Y, Sawa H, Tamura M, Kinoshita M, Oshima K, Kobayashi A, Bun K, Kobayashi H (1993) Metallic and superconducting salts based on an unsymmetrical π-donor dimethyl(ethylenedithio)tetraselenafulvalene (DMET-TSeF). Synth Met 61:199–206
262. Imakubo T, Tajima N, Tamura M, Kato R, Nishio Y, Kajita K (2002) A supramolecular superconductor θ-(DIETS)$_2$[Au(CN)$_4$]. J Mater Chem 12:159–161
263. Lyubovskaya RN, Zhilyaeva EI, Torunova SA, Mousdis GA, Papavassiliou GC, Perenboom JAAJ, Pesotskii SI, Lyubovskii RB (2004) New ambient pressure organic superconductor with $T_c = 8.1$ K: (EDT-TTF)$_4$Hg$_{3-\delta}$I$_8$. J Phys IV 114:463–466
264. Papavassiliou GC, Mousdis GA, Zambounis JS, Terzis A, Hountas A, Hilti B, Mayer CW, Pfeiffer J (1988) Low temperature measurements of the electrical conductivities of some charge transfer salts with the asymmetric donors MDT-TTF, EDT-TTF and EDT-DSDTF. (MDT-TTF)$_2$AuI$_2$, a new superconductor ($T_c = 3.5$ K at ambient pressure). Synth Met B27:379–383
265. Takahashi T, Kobayashi Y, Nakamura T, Kanoda K, Hilti B, Zambounis JS (1994) Symmetry of the order parameter in organic superconductors: (MDT-TTF)$_2$AuI$_2$ vs. (TMTSF)$_2$ClO$_4$. Physica C 235–240:2461–2462
266. Takimiya K, Takamori A, Aso Y, Otsubo T, Kawamoto T, Mori T (2003) Organic superconductors based on a new electron donor, methylenedithio-diselenadithiafulvalene (MDT-ST). Chem Mater 15:1225–1227
267. Kawamoto T, Mori T, Enomoto K, Koike T, Terashima T, Uji S, Takamori A, Takimiya K, Otsubo T (2006) Fermi surface of the organic superconductor (MDT-ST)(I$_3$)$_{0.417}$ reconstructed by incommensurate potential. Phys Rev B73: 024503/1-5
268. Takimiya K, Kodani M, Niihara N, Aso Y, Otsubo T, Bando Y, Kawamoto T, Mori T (2004) Pressure-induced superconductivity in (MDT-TS)(AuI$_2$)$_{0.441}$ [MDT-TS = 5H-2-(1,3-diselenol-2-ylidene)-1,3,4,6-tetrathiapentalene]: a new organic superconductor possessing an incommensurate anion lattice. Chem Mater 16:5120–5123
269. Takimiya K, Kataoka Y, Aso Y, Otsubo T, Fukuoka H, Yamanaka S (2001) Quasi one-dimensional organic superconductor MDT-TSF·AuI$_2$ with $T_c = 4.5$ K at ambient pressure. Angew Chem Int Ed 40:1122–1125
270. Kawamoto T, Mori T, Konoike T, Enomoto K, Terashima T, Uji S, Kitagawa H, Takimiya K, Otsubo T (2006) Charge transfer degree and superconductivity of the incommensurate organic superconductor (MDT-TSF)(I$_3$)$_{0.422}$. Phys Rev B 73:094513/1-8

271. Kawamoto T, Mori T, Terakura C, Terashima T, Uji S, Takimiya K, Aso Y, Otsubo T (2003) Incommensurate anion potential effect on the electronic states of the organic superconductor (MDT-TSF)$(AuI_2)_{0.436}$. Phys Rev B 67:020508/1-4
272. Kawamoto T, Mori T, Terakura C, Terashima T, Uji S, Tajima H, Takimiya K, Aso Y, Otsubo T (2003) Electronic state anisotropy and the Fermi surface topology of the incommensurate organic superconducting crystal (MDT-TSF)$(AuI_2)_{0.436}$. Eur Phys J B 36:161–167
273. Kodani M, Takamori A, Takimiya K, Aso Y, Otsubo T (2002) Novel conductive radical cation salts based on methylenediselenotetraselenafulvalene (MDSe-TSF): a sign of superconductivity in κ-$(MDSe\text{-}TSF)_2Br$ below 4 K. J Solid State Chem 168:582–589
274. Yamada J, Watanabe M, Akutsu H, Nakatsuji S, Nishikawa H, Ikemoto I, Kikuchi K (2001) New organic superconductors β-$(BDA\text{-}TTP)_2X$ [BDA-TTP = 2,5-bis(1,3-dithian-2-ylidene)-1,3,4,6-tetrathiapentalene; $X^- = SbF_6^-$, AsF_6^-, and PF_6^-]. J Am Chem Soc 123: 4174–4180
275. Ito H, Ishihara T, Tanaka H, Kuroda S, Suzuki T, Onari S, Tanaka Y, Yamada J, Kikuchi K (2008) Roles of spin fluctuation and frustration in the superconductivity of β-$(BDA\text{-}TTP)_2X$ ($X = SbF_6$, AsF_6) under uniaxial compression. Phys Rev B 78:172506/1-4
276. Shimojo Y, Ishiguro T, Toita T, Yamada J (2002) Superconductivity of layered organic compound β-$(BDA\text{-}TTP)_2SbF_6$, where BDA-TTP is 2,5-bis(1,3-dithian-2-ylidene)-1,3,4,6-tetrathiapentalene. J Phys Soc Jpn 71:717–720
277. Nomura K, Muraoka R, Matsunaga N, Ichimura K, Yamada J (2009) Anisotropic superconductivity in β-$(BDA\text{-}TTP)_2SbF_6$: STM spectroscopy. Physica B 404:562–564
278. Misaki Y, Higuchi N, Fujiwara H, Yamabe T, Mori T, Mori H, Tanaka S (1995) (DTEDT) $[Au(CN)_2]_{0.4}$: an organic superconductor based on the novel π-electron framework of vinylogous bis-fused tetrathiafulvalene. Angew Chem Int Ed Engl 34:1222–1225
279. Tanaka H, Ojima E, Fujiwara H, Nakazawa Y, Kobayashi H, Kobayashi A (2000) A new κ-type organic superconductor based on BETS molecules, κ-$(BETS)_2GaBr_4$ [BETS = bis (ethylenedithio)tetraselenafulvalene]. J Mater Chem 10:245–247
280. Gritsenko V, Tanaka H, Kobayashi H, Kobayashi A (2001) A new molecular superconductor, κ-$(BETS)_2TlCl_4$ [BETS = bis(ethylenedithio)tetraselenafulvalene]. J Mater Chem 11: 2410–2411
281. Konoike T, Uji S, Terashima T, Nishimura M, Yasuzuka S, Enomoto K, Fujiwara H, Zhang B, Kobayashi H (2004) Magnetic-field-induced superconductivity in the antiferromagnetic organic superconductor κ-$(BETS)_2FeBr_4$. Phys Rev B70:094514/1-5
282. Brossard L, Clerac R, Coulon C, Tokumoto M, Ziman T, Petrov DK, Laukhin VN, Naughton MJ, Audouard A, Goze F, Kobayashi A, Kobayashi H, Cassoux P (1998) Interplay between chains of S = 5/2 localised spins and two-dimensional sheets of organic donors in the synthetically built magnetic multilayer λ-$(BETS)_2FeCl_4$. Eur Phys J B 1:439–452
283. Kobayashi H, Tomita H, Naito T, Kobayashi A, Sakai F, Watanabe T, Cassoux P (1996) New BETS conductors with magnetic anions (BETS = bis(ethylenedithio)tetraselenafulvalene). J Am Chem Soc 118:368–377
284. Kobayashi H, Udagawa T, Tomita H, Bun K, Naito T, Kobayashi A (1993) A new organic superconductor, λ-$(BEDT\text{-}TSF)_2GaCl_4$. Chem Lett 22:1559–1562
285. Uji S, Shinagawa H, Terashima T, Yakabe T, Terai Y, Tokumoto M, Kobayashi A, Tanaka H, Kobayashi H (2001) Magnetic-field-induced superconductivity in a two-dimensional organic conductor. Nature 410:908–910
286. Matsui H, Tsuchiya H, Suzuki T, Negishi E, Toyota N (2003) Relaxor ferroelectric behavior and collective modes in the π-d correlated anomalous metal λ-$(BEDT\text{-}TSF)_2FeCl_4$. Phys Rev B68:155105/1-10
287. Toyota N, Abe N, Matsui H, Negishi E, Ishizaki Y, Tsuchiya H, Uozaki H, Endo S (2002) Nonlinear electrical transport in λ-$(BEDT\text{-}TSF)_2FeCl_4$. Phys Rev B66:033201/1-4
288. Kobayashi H, Sato A, Arai E, Akutsu H, Kobayashi A, Cassoux P (1997) Superconductor-to-insulator transition in an organic metal incorporating magnetic anions: λ-$(BETS)_2(Fe_xGa_{1-x})$

Cl_4 [BETS = bis(ethylenedithio)tetraselenafulvalene; x ≈ 0.55 and 0.43]. J Am Chem Soc 119:12392–12393
289. Fujiwara E, Fujiwara H, Kobayashi H, Otsuka T, Kobayashi A (2002) A series of organic conductors, κ-$(BETS)_2FeBr_xCl_{4-x}$ ($0 \leq x \leq 4$), exhibiting successive antiferromagnetic and superconducting transitions. Adv Mater 14:1376–1379
290. Meul HW, Rossel C, Decroux M, Fischer Ø, Remenyi G, Briggs A (1984) Observation of magnetic-field-induced superconductivity. Phys Rev Lett 53:497–500
291. Lin CL, Teter J, Crow JE, Mihalisin T, Brooks J, Abou-Aly AI, Stewart GR (1985) Observation of magnetic-field-induced superconductivity in a heavy-Fermion antiferromagnet: $CePb_3$. Phys Rev Lett 54:2541–2544
292. Clark K, Hassanien A, Khan S, Braun KF, Tanaka H, Hla SW (2010) Superconductivity in just four pairs of $(BETS)_2GaCl_4$ molecules. Nat Nanotechnol 5:261–265
293. Bousseau L, Valade L, Legros JP, Cassoux P, Garbauskas M, Interrante LV (1986) Highly conducting charge-transfer compounds of tetrathiafulvalene and transition metal-"dmit" complexes. J Am Chem Soc 108:1908–1916
294. Mitsuhashi R, Suzuki Y, Yamanari Y, Mitamura H, Kambe T, Ikeda N, Okamoto H, Fujiwara A, Yamaji M, Kawasaki N, Maniwa Y, Kubozono Y (2010) Superconductivity in alkali-metal-doped picene. Nature 464:76–79
295. Tajima H, Inokuchi M, Kobayashi A, Ohta T, Kato R, Kobayashi H, Kuroda H (1993) First ambient-pressure superconductor based on $Ni(dmit)_2$, α-EDT-TTF$[Ni(dmit)_2]$. Chem Lett 22:1235–1238
296. Miura YF, Horikiri M, Saito SH, Sugi M (2000) Evidence for superconductivity in Langmuir-Blodgett films of ditetradecyldimethylammonium-$Au(dmit)_2$. Solid State Commun 113: 603–605
297. Avdeev VV, Zharikov OV, Nalimova VA, Pal'nichenko AV, Semenenko KN (1986) Superconductivity of the layered potassium-graphite compounds C_6K and C_4K. Pis'ma Zh Eksp Teor Fiz 43:376–378
298. Krätschmer W, Lamb LD, Fostiropoulos K, Huffman DR (1990) Solid C_{60}: a new form of carbon. Nature 347:354–358
299. Haddon RC, Hebard AF, Rosseinsky MJ, Murphy DW, Duclos SJ, Lyons KB, Miller B, Rosamilia JM, Fleming RM, Kortan AR, Glarum SH, Makhija AV, Muller AJ, Eick RH, Zahurak SM, Tycko R, Dabbagh G, Thiel FA (1991) Conducting films of C_{60} and C_{70} by alkali-metal doping. Nature 350:320–322
300. Stephens PW, Mihaly L, Lee PL, Whetten RL, Huang SM, Kaner R, Diederich F, Holczer K (1991) Structure of single-phase superconducting K_3C_{60}. Nature 351:632–634
301. Rosseinsky MJ, Ramirez AP, Glarum SH, Murphy DW, Haddon RC, Hebard AF, Palstra TTM, Kortan AR, Zahurak SM, Makhija AV (1991) Superconductivity at 28 K in Rb_xC_{60}. Phys Rev Lett 66:2830–2832
302. Tanigaki K, Ebbesen TW, Saito S, Mizuki J, Tsai JS, Kubo Y, Kuroshima S (1991) Superconductivity at 33 K in $Cs_xRb_yC_{60}$. Nature 352:222–223
303. Fleming RM, Ramirez AP, Rosseinsky MJ, Murphy DW, Haddon RC, Zahurak SM, Makhija AV (1991) Relation of structure and superconducting transition temperatures in A_3C_{60}. Nature 352:787–788
304. Stenger VA, Pennington CH, Buffinger DR, Ziebarth RP (1995) Nuclear magnetic resonance of A_3C_{60} superconductors. Phys Rev Lett 74:1649–1652
305. Kiefl RF, MacFarlane WA, Chow KH, Dunsiger S, Duty TL, Johnston TMS, Schneider JW, Sonier J, Brard L, Strongin RM, Fischer JE, Smith AB III (1993) Coherence peak and superconducting energy gap in Rb_3C_{60} observed by muon spin relaxation. Phys Rev Lett 70:3987–3990
306. Chen CC, Lieber CM (1992) Synthesis of pure $^{13}C_{60}$ and determination of the isotope effect for fullerene superconductors. J Am Chem Soc 114:3141–3142
307. Takenobu T, Muro T, Iwasa Y, Mitani T (2000) Antiferromagnetism and phase diagram in ammoniated alkali fulleride salts. Phys Rev Lett 85:381–384

308. Dahlke P, Denning MS, Henry PF, Rosseinsky MJ (2000) Superconductivity in expanded fcc C_{60}^{3-} fullerides. J Am Chem Soc 122:12352–12361
309. Yildirim T, Barbedette L, Fischer JE, Lin CL, Robert J, Petit P, Palstra TTM (1996) T_c vs carrier concentration in cubic fulleride superconductors. Phys Rev Lett 77:167–170
310. Kosaka M, Tanigaki K, Prassides K, Margadonna S, Lappas A, Brown CM, Fitch AN (1999) Superconductivity in Li_xCsC_{60} fullerides. Phys Rev B59:R6628–R6630
311. Palstra TTM, Zhou O, Iwasa Y, Sulewski PE, Fleming RM, Zegarski BR (1995) Superconductivity at 40K in cesium doped C_{60}. Solid State Commun 93:327–330
312. Ganin AY, Takabayashi Y, Khimyak YZ, Margadonna S, Tamai A, Rosseinsky MJ, Prassides K (2008) Bulk superconductivity at 38 K in a molecular system. Nat Mater 7:367–371
313. Takabayashi Y, Ganin AY, Jeglič P, Arčon D, Takano T, Iwasa Y, Ohishi Y, Takata M, Takeshita N, Prassides K, Rosseinsky MJ (2009) The disorder-free non-BCS superconductor Cs_3C_{60} emerges from an antiferromagnetic insulator parent state. Science 323:1585–1590
314. Ganin AY, Takabayashi Y, Jeglič P, Arčon D, Potočnik A, Baker PJ, Ohishi Y, McDonald MT, Tzirakis MD, McLennan A, Darling GR, Takata M, Rosseinsky MJ, Prassides K (2010) Polymorphism control of superconductivity and magnetism in Cs_3C_{60} close to the Mott transition. Nature 466:221–225
315. Özdas E, Kortan AR, Kopylov N, Ramirez AP, Siegrist T, Rabe KM, Bair HE, Schuppler S, Citrin PH (1995) Superconductivity and cation-vacancy ordering in the rare-earth fulleride $Yb_{2.75}C_{60}$. Nature 375:126–129
316. Chen XH, Roth G (1995) Superconductivity at 8 K in samarium-doped C_{60}. Phys Rev B52: 15534–15536
317. Baenitz M, Heinze M, Lüders K, Werner H, Schlögl R, Weiden M, Sparn G, Steglich F (1995) Superconductivity of Ba doped C_{60} – susceptibility results and upper critical field. Solid State Commun 96:539–544
318. Brown CM, Taga S, Gogia B, Kordatos K, Margadonna S, Prassides K, Iwasa Y, Tanigaki K, Fitch AN, Pattison P (1999) Structural and electronic properties of the noncubic superconducting fullerides $A_4'C_{60}$ (A' = Ba, Sr). Phys Rev Lett 83:2258–2261
319. Ksari-Habiles Y, Claves D, Chouteau G, Touzain Ph, Cl J, Oddou JL, Stepanov A (1997) Superexchange and magnetic relaxation in novel Eu-doped C_{60} phases. J Phys Chem Solids 58:1771–1778
320. Ishii K, Fujiwara A, Suematsu H, Kubozono Y (2002) Ferromagnetism and giant magnetoresistance in the rare-earth fullerides $Eu_{6-x}Sr_xC_{60}$. Phys Rev B65:134431/1-6
321. Maruyama Y, Motohashi S, Sakai N, Watanabe K, Suzuki K, Ogata H, Kubozono Y (2002) Possible competition of superconductivity and ferromagnetism in Ce_xC_{60} compounds. Solid State Commun 123:229–233
322. Wang P, Metzger RM, Bandow S, Maruyama Y (1993) Superconductivity in Langmuir-Blodgett multilayers of fullerene (C_{60}) doped with potassium. J Phys Chem 97:2926–2927
323. Ikegami K, Kuroda S, Matsumoto M, Nakamura T (1995) Alkali-metal doping of Langmuir-Blodgett films of C_{60} studied by ESR. Jpn J Appl Phys 34:L1227–L1229
324. Little WA (1964) Possibility of synthesizing an organic superconductor. Phys Rev A134: 1416–1424
325. Greene RL, Street GB, Suter LJ (1975) Superconductivity in polysulfur nitride $(SN)_x$. Phys Rev Lett 34:577–579
326. Labes MM, Love P, Nichols LF (1979) Polysulfur nitride – a metallic, superconducting polymer. Chem Rev 79:1–15
327. Banister AJ, Gorrell IB (1998) Poly(sulfur nitride): the first polymeric metal. Adv Mater 10: 1415–1429
328. Kawamura H, Shirotani I, Tachikawa K (1984) Anomalous superconductivity in black phosphorus under high pressures. Solid State Commun 49:879–881
329. Lee K, Cho S, Park SH, Heeger AJ, Lee C-W, Lee S-H (2006) Metallic transport in polyaniline. Nature 441:65–68

330. Sato K, Yamaura M, Hagiwara T, Murata K, Tokumoto M (1991) Study on the electrical conduction mechanism of polypyrrole films. Synth Met 40:35–48
331. Aleshin A, Kiebooms R, Menon R, Heeger AJ (1997) Electronic transport in doped poly (3,4-ethylenedioxythiophene) near the metal-insulator transition. Synth Met 90:61–68
332. Wolf AA, Halpern EH (1976) On a class of organic superconductors: a summary of findings. Proc IEEE 64:357–359
333. Langer J (1978) Unusual properties of the aniline black: does the superconductivity exist at room temperature? Solid State Commun 26:839–844
334. Enikolopyan NS, Grigorov LN, Smirnova SG (1989) Possible superconductivity near 300 K in oxidized polypropylene. JETP Lett 49:371–375
335. Lyubchenko LS, Stepanov SV, Lyubchenko ML, Sherle AI, Epstein VP, Dadali AA, Malinsky J (1992) Low-field microwave absorption in high-T_c superconductors based on olygo- and polyphthalocyanines. Phys Lett A162:69–78
336. Iwasa Y, Takenobu T (2003) Superconductivity, Mott-Hubbard states, and molecular orbital order in intercalated fullerides. J Phys Condens Matter 15:R495–R519
337. Chu CW (2009) High-temperature superconductivity: alive and kicking. Nat Phys 5:787–789
338. Anderson PW (1973) Resonating valence bonds: a new kind of insulator? Mater Res Bull 8: 153–160
339. Ohmichi E, Ito H, Ishiguro T, Komatsu T, Saito G (1997) Angle-dependent magnetoresistance in the organic superconductor $(BEDT\text{-}TTF)_2Cu_2(CN)_3$ under pressure. J Phys Soc Jpn 66:310–311
340. Ohmichi E, Ito H, Ishiguro T, Komatsu T, Saito G (1998) Pressure dependence of magnetoresistance and Fermi surface of $(BEDT\text{-}TTF)_2Cu_2(CN)_3$. Rev High Pressure Sci Technol 7: 523–525
341. Ohmichi E, Ito H, Ishiguro T, Saito G (1998) Shubnikov-de Haas oscillation with unusual angle dependence in the organic superconductor $\kappa\text{-}(BEDT\text{-}TTF)_2Cu_2(CN)_3$. Phys Rev B57:7481–7484
342. Shimizu Y, Miyagawa K, Kanoda K, Maesato M, Saito G (2003) Spin liquid state in an organic Mott insulator with a triangular lattice. Phys Rev Lett 91:107001/1-4. $\kappa\text{-}(ET)_2Cu_2(CN)_3$ was first prepared and reported by some groups including [206–209]
343. McKenzie RH (1998) A strongly correlated electron model for the layered organic superconductors $\kappa\text{-}(BEDT\text{-}TTF)_2X$. Comments. Condens Matter Phys 18:309–337
344. Kino H, Fukuyama H (1995) Electronic states of conducting organic $\kappa\text{-}(BEDT\text{-}TTF)_2X$. J Phys Soc Jpn 64:2726–2729
345. Imamura Y, Ten-no S, Yonemitsu K, Tanimura Y (1999) Structures and electronic phases of the bis(ethylenedithio)tetrathiafulvalene (BEDT-TTF) clusters and κ-(BEDT-TTF) salts: a theoretical study based on ab initio molecular orbital methods. J Chem Phys 111:5986–5994. Though the previous density-function theory (DFT) calculations gave a little higher t'/t than those by the extended Hückel method [345, 346], the recent DFT calculations using a generalized-gradient-approximation gave the smaller t'/t (~0.8) [347, 348].
346. Kawakami T, Taniguchi T, Kitagawa Y, Takano Y, Nagao H, Yamaguchi K (2002) Theoretical investigation of magnetic parameters in two-dimensional sheets of pure organic BEDT-TTF and BETS molecules by using ab initio MO and DFT methods. Mol Phys 100: 2641–2652
347. Kandpal HC, Opahle I, Zhang Y, Jeschke HO, Valenti R (2009) Revision of model parameters for κ-type charge transfer salts: an ab initio study. Phys Rev Lett 103:067004/1-4
348. Nakamura K, Yoshimoto Y, Kosugi T, Arita R, Imada M (2009) Ab initio derivation of low-energy model for κ-ET Type organic conductors. J Phys Soc Jpn 78:083710/1-4
349. Shimizu Y, Maesato M, Saito G, Miyagawa K, Kanoda K (2003) Magnetic properties of $\kappa\text{-}(ET)_2Cu_2(CN)_3$. Synth Met 137:1247–1248
350. Shimizu Y, Maesato M, Saito G, Drozdova O, Ouahab L (2003) Transport properties of a Mott insulator $\kappa\text{-}(ET)_2Cu_2(CN)_3$ under the uniaxial strain. Synth Met 133–134:225–226
351. Shimizu Y, Miyagawa K, Oda K, Kanoda K, Maesato M, Saito G (2004) ^{1}H-NMRstudy of Mott insulator $\kappa\text{-}(ET)_2Cu_2(CN)_3$ with isotropic triangular lattice. J Phys IV 114:377–378

352. Shimizu Y, Kurosaki Y, Miyagawa K, Kanoda K, Maesato M, Saito G (2005) NMR study of the spin-liquid state and Mott transition in the spin-frustrated organic system, κ-$(ET)_2Cu_2(CN)_3$. Synth Met 152:393–396
353. Shimizu Y, Miyagawa K, Kanoda K, Maesato M, Saito G (2005) Spin liquid in a spin-frustrated organic Mott insulator. Prog Theor Phys Suppl 159:52–60
354. Kurosaki Y, Shimizu Y, Miyagawa K, Kanoda K, Saito G (2005) Mott transition from spin liquid in the spin-frustrated organic conductor κ-$(ET)_2Cu_2(CN)_3$. Phys Rev Lett 95: 177001/1-4
355. Kézsmárki I, Shimizu Y, Mihály G, Tokura Y, Kanoda K, Saito G (2006) Depressed charge gap in the triangular-lattice Mott insulator κ-$(ET)_2Cu_2(CN)_3$. Phys Rev B74:201101/1-4
356. Ohira S, Shimizu Y, Kanoda K, Saito G (2006) Spin liquid state in κ-(BEDT-TTF)$_2$Cu$_2$(CN)$_3$ studied by muon spin relaxation method. J Low Temp Phys 142:153–158
357. Shimizu Y, Miyagawa K, Kanoda K, Maesato M, Saito G (2006) Emergence of inhomogeneous moments from spin liquid in the triangular-lattice Mott insulator κ-$(ET)_2Cu_2(CN)_3$. Phys Rev B 73:140407/1-4
358. Yamashita S, Nakazawa Y, Oguni M, Oshima Y, Nojiri H, Shimizu Y, Miyagawa K, Kanoda K (2008) Thermodynamic properties of a spin-1/2 spin-liquid state in a κ-type organic salt. Nat Phys 4:459–462
359. Yamashita M, Nakata N, Kasahara Y, Sasaki T, Yoneyama N, Kobayashi N, Fujimoto S, Shibauchi T, Matsuda Y (2009) Thermal-transport measurements in a quantum spin-liquid state of the frustrated triangular magnet κ-(BEDT-TTF)$_2$Cu$_2$(CN)$_3$. Nat Phys 5:44–47
360. Manna RS, de Souza M, Bruhl A, Schulueter JA, Lang M (2010) Lattice effects and entropy release at the low-temperature phase transition in the spin-liquid candidate κ-(BEDT-TTF)$_2$Cu$_2$(CN)$_3$. Phys Rev Lett 104:016403/1-4
361. Maesato M, Shimizu Y, Ishikawa T, Saito G (2003) Anisotropy in the superconducting transition temperature of κ-(BEDT-TTF)$_2$X. Synth Met 137:1243–1244
362. Komatsu T, Matsukawa N, Inoue T, Saito G (1996) Realization of superconductivity at ambient pressure by band-filling control in κ-(BEDT-TTF)$_2$Cu$_2$(CN)$_3$. J Phys Soc Jpn 65:1340–1354
363. Maesato M, Shimizu Y, Ishikawa T, Saito G, Miyagawa K, Kanoda K (2004) Spin-liquid behavior and superconductivity in κ-(BEDT-TTF)$_2$X: the role of uniaxial strain. J Phys IV 114:227–232
364. Balents L (2010) Spin liquids in frustrated magnets. Nature 464:199–208
365. Itou T, Oyamada A, Maegawa S, Tamura M, Kato R (2007) Spin-liquid state in an organic spin-1/2 system on a triangular lattice, $EtMe_3Sb[Pd(dmit)_2]_2$. J Phys Condens Matter 19: 145247/1-5
366. Komatsu T, Saito G (1996) Carrier-doping in a Mott-insulator κ-(BEDT-TTF)$_2$Cu$_2$(CN)$_3$. Mol Cryst Liq Cryst 285:51–56
367. Komatsu T, Kojima N, Saito G (1997) Ambient-pressure superconductivity of κ'-(BEDT-TTF)$_2$Cu$_2$(CN)$_3$ realized by a carrier-doping into a Mott-insulating state. Synth Met 85: 1519–1520
368. Drozdova O, Saito G, Yamochi H, Ookubo K, Yakushi K, Uruichi M, Ouahab L (2001) Composition and structure of the anion layer in the organic superconductor κ'-$(ET)_2$ $Cu_2(CN)_3$: optical study. Inorg Chem 40:3265–3266
369. Saito G, Ookubo K, Drozdova O, Yakushi K (2002) Tuning of critical temperature in an ET organic superconductor. Mol Cryst Liq Cryst 380:23–27
370. Yamochi H, Nakamura T, Komatsu T, Matsukawa N, Inoue T, Saito G, Mori T, Kusunoki M, Sakaguchi K (1992) Crystal and electronic structures of the organic superconductors, κ-(BEDT-TTF)$_2$Cu(CN)[N(CN)$_2$] and κ'-(BEDT-TTF)$_2$Cu$_2$(CN)$_3$. Solid State Commun 82: 101–105
371. Shimizu Y, Kasahara H, Furuta T, Miyagawa K, Kanoda K, Maesato M, Saito G (2010) Pressure-induced superconductivity and Mott transition in spin-liquid κ-$(ET)_2Cu_2(CN)_3$ probed by ^{13}C NMR. Phys Rev B 81:224508/1-5

Top Curr Chem (2012) 312: 127–174
DOI: 10.1007/128_2011_176

Published online: 6 September 2011

Fullerenes, Carbon Nanotubes, and Graphene for Molecular Electronics

Julio R. Pinzón, Adrián Villalta-Cerdas, and Luis Echegoyen

Abstract With the constant growing complexity of electronic devices, the top-down approach used with silicon based technology is facing both technological and physical challenges. Carbon based nanomaterials are good candidates to be used in the construction of electronic circuitry using a bottom-up approach, because they have semiconductor properties and dimensions within the required physical limit to establish electrical connections. The unique electronic properties of fullerenes for example, have allowed the construction of molecular rectifiers and transistors that can operate with more than two logical states. Carbon nanotubes have shown their potential to be used in the construction of molecular wires and FET transistors that can operate in the THz frequency range. On the other hand, graphene is not only the most promising material for replacing ITO in the construction of transparent electrodes but it has also shown quantum Hall effect and conductance properties that depend on the edges or chemical doping. The purpose of this review is to present recent developments on the utilization carbon nanomaterials in molecular electronics.

Keywords Bottom-up · Carbon nanomaterials · Molecular conductance · Molecular electronics · Unimolecular electronic devices

Contents

J.R. Pinzón, A. Villalta-Cerdas, and L. Echegoyen (✉)
Department of Chemistry, University of Texas at El Paso, El Paso, TX, USA
e-mail: echegoyen@nsf.gov

1 Introduction

With the constant growing complexity of electronic devices, the top-down approach used with silicon based technology is facing both technological and physical challenges [1–3]. In contrast, constructing circuitry from small pieces using bottom-up approaches is gaining considerable attention [4]. Single molecules are ideal candidates to be used as electronic devices because they can be prepared using well known and easily reproducible synthetic procedures. In 2003, Tao and collaborators measured single molecule conductances for several organic compounds between a gold scanning tunneling microscope (STM) tip and a gold surface [5]. Conductance histograms showed peaks at integer values of 1/100 of conductance quantization G_0 (=$2e^2/h$ = 77 μS), ascribed to the number of molecules forming a stable junction between the electrodes (see Fig. 1a).

In 2005, the same group reported the use of an organic molecule as a single molecule transistor (see Fig. 1b), in which the current through the molecule was modulated by a gate potential (V_g) applied at room temperature [6]. As the gate potential was reduced the current increased, resulting in a huge increase (~500 times vs $V_g = 0$ V) at −0.65 V, corresponding to the reduction of the organic linker (see Fig. 1c). Control experiments using alkanedithiol showed no current dependence upon applied gate voltage (see Fig. 1d), further proving the effectiveness of the transistor behavior in these experiments.

Carbon based nanomaterials [7] are strong candidates to be used as building blocks for molecular electronics, because they have semiconductor properties and dimensions within the required physical limit to establish electrical connections [8]. Many researchers have studied and measured single molecule conductance and its dependence on intrinsic chemical properties. For example, in 2006 Venkataraman et al. demonstrated that conductances for a series of biphenyl compounds decreases when the molecular twist angle between the phenyl groups (θ) increases (see Fig. 2) [9]. This correlates with the predicted transport through p-conjugated biphenyl systems considering the $\cos^2\theta$.

Therefore, single molecule conductance measurements provide experimental evidence for theoretically expected behavior which cannot be determined by any other means but single molecule analysis.

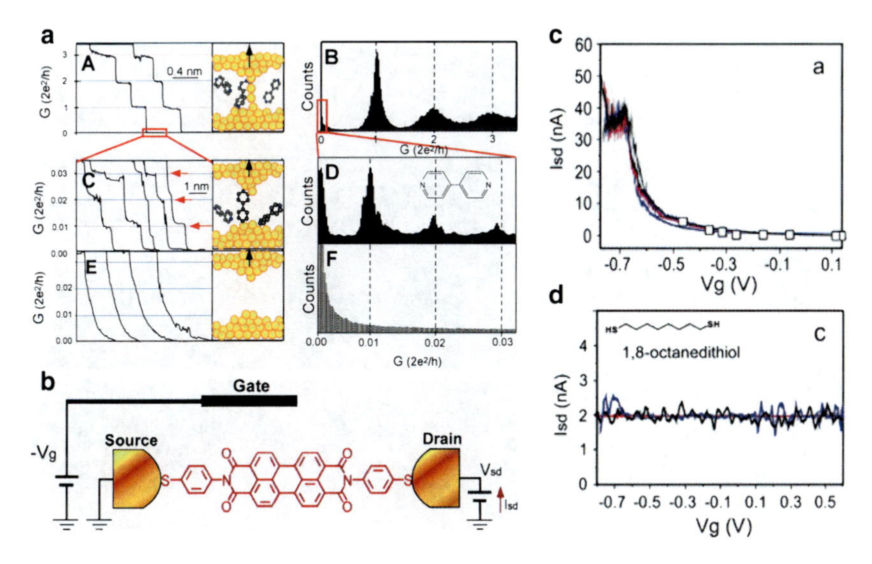

Fig. 1 (**a**) Single molecule conductance experiments of bipyridine. (Reprinted with permission from [5]. (**b, c**) Single molecule transistor experiments using an organic molecule. (Reprinted with permission from [6])

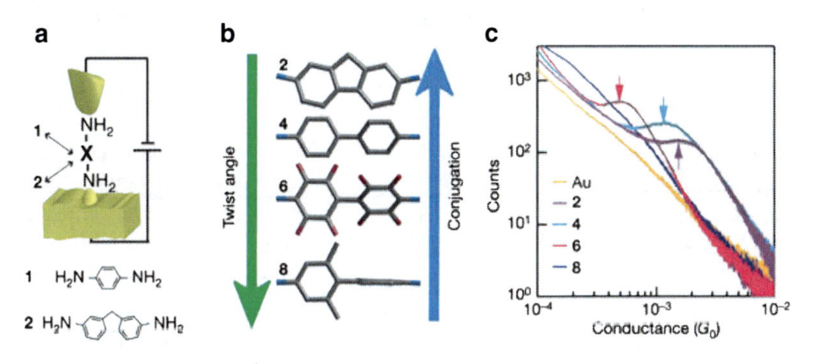

Fig. 2 (**a**) Schematic diagram of single molecule junction conductance measurement. (**b**) Structures of a subset of the biphenyl series studied, shown in order of increasing twist angle (θ) or decreasing conjugation. (**c**) Biphenyl junction conductance as a function of molecular twist angle. (Reprinted with permission from [9])

The purpose of this review is to present recent developments on the utilization of fullerenes, carbon nanotubes, and graphene in molecular electronics.

1.1 Types and Shapes of Carbon Nanomaterials

After the discovery of fullerenes [10] the family of carbon based nanostructures has been constantly growing (see Fig. 3). Fullerene discovery was followed by

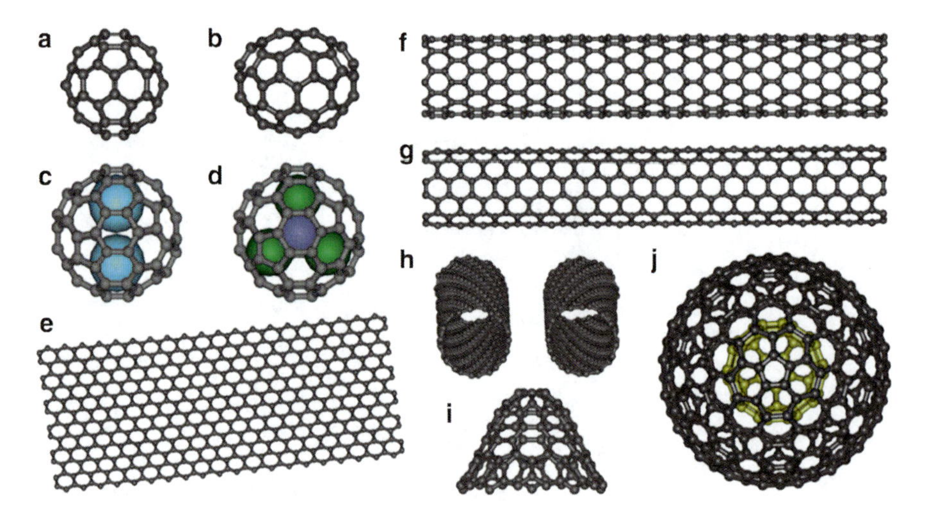

Fig. 3 Representative examples of carbon nanomaterials. Empty cage fullerenes (**a**) C_{60}, (**b**) C_{70}; endohedral fullerenes (**c**) $La_2@I_h$-C_{80}, (**d**) $Lu_3N@I_h$-C_{80}, (**e**) graphene sheet, (**f**) zig-zag single wall carbon nanotube, (**g**) arm chair single wall carbon nanotube, (**h**) chiral carbon nanotubes, (**i**) carbon nanohorn, (**j**) carbon nanoonion

multiwall [11] (MWCNTs) and single wall carbon nanotubes (SWCNTs) [12, 13]. "Onion-like" structures of concentric multishell fullerenes were observed in the carbon soot resulting from the resistive evaporation of graphite [14, 15], and they are also formed upon strong electron beam irradiation of amorphous carbon and carbon nanotubes [16], by high temperature annealing of nanodiamonds [17], or by arc discharge under water [18], organic solvents [19], or liquid nitrogen [20]. Other striking nanostructures are the so-called carbon nanohorns (CNHs) which result from the CO_2 laser ablation of carbon at room temperature without a metal catalyst [21]. Despite the fact that endohedral metallofullerenes were observed right after the discovery of fullerenes via mass spectrometry [22], their endohedral nature was only confirmed a few years later [23]. Atom clusters can also be confined inside fullerenes. Perhaps the most important family within this group is the trimetallic nitride endohedral metallofullerenes (TNT-EMFs) [24, 25]. However, clusters containing metal oxides [26], metal carbides [27], or metal sulfides [28, 29] can also be encapsulated inside the fullerene cages.

2 Fullerenes

2.1 Fullerene Preparation

Fullerenes were detected for the first time upon laser-induced vaporization of graphite [10]. However, the first preparative method involved the vaporization of graphite by arc discharge (see Fig. 4) [30–33]. Today, the sooting flame from a

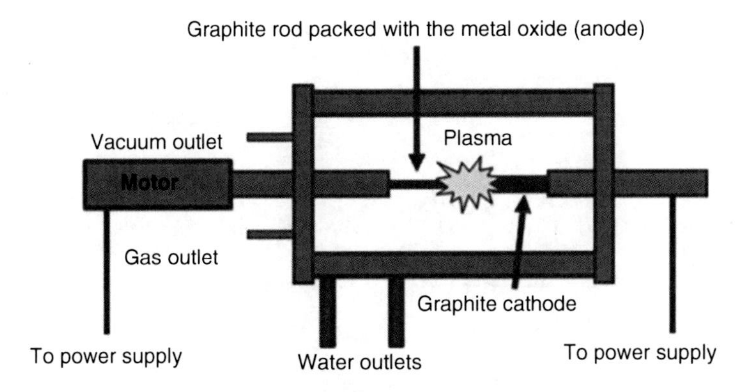

Fig. 4 Arc discharge reactor for the production of fullerenes. (Reprinted with permission from [25])

hydrocarbon feedstock is the most efficient method for the preparation of empty cage fullerenes [34–38]. This is the method used in the largest fullerene production facility in the world located in Japan [39]. The flame is produced by burning part of the hydrocarbon; therefore, the reaction temperature depends on the hydrocarbon to oxygen ratio and it is usually lower than 2,000 K [34]. The main advantage of this method is that it is a continuous process with overall low energy consumption. However, the flame introduces limitations to the scope of this method; therefore, the arcing process is preferred in lab scale preparation of fullerenes. In this process, high purity graphite rods are evaporated in electrically generated plasmas by either alternating current AC [32, 40] or direct current DC discharge [41, 42]. The distance gap between the electrodes not only affects the electrical characteristics of the plasma, but also the radiation level and the heat exchange between the plasma and the surroundings [43].

Other parameters such as the metals added, plasma gas, and temperature become independent and can be adjusted freely, thus making possible the production of a large variety of carbon nanomaterials by using the same arc discharge reactor [44]. Powdered metals or metal oxides can be packed in the graphite rods and can lead to the formation of endohedral metallofullerenes [45–47]. If, simultaneously, the arcing atmosphere is charged with reactive gases, different clusters can be incorporated inside the fullerene cages [48–51]. Salts or organic molecules can be used as packing materials as well, further diversifying the products that can be obtained directly from the arcing process [26, 29, 52].

2.2 *Redox Properties of Fullerenes*

Early theoretical calculations established that C_{60} has a low energy LUMO orbital that is triply degenerate and hence capable of accepting six electrons upon reduction [53]. However, the electrochemical detection of the C_{60} hexaanion was only

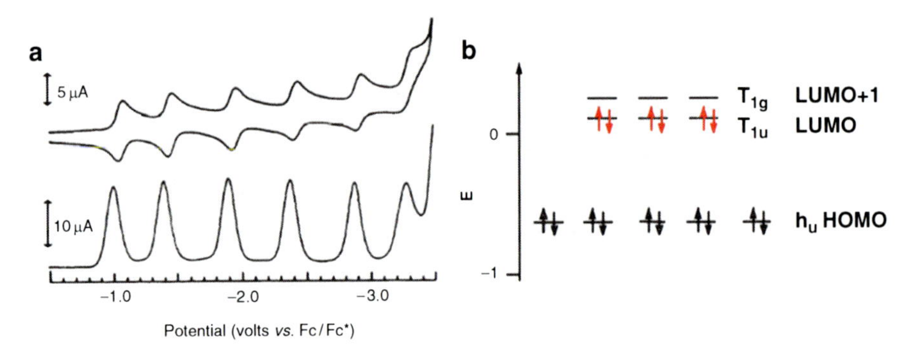

Fig. 5 (**a**) Electrochemistry of C_{60} fullerene, cyclic voltammetry (*top*) and differential pulse voltammetry (*bottom*) (Reprinted with permission from [54]). (**b**) Schematic representation of HOMO and LUMO orbitals after addition of six electrons (*red arrows*) to the fullerene

possible when the solvent potential window was expanded by running the electrochemical experiments using a toluene/acetonitrile 5:1 v/v solvent mixture at −10 °C (see Fig. 5) [54].

In the case of C_{70}, it was predicted that the LUMO orbital was doubly degenerate and thus capable of accepting four electrons; however, the energy difference between the LUMO and the LUMO + 1 orbital is very small and, as in the case of C_{60}, six reversible reduction processes were observed [54]. On the oxidation scan, using 1,1,2,2-tetrachloroethane as the solvent, one chemically reversible oxidation was observed for C_{60} whereas two oxidative processes were observed for C_{70} [55] and, by calculating the energy difference between the first reduction and the first oxidation processes, the HOMO–LUMO gaps in solution for C_{60} and C_{70} were established at 2.32 eV and 2.22 eV respectively. These findings also show the rich redox chemistry of C_{60} and C_{70} [56]. The redox properties of larger empty cage fullerenes are equally rich and it has become an important technique for differentiating among different isomers [57].

The electrochemical properties of TNT-EMFs, $M_3N@C_{2n}$ ($n > 39$) differ from those of the empty cage fullerenes (see Fig. 6) due to the interaction of the metal cluster with the carbon cage and because the structure of these carbon cages are generally different. As a consequence, the reductive processes are electrochemically irreversible but chemically reversible. The oxidative processes occur at lower potentials because the HOMO orbital is mainly localized on the trimetallic nitride clusters and the HOMO–LUMO gaps in solution are smaller [25, 58]. The endohedral metallofullerenes $M@C_{2n}$ show similar behavior but even smaller HOMO–LUMO gaps [59].

2.3 *Electronic Transport Properties of Single Fullerene Molecules*

In order to understand the electronic transport properties of single fullerene molecules, and thus to explore the possibility of device application, it is very important to

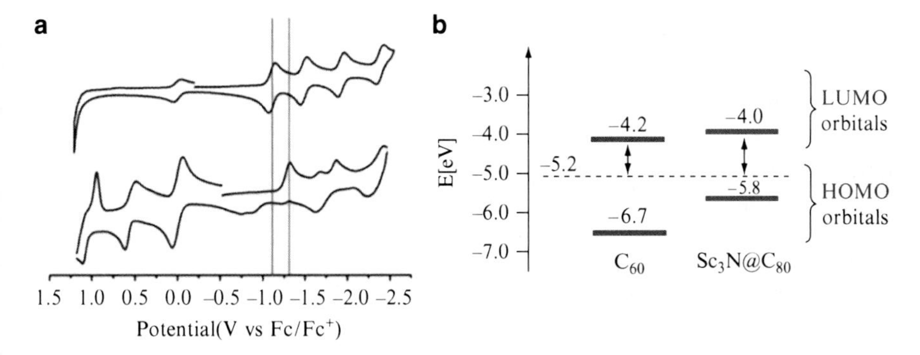

Fig. 6 (**a**) Cyclic voltammograms C_{60} (*top*) and $Sc_3N@I_h$-C_{80} (*bottom*) in a 0.050 M solution of $TBAPF_6$ in *o*-DCB. (**b**) Schematic energy diagram of HOMO-LUMO gaps difference between C_{60} and $Sc_3N@I_h$-C_{80}

study the current–voltage (I–V) characteristics of single fullerene molecules connected to two electrodes [60]. The charge transport through single C_{60} molecules has been studied experimentally using STM. C_{60} molecules have been studied using two different approaches. In the first approach, C_{60} molecules were deposited on a metal surface and a single molecule was contacted with an STM tip, thus creating a single barrier tunnel junction. C_{60} fullerene has a band gap above 2 V and thus it behaves as an insulator at room temperature, but it has a low lying energy LUMO level and, thus, when a C_{60} molecule is contacted with metal electrodes, charge transfer can take place and conductance can be established through the LUMO orbital. Under these conditions the I–V characteristics of single fullerene molecules deposited on top of Au electrodes are linear at voltages lower than the HOMO–-LUMO gap [61, 62]. Low temperature studies of C_{60} molecules absorbed on Cu(100) showed a sudden change of conductance from $\approx 0.25\ G_0$ to G_0 ($G_0 = 2e^2/h$) when the conductance regime changed from tunneling to contact and the increase on the conductance is attributed to the formation of a chemical bond between the fullerene and the STM tip [63]. Theoretical studies have established that the number of contacts between the tip and the atoms in the fullerene cage affects the conductance due to interference effects [64]. Other theoretical studies have predicted that the charge transport through endohedral fullerenes differs from that of empty cages depending on the nature of the encapsulated atom because the main channel for the charge transport could be the fullerene cage or the encapsulated atom [65]. This difference could be useful for controlling the current in future nano-sized devices.

In the second experimental approach, the C_{60} molecules are deposited on top of an insulating self-assembled monolayer, thus creating a double barrier tunnel junction connected in series and sharing an electrode [66, 67]. Under these conditions current steps in the I–V graph are observed, because when a potential is applied the capacitances of each junction has to be charged to a threshold potential before an electron can tunnel through the junction and when it is favorable for an electron to sit in the middle electrode the amount of current that flows through the junctions increases [68].

2.4 Fullerene Based Unimolecular Devices

2.4.1 Molecular Wires and Donor–Acceptor Systems

Fullerenes have unique properties that make them good candidates as electron acceptors in the construction of photo induced charge transfer systems. Due to their low energy and triply degenerate LUMO orbital (see Fig. 5b), they can accept electrons easily [69]. Another important property is that their reorganization energy upon reduction is very low because of their rigid spherical geometry and electron delocalization through the whole molecule [70]. According to Marcus' theory of electron transfer [71–73] the low reorganization energy favors charge separation while slowing down the charge recombination, thus leading to long lived charge separated states. Therefore, many fullerene and fullerene derivatives have been studied as electron acceptor materials as has the coupling through covalent bonding of the electron donor moieties and the fullerene to produce molecular dyads (see Fig. 7).

The bridge that links the donor and the acceptor plays a vital role in several key aspects. For example, they eliminate diffusion as the rate determining step for the charge transfer process. The chemical nature and length of the bridge is more important than the donor–acceptor (D–A) separation, orientation, overlap, and topology [74]. Three different scenarios can appear; the first is when the donor and acceptor moieties are connected by an insulating bridge (see Fig. 8a). A rigid bridge is more advantageous because it prevents undesired rearrangements or conformers whereas a flexible one may lead to different behaviors and the chemical nature of the bridge affects the conductance or charge transport behavior. For those donor-bridge-acceptor architectures, the efficiency of the charge transfer process decreases exponentially with the distance between the donor and the acceptor [75, 76] but longer distances produce longer lived charge separated states.

A different architecture creates molecular redox gradients in which several short distance charge separation processes can occur instead of a single long distance transfer [77, 78] (see Fig. 8b). This strategy allows the formation of extremely long lived radical ion pairs separated by distances up to 50 Å compared to the maximum 20 Å observed in donor–acceptor systems connected by an aliphatic bridge; however, some energy is lost in each redox step.

Finally, another alternative is to connect the donor and the acceptor using highly conjugated molecules that act as molecular wires (see Fig. 8c). In that case the charge separation process is dominated by the degree of overlap between the donor, the acceptor, and the linker orbitals. In such systems, the charge separation usually takes place through the LUMO of the bridges via a super-exchange mechanism and/or a hopping mechanism, depending on their relative LUMO energy levels. In contrast, the charge recombination of the radical ion pairs usually takes place through the HOMO of the bridges [79]. For example, olefin conjugated systems such as *p*-phenyleneethynylene facilitate the charge transfer process because the energies of the C_{60} HOMOs match those of the bridge, which facilitates electron/

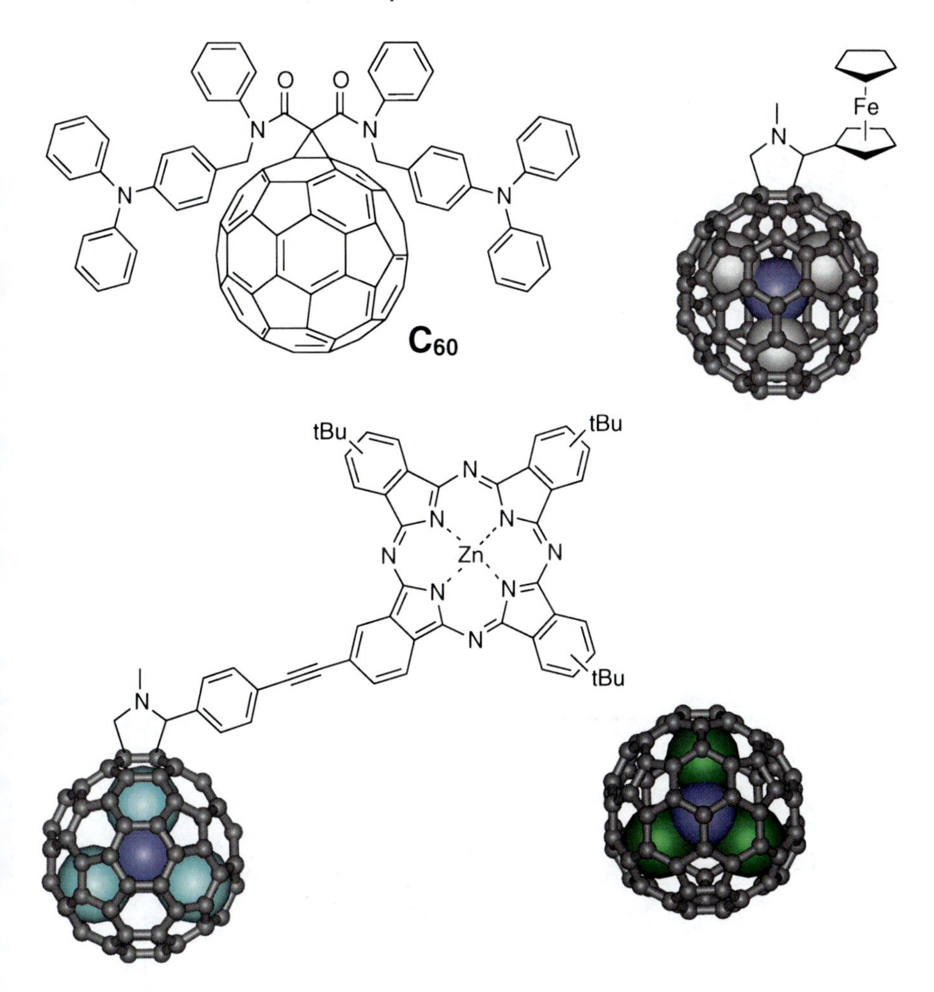

Fig. 7 Molecular dyads based on fullerene derivatives

hole injection into the wire [80]. Polythiophene oligomers [80], on the other hand [81], can participate in the charge separation and charge recombination processes and the radical cations produced upon photoexcitation are delocalized not only in the donor but also in the linker.

The conjugation in the molecular wire may be disrupted or modulated to create systems with different properties. For example, a porphyrin C_{60} donor–acceptor system linked with a conjugated binaphthyl unit, has a preference for the atropisomer where the fullerene unit is closer to the porphyrin system, thus increasing the through space interactions [82]. The charge transfer process on a dyad containing a crown ether in the linker structure can be modulated by complexation/decomplexation of sodium cations [83] but even more interesting is the construction of supramolecular systems where the donor and acceptor moieties are

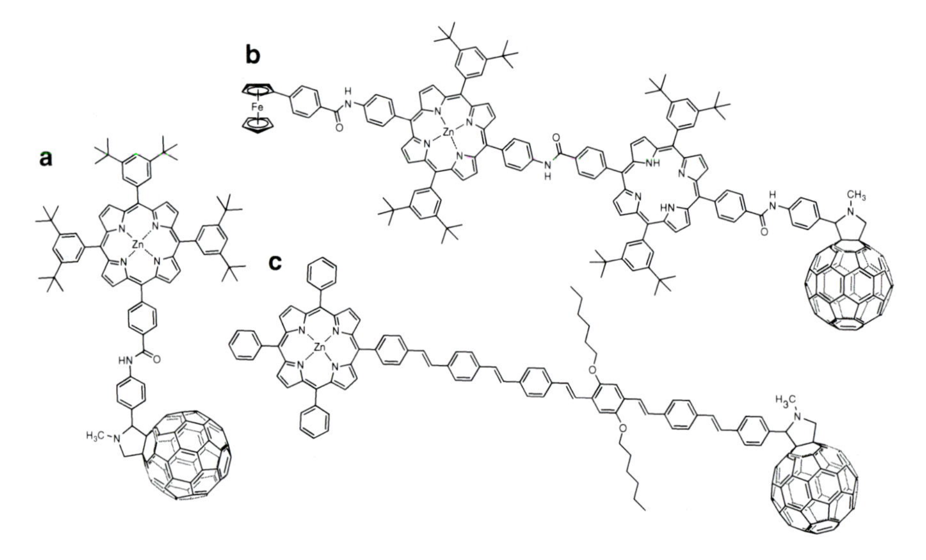

Fig. 8 Bridge type connecting the donor and acceptor moieties. (**a**) Insulating bridge, (**b**) Redox gradient bridge, and (**c**) highly conjugated bridge

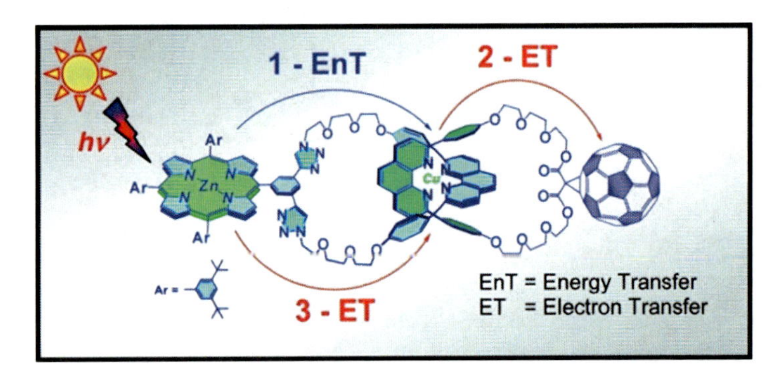

Fig. 9 Example of a supramolecular donor–acceptor system. (Reprinted with permission from [85])

connected by selective metal complexation [84], catenanes [85], or π–π interactions [86, 87] (see Fig. 9).

2.4.2 Rectifiers

Rectification phenomena through molecules is attributed to three different effects. The first is due to Schottky barriers because a surface dipole is formed at the organic/metal interface. The second effect occurs when the LUMO orbital, which has to be accessed during the conduction process, is placed unsymmetrically between two metal electrodes and the third one occurs when there is electron transfer between the HOMO and LUMO orbitals unsymmetrically located within a molecule [88, 89]. This last phenomenon is really the true molecular rectification and it was originally proposed by Aviram and Ratner [90]. Theoretical comparative studies between N doped and P doped fullerenes have shown that these materials can be used to prepare a classical N–P junction and therefore a diode [91, 92].

Two different rectification behaviors on a Langmuir–Blodgett monolayer of dimethylanilino-aza-[C_{60}]-fullerene sandwiched between gold electrodes was observed. At low potentials a moderate molecular rectification ratio of ~2 can be attained [93]. However, if the device is operated at higher potentials, it can reach a rectification ratio up to 20,000 at 1.5 V. This observation is presumably due to the presence of defects or stalagmitic filaments of gold growing from the bottom electrode under positive bias which are broken when a negative potential is applied [94]. A detailed study with a similar molecule *N*-3-γ-Pyridyl-aza[C_{60}]fulleroid showed that both the molecular rectification and the geometric asymmetry of the junction contribute to the rectifying effect [95].

Rectification ratios between 87 and 158 were observed in Langmuir–Blodgett monolayer films of C_{60}-didodecyloxybenzene dyad at 3.0 V [96]. Supramolecularly connected ZnTTP and a C_{60} fulleropyrrolidine donor–acceptor system was prepared over a gold(111) surface. This device shows molecular rectification with high tunneling currents at positive substrate voltages [97]. The amphiphilic fullerene derivative 1,4,11,15,30-pentakis(4-hydroxyphenyl)-*2H*-1,2,4,11,15,30-hexahydro-[60]fullerene (see Fig. 10) was used to create a stable Langmuir–Blodgett film on water with the hydrophobic fullerenes pointing up. This film was transferred to a gold electrode and asymmetric rectification behavior was

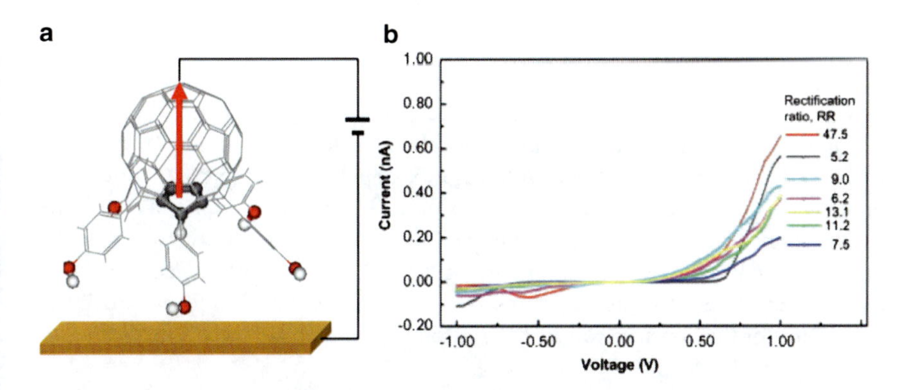

Fig. 10 (**a**) Schematic representation of the I–V measurement of the C_{60} derivative based molecular rectifier. (**b**) I–V characteristics of the Langmuir–Blodgett film of the fullerene pentapod showing current rectification at bias voltage of ±1.0 V measured at different junction positions, with high rectification ratios (RR) registered at each junction. (Reprinted with permission from [98])

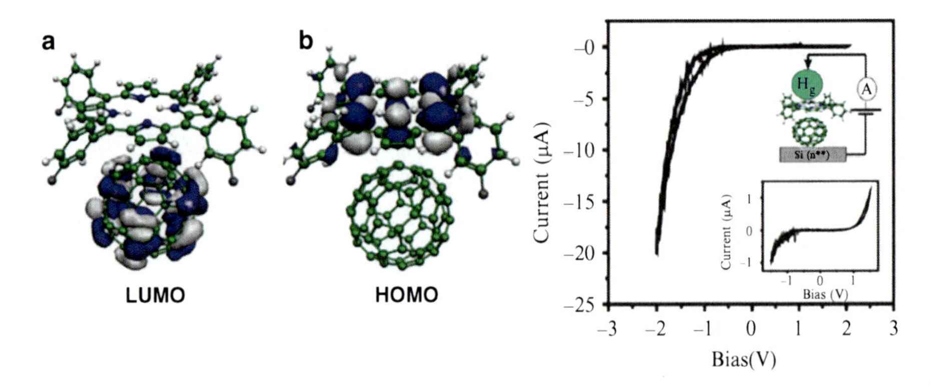

Fig. 11 (**a**) Molecular orbital location in a C_{60} based molecular rectifier. (**b**) Typical current–voltage characteristics recorded for Hg/TFPP/C_{60}/Si(n^{++}) structure (schematic shown in the *upper inset*). The *lower inset* shows a symmetric I–V curve recorded for the Hg/C_{60}/Si(n^{++}) structure. (Reprinted with permission from [99])

observed in the range of 1.0–2.0 V. Theoretical calculations demonstrate that both the HOMO and LUMO orbitals are located on the fullerene cage, in contrast to other unimolecular diodes prepared with fullerenes [98].

The fabrication of diodes on silicon substrates was demonstrated using the supramolecular interactions between a 5,10,15,20-tetra(3-fluorophenyl)porphyrin and C_{60} fullerene with a rectification ratio of ~1,500 (see Fig. 11). The rectifying behavior is explained by theoretical calculations which show that the LUMO orbital is located mainly on the fullerene whereas the HOMO orbital is located on the porphyrin moiety [99].

2.4.3 Transistors

The first device fabricated with C_{60} that can be considered a transistor because it performed voltage amplification is an electromechanical amplifier where a single C_{60} molecule was pressed with an STM tip connected to a piezoelectric actuator (see Fig. 12) [100]. The input voltage was applied to the piezoelectric which pressed the C_{60} molecule against the copper(100) surface where it was deposited, thus changing the conductance characteristics by approximately two orders of magnitude, corresponding to deformations around 0.2 nm. This device can tolerate currents up to several microamperes but, more important, this result proved the feasibility for the construction and operation of single molecule C_{60} based transistors (see Fig. 13) [101].

Single molecule C_{60} based transistors were also fabricated by depositing a diluted C_{60} toluene solution onto a pair of gold electrodes. The whole structure was built on an insulating SiO_2 layer on top of a doped Si wafer that was used as the gate electrode to modulate the electrostatic potential of the C_{60} molecule trapped in the middle of the junction (see Fig. 14). The observed I–V characteristics were

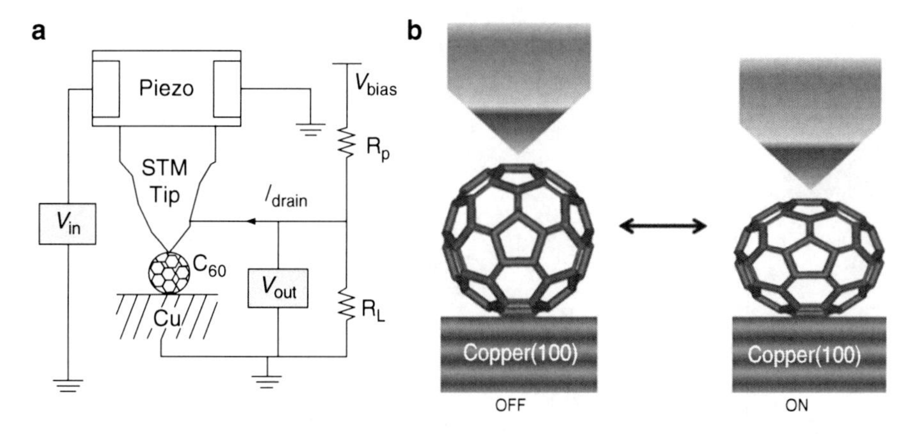

Fig. 12 (**a**) Schematic diagram of single molecule C_{60} based electromechanical amplifier. (**b**) Schematic representation of the on/off states. (Reprinted with permission from [100])

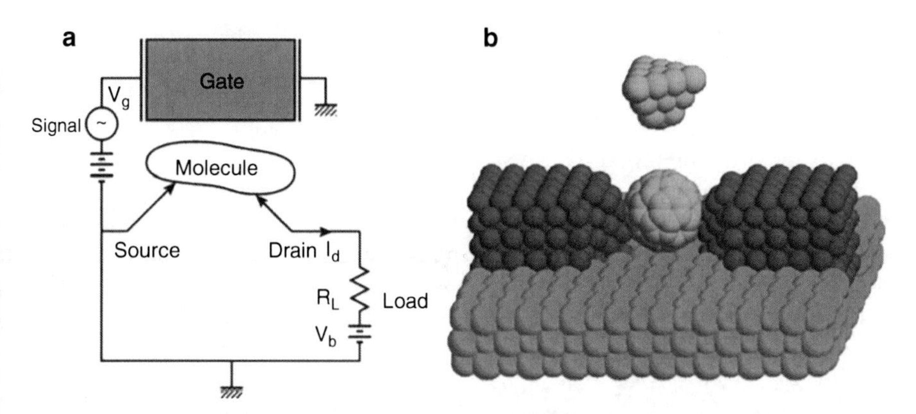

Fig. 13 (**a**) Schematic of an electromechanical single molecule transistor. (**b**) Schematic of the planar version of the C_{60} amplifier. (Reprinted with permission from [101])

explained by the nanomechanical oscillations of the C_{60} molecule against the gold electrode surface with a frequency of ~1.2 THz [102].

C_{60} single molecules between electrodes fabricated by electromigration [103] to create single C_{60} molecule transistors were measured at a temperature of 40 mK. The results showed the coexistence and competition of the effects of Coulomb repulsion, Kondo correlations and superconductivity [104]. The Kondo effect had been previously observed in similar devices [105]. Recently a SAM of a tricarboxylic acid fullerene derivative was used to fabricate a transistor. The SAM was created by allowing the fullerene compound to self assemble on top of an Al_2O_3 layer just above the aluminum drain electrode; the source lead was created by

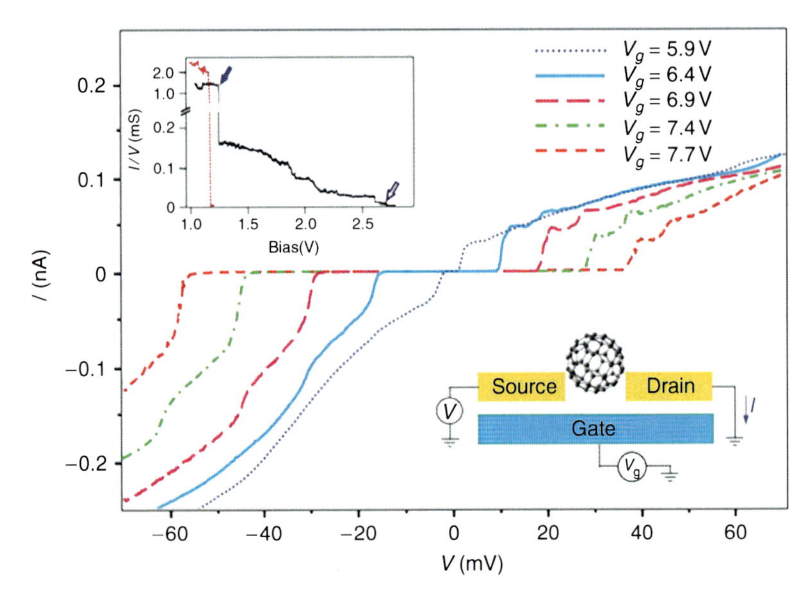

Fig. 14 Current–voltage (I–V) curves for a single fullerene transistor at different applied gate voltages. (Reprinted with permission from [102])

evaporation of Pd on top of the SAM whereas the gate electrode was lithographically defined as a vertical rod of Ti metal covered with TiO_2 acting as the gate oxide layer. The resulting device has a good current dependence on the gate voltage and the phenomenon was attributed to the strong coupling with the gate electrode which is separated by an oxide dielectric layer with dimensions close to the molecular length. The device also shows asymmetric conduction behavior because the bottom covalent contact is different from the purely electrostatic metal at the fullerene top contact. A gate voltage dependent hysteresis that decreases when the gate voltage is increased was also observed. Hence this transistor can be operated both by voltage driven switching and voltage controlled hysteresis. The first case can be applied in regular switching applications whereas the second may be of interest for the fabrication of memory devices [106].

2.5 Conclusions and Future Directions

Fullerene molecules show potential for the construction of nano-sized electronic devices because of their easily accessible and degenerate LUMO orbitals, their spherical shape that makes the self assembling process highly predictable and geometry independent, the possibility of establishing multiple conductance channels through a single molecule to avoid the intrinsic high quantum resistance values, and their good tolerance to relatively high temperatures when compared to ordinary organic compounds. In contrast to other carbon nanomaterials, fullerenes have a precise and well defined structure and can be isolated in high purity using well established chemical methods. The diversity is increasing constantly with the discovery of new fullerene families that also bring new properties.

Chemical functionalization can provide additional modulation of the physical and electronic properties. For example, potential-controlled STM conductance measurements of a *trans-2*-C_{60} derivative immobilized on gold electrodes by amino-gold linkages showed that the tunneling current changes reproducibly upon first and second reduction/reoxidation reactions (see Fig. 15). The molecular conductance is higher in the order C_{60} dianion > C_{60} anion > C_{60} neutral, showing the potential of using the eight different redox states of C_{60} for constructing molecular switching devices [107]. Similar behavior can be expected for *trans-1*-C_{60} derivatives [108–110]. However, further studies are required to incorporate fullerene or fullerene derivatives into more complex electronic devices such as memories or logic gates based on rectifiers or single molecule transistors.

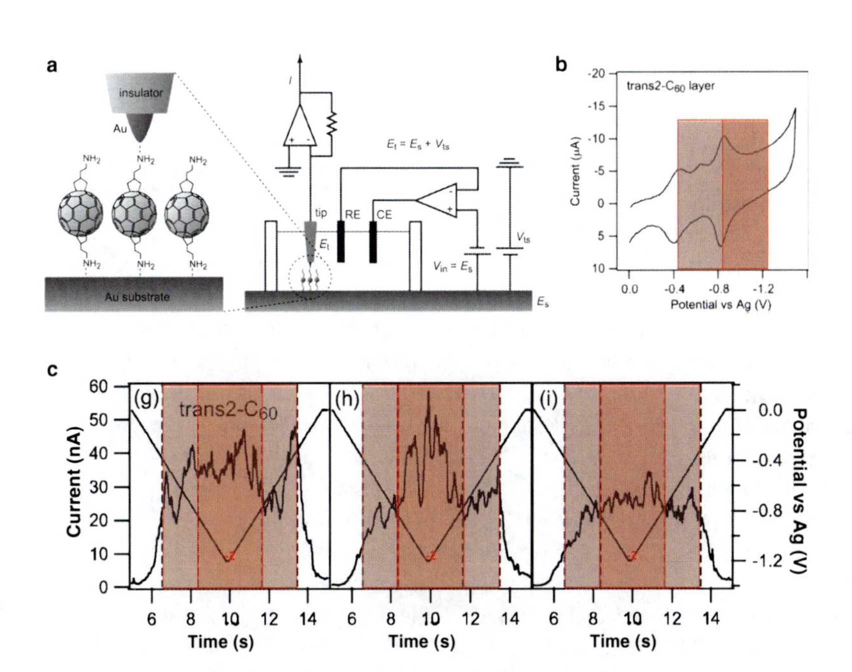

Fig. 15 (**a**) Schematic illustration of the potential-controlled STM measurement. (**b**) Cyclic voltammetry of the C_{60}-modified substrate in a 0.1 M $TBAPF_6$ DMF solution. (**c**) Representative current–time curves upon potential sweep for the bare gold surface *trans*-2-C_{60} in a 0.1 M $TBAPF_6$ DMF solution. (Reprinted with permission from [107])

3 Carbon Nanotubes

Multiwall carbon nanotubes were discovered in the cathodic deposit formed during the preparation of fullerenes using arcing techniques [11]. The formation of SWCNTs only occurs if the arcing process is made in the presence of a metal catalyst [12, 13]. Since their discovery, many applications have been suggested for nanotubes based on their unique properties. The tensile strength, for example, far exceeds that of steel [111]. However, the electronic properties are even more interesting. SWCNTs can be described as a graphene sheet rolled up to form a tube. Since a graphene sheet has a honeycomb arrangement of carbon atoms, different structures may result upon rolling up the graphene sheet. However, every possible nanotube that results can be differentiated using Hamada's notation [112].

In that nomenclature system, the center of a hexagon is chosen as the origin (0,0) and then it is superimposed with the center (m,n) of another hexagon to form the nanotube. There are three types of carbon nanotubes. If the graphene sheet is rolled in the direction of the axis, it will produce either an armchair nanotube ($m = n$) or a zig-zag nanotube ($m = 0$). On the other hand, if the graphene sheet is rolled in any other (m,n) direction it will produce a chiral nanotube and the chirality will depend on whether the sheet is rolled upwards or backwards.

3.1 *CNT Preparation and Purification*

Since the discovery of SWCNTs prepared by the arc discharge of graphite rods doped with a metal catalyst [12, 13], considerable effort has been directed to developing other methods for the production of large amounts of nanotube materials; most methods are based on three primary approaches: electric arc discharge, laser ablation, and chemical vapor deposition (CVD) [113]. However, the preparation of CNTs in flames is also gaining attention. In the arc discharge method CNTs are prepared in a similar fashion to the preparation of fullerenes [11]; a doped graphite anode is evaporated in an inert 500–600 mbar He atmosphere with currents between 50 and 100 A. Carbon nanotubes are only formed in the presence of a metallic catalyst and most of them accumulate as a growing deposit formed in the cathode. Diverse metal catalysts have been used but the highest yield has been achieved with a mixture of 1 wt% Y and 4.2 wt% Ni [114].

In the laser ablation technique, the apparatus consists of a tube furnace operating at 1,200 °C under argon where a graphite target impregnated with a metal catalyst is ablated with a high power laser [115]. This method displays a high conversion of graphite into CNTs. The CVD is perhaps the most interesting method because it allows the controlled growth of SWCNTs along a surface. The direct growth of both MWCNTs [116] and SWCNTs by CVD has been observed by field emission microscopy (FEM) and it showed that the nanotube rotates during the growing process [117] (see Fig. 16). The metal catalyst is first deposited on a surface where the SWCNTs will grow by simple spin coating or ultrahigh vacuum deposition. The

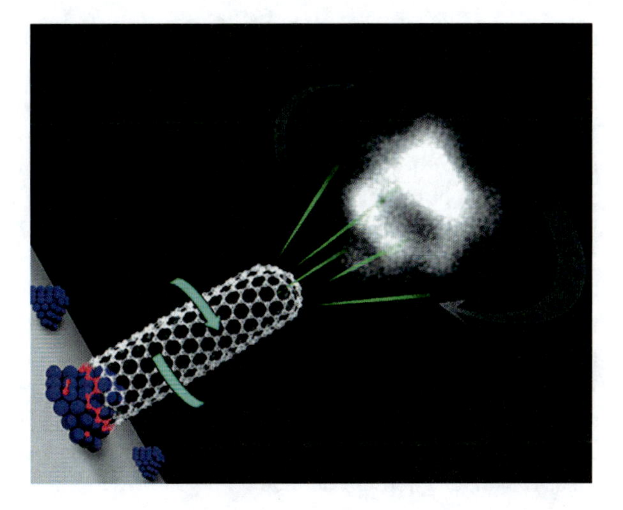

Fig. 16 Growing of a SWCNT on a catalyst particle by chemical vapor deposition. (Reprinted with permission from [117])

most typical carbon feedstock used is C_2H_2/NH_3 or CH_4/H_2 among numerous other source gases, while nickel, iron, and cobalt, or alloys of those metals are used as catalysts [118]. Recently, the preparation of MWCNTs in a rotating counterflow diffusion flame using nickel nitrate coated nickel substrates was reported. A mixture of 86% nitrogen and 14% ethylene was used as the fuel, with air as the oxidant [119]. A similar method uses a V-type pyrolysis flame using CO as the carbon source [120].

Independent of the preparation method, SWCNTs are a complex mixture containing tubes with different length, diameter, chirality, and electronic properties. Additionally they are grouped in insoluble tube bundles or tube ropes. SWCNT samples are usually contaminated with metals, fullerenes, carbon nano-onions, and related carbon nanomaterials. One of the most significant obstacles to the application of SWCNTs in nanoelectronics has been their separation according to their length, diameter, and chirality. The production of highly homogenous samples or the purification of SWCNTs is still an open field and is a big obstacle towards practical applications in nanoelectronics [121]. CNT purification methods have been extensively reviewed in the literature [113, 122–125]; however, some remarkable methods for the separation of nanotubes exploit their different electrical properties [126, 127], their difference in density [128, 129], and their selective chirality driven DNA wrapping [130–135] that allows their separation by high performance liquid chromatography methods.

3.2 Electrochemical Properties of CNT

Knowing the electrochemical properties of CNTs is important for the rational design of electronic devices; however, the study of the electrochemical properties of carbon nanotubes as individual entities or individual molecules has encountered diverse problems. The extremely low solubility of CNT samples, the ionic strength of the electrolyte solutions, the interference of surfactants, and the limited electrochemical window of the aqueous medium have prevented the study of the electrochemical properties of CNTs in solution [136]. The electrochemical properties of CNTs supported on metal electrodes was studied in aqueous [137] and organic electrolytes [138] and it was observed that CNTs do not have discrete reduction peaks but a constantly growing current on the cathodic scan, which reflects the presence of a complex mixture of different CNTs. The first electrochemical studies in solution were made on soluble samples of pyrrolidine functionalized CNTs. Those derivatives showed irreversible reductive behavior that was attributed to the decomposition of the functionalized nanotubes on the surface of the electrode; theoretical calculations suggested that CNT functionalization significantly affects the energy of the low lying electronic states of the nanotubes [139].

Pyrrolidine functionalized CNTs having ferrocenyl groups showed a reversible oxidation step associated with the oxidation of the ferrocene moieties, no discrete electrochemical steps associated with single electron transfer events, but a continuous increase of the current with the increase of the negative potential, indicative of multiple electron transfer events starting at -0.5 V relative to a pseudo reference silver electrode [140], similar to what was observed for the reductive scan of CNTs supported on metal electrodes. Attempts to study the redox properties of CNTs in solution were made using UV–Vis/NIR spectroscopy to monitor electron transfer between (6,5)-enriched SWCNTs and K_2IrCl_6 during a titration experiment that revealed a reduction potential of approximately 800 mV vs NHE [141]. The reduction of nanotubes with alkali metals produces polyionic salts that are soluble without using sonication, surfactants or chemical functionalization [142]. By using this technique, SWCNTs were solubilized and their electrochemical properties studied. The CV experimental results were checked for consistency with the electronic spectra as a function of the potential and the exciton binding energy [143, 144] of the nanotubes and the average reduction and oxidation potentials for the CNTs were confidently established (see Fig. 17) [136].

3.3 Charge Transport Properties of CNTs

The conductive properties of SWCNTs were predicted to depend on the helicity and the diameter of the nanotube [112, 145]. Nanotubes can behave either as metals or semiconductors depending upon how the tube is rolled up. The armchair nanotubes are metallic whereas the rest of them are semiconductive. The conductance through carbon nanotube junctions is highly dependent on the CNT/metal contact [146]. The first measurement of conductance on CNTs was made on a metallic nanotube connected between two Pt electrodes on top of a Si/SiO_2 substrate and it was observed that individual metallic SWCNTs behave as quantum wires [147]. A third electrode placed nearby was used as a gate electrode, but the conductance had a minor dependence on the gate voltage for metallic nanotubes at room temperature. The conductance of metallic nanotubes surpasses the best known metals because the

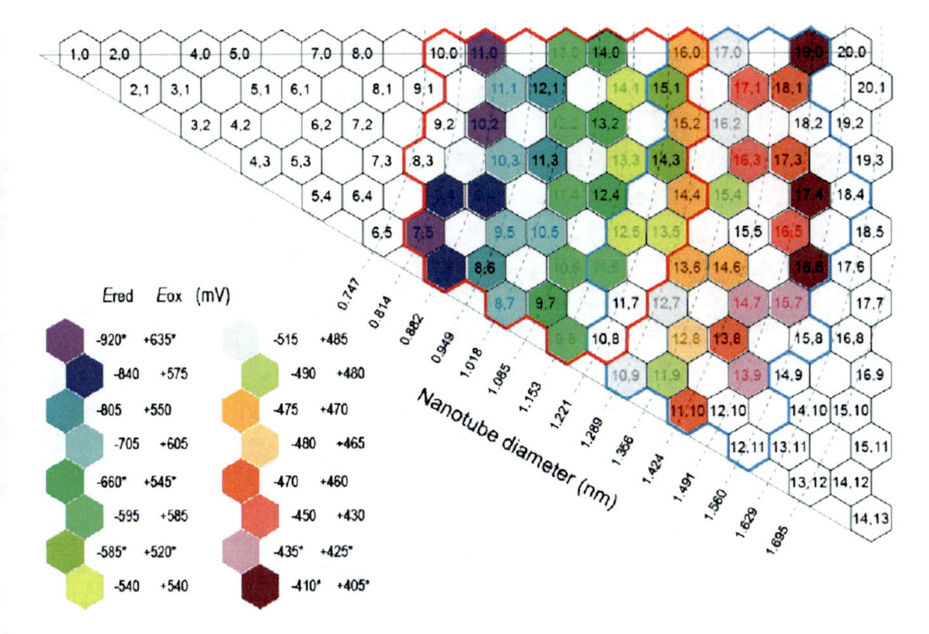

Fig. 17 Redox potentials of SWCNTs as a function of their chirality. (Reprinted with permission from [136])

electron transport is ballistic [148] and they are able to carry currents with densities that exceed 2.5×10^9 A/cm^2 for a 1 nm diameter nanotube [149].

The conductance of MWCNTs is quantized. The experimental setup to measure the conducting properties involved the replacement of an STM tip with a nanotube fiber that was lowered into a liquid metal to establish the electrical contact. The conductance value observed corresponded to one unit of quantum conductance ($G_0 = 2e^2/h = 12.9\ k\Omega^{-1}$). This value may reflect the conductance of the external tube because, for energetic reasons, the different layers are electrically insulated [150]. Finally, the conductance of semiconductor nanotubes depends on the voltage applied to the gate electrode; their band gap is a function of their diameter and helicity [145] and the ON/OFF ratio of the transistors fabricated with semiconductor nanotubes is typically 10^5 at room temperature and can be as high as 10^7 at extremely low temperatures [151].

3.4 CNT Based Devices

3.4.1 CNT as Contacts

Construction of devices based on molecular electronics will require connections at the molecular level. One of the biggest challenges is to construct wires or electrodes

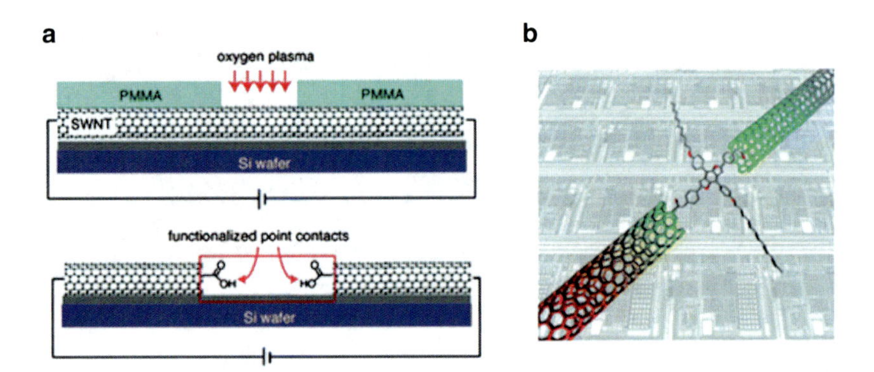

Fig. 18 (**a**) Fine cutting of SWNT with oxygen plasma introduced through an opening in a window of PMMA defined with e-beam lithography. (**b**) Schematic demonstration of holistic construction of a single molecule circuit. (Reprinted with permission from [153])

to connect molecules. Carbon nanotubes are 1D ballistic electron conductors so they can transport electrons essentially without heat dissipation; thus these properties can be used for the construction of true molecular devices. There are two different methods to cut SWCNTs and create nanogaps with ~2 nm length. In the first method a bias voltage is applied to a nanotube connected to two electrodes until an electrical breakdown occurs [152].

In the other method the carbon nanotubes are spin coated and covered with a polymethyl methacrylate PMMA mask where a small window is defined using ultrahigh resolution electron beam lithography; then the carbon nanotubes are cut through the window using oxygen plasma ion etching (see Fig. 18) [153]. Carboxylic acid groups are normally produced at the points where the nanotubes are cut. They are easily converted into acyl halides that after reacting with amines yield amide linkages that are then used to connect single-walled carbon nanotubes covalently to other molecules. Theoretical calculations have shown that CNTs are well suited for establishing these connections because of the good match between the energy levels of the molecules responsible for the conduction and the Fermi level of the nanotubes [154, 155]. Cutting nanotubes connected to electrodes and rejoining the ends with conductive molecules provides a new generation of nanoscale devices [156] with applications including selective ion detection [157] and reversible photo driven switches [158] among others. CNTs also have the potential to be used as atomic force microscopy (AFM) tips but they are still not widely adopted [159].

3.4.2 CNT Transistors

Semiconductive SWCNT FET transistors capable of operating at room temperature were constructed several years ago and their operation in the terahertz frequency range was predicted [160]. These early devices had *p*-type characteristics (hole conductors); however, this behavior is not an intrinsic property of the nanotubes and the *p*-type nanotube based field-effect transistors (FETs) can be switched to *n*-type simply by annealing them under vacuum [161] or by chemical doping with alkali metals [162] or other reducing agents [163, 164]. However, these doping strategies are not completely reliable and make the fabrication process of complex devices a big challenge. This problem was solved by choosing the right materials for the electrodes, Pd for *p*-type and Sc for *n*-type carbon nanotube FETs [165]. A disadvantage of this method is that Sc is ~5 times more expensive than gold, but Sc can be substituted by Y which is ~1,000 times cheaper and gives the same effect [166].

As an alternative, top gated devices contacted with TiC show ambipolar behavior and better stability [167, 168]. Ambipolar transistors, however, cannot be used to reproduce the current CMOS technology where both *p*-type and *n*-type transistors are used because the combination of the two has superior performance and lower power consumption than devices built with only one type of transistor. An alternative solution was recently presented and consists of selectively applying negative or positive gate voltages to CNT ambipolar transistors, which make them behave as *p*-type or *n*-type respectively [169].

Other alternatives for the construction of CNT based FETs have been explored. For example, carbon nanotube branches with Y shape can be used directly as transistors where the modulation of the current from an ON to an OFF state is presumably mediated by the defects and the morphology of the junction (see Fig. 19) [170, 171]. Carbon nanotube based FETs can be gated by an electrode immersed in a solution, or by charged molecules in solution (proteins, DNA, etc.) which opens a huge field of applications in sensors [172–176] (see Fig. 20). Their ability to operate under biological conditions allows their direct use or integration into biological systems [177].

CNT based FETs can outperform the current FET technologies in many ways; however, one of the most interesting properties of carbon nanotubes is the ballistic transport of electrons [178], which opens the possibility of constructing FETs that can operate at extremely high frequencies, making them suitable for the next generation electronic devices. Operation of SWCNT transistors has been demonstrated at microwave frequencies (see Fig. 21) [179] and more recently the operation of an SWCNT transistor in the terahertz frequency range was demonstrated [148].

3.5 Integration into ICs and Future Direction

Fundamental studies of CNTs reveal extraordinary properties at room temperature including mobilities that exceed all of the known semiconductors [180] and large current carrying capacities. Despite the report of the construction and operation of logic circuits using SWCNT transistors [151], the construction of complex devices that use single nanotube transistors needs further development because of the very difficult problem of synthesizing and precisely positioning a large number of

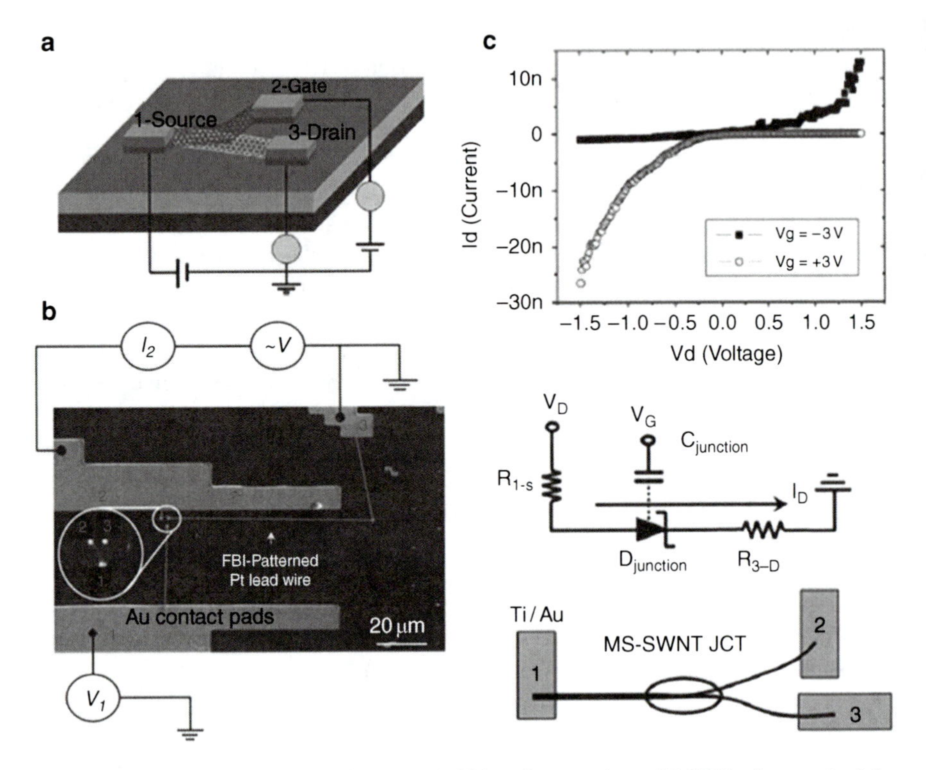

Fig. 19 (**a**) The device schematic for a pseudo Y-junction transistor. (**b**) SEM micrograph of the overall circuit arrangement used in the measurement of the electrical characteristics, with Au contact pads and an FIB-patterned Pt wire contacting the Au pads and the Y-junction. (**c**) The ambipolar I–V curves resemble that of an *n*-type semiconductor at a positive gate potential, and a *p*-type semiconductor at a negative gate potential (*top*), and the equivalent circuit for a pseudo Y junction SWNT device (*bottom*). (Reprinted with permission from [170, 171])

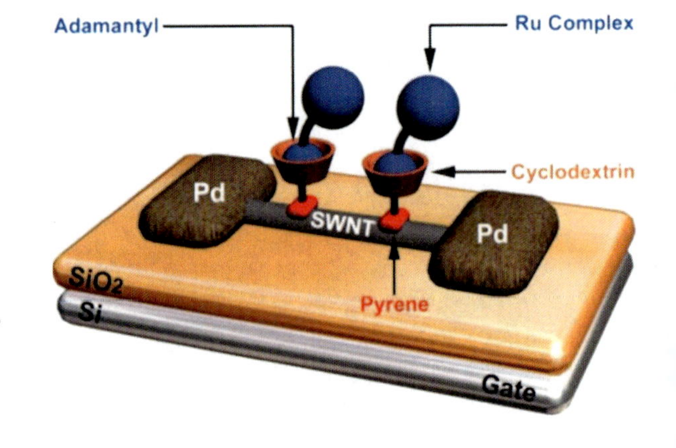

Fig. 20 Example of a CNT based FET used as sensor. (Reprinted with permission from [176])

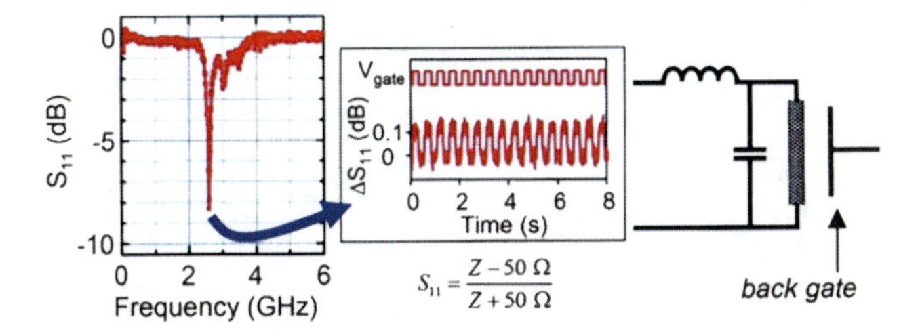

Fig. 21 Operation of a SWCNT based transistor in the microwave frequency range. (Reprinted with permission from [179])

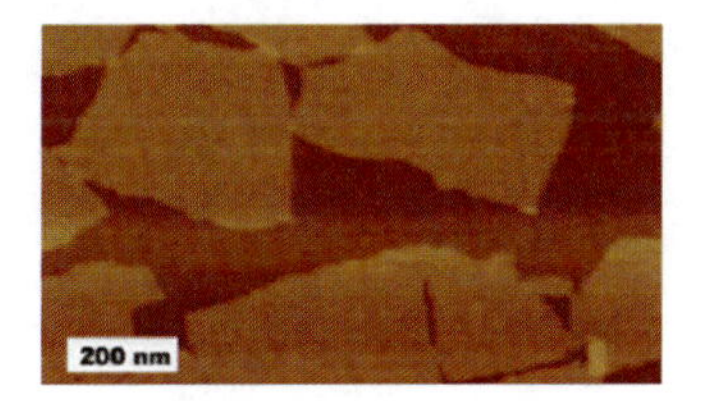

Fig. 22 Atomic force microscopy (AFM) image of graphene flakes. (Reprinted with permission from [182])

individual electrically and geometrically uniform nanotubes, but some research groups are already making progress in that direction [181] and the future looks very promising. The basic principles of fabrication and operation are already understood but much of the progress in the near future will rely on the chemist's ability to find methods that allow either selective synthesis in high scale or highly reliable methods of purification or a combination of the two.

4 Graphene

4.1 Introduction

Graphite, the most common allotropic form of carbon found in nature, is composed of stacked two-dimensional flat sheets of honeycomb-like carbon hexagons. Each of these sheets is a single carbon atom thick. Recently the separation, isolation, and synthesis of these individual sheets was reported and the product is known as graphene (see Fig. 22) [182, 183]. Its planar shape and chemical structure grant it a collection of properties not found in other materials. In the field of molecular electronics, graphene has found a realm of potential applications due to interesting properties such as high current density [183–185], quantum Hall effect (QHE) behavior [186–192], high electron mobility [193–196], high optical transparency [197, 198], chemical stability and inertness, good mechanical resistance [199], and differential electrical behavior depending on its edge structure (metallic or semiconductor), among others [200–203].

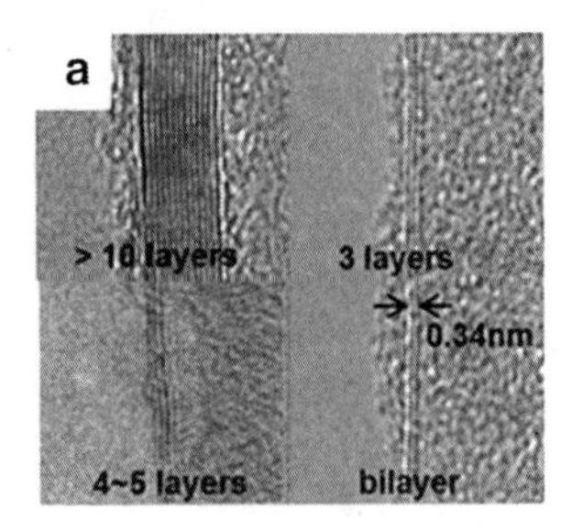

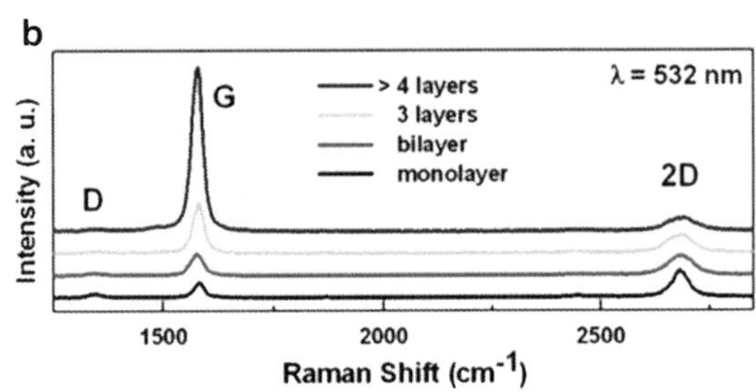

Fig. 23 Different layer thicknesses of graphene films. (**a**) TEM image and (**b**) Raman spectra (increasing number of layers from *bottom* to *top*). (Reprinted with permission from [193])

Isolation of high quality graphene layers in 2D crystal lattices with homogeneous size and layer count (see Fig. 23), as well as in good amounts, continues to be a challenge. However, its fascinating properties have resulted in the design of multiple synthetic routes for its preparation [193, 204, 205]. From the top-down synthetic point of view, the isolation of graphene sheets (single and few layers) has been possible by mechanical exfoliation of graphite with high quality crystallites [183, 206–209]. In addition, extensive work has been performed by means of mechanical intercalation in graphite to exfoliate it later in single sheets, as well as chemical modification to ease the exfoliation (e.g., graphite oxide) to later remove the functional groups and recover the carbon based sheets [193]. Although good quantities are obtained, the chemical methods show low selectivity for 2D crystal size, quality, and layer count. In contrast, bottom-up methods (e.g., total organic syntheses, CVD, and reduction of silicon carbide) promise good control of the number and size of the deposited layers [200, 201, 204]. However they are far from being scalable for large quantities with uniform single layers.

4.2 *Properties*

Many of the interesting properties of graphene have been measured on single layer samples; however, the use of bi-layer and multi layers (three to ten layers) have been considered in order to access those properties in adequate material quantities, and have shown significant differences depending on the number of layers [193, 204].

4.2.1 Single- and Bi-Layer Graphene

Single layer graphene is thermodynamically unstable [204, 210]; because of this, it has been found to be rough and not entirely flat. Ripples are observed by STM and TEM analysis [211, 212]. However, this distortion of its expected planar geometry does not interfere with its properties. Graphene is a strong material in terms of physical–mechanical properties with a Young's modulus of 1 TPa [213, 214], an intrinsic strength of 130 GPa [214], and thermal conductivity of 5,300 $Wm^{-1}K^{-1}$ with high thermal stability [215]. The high electron mobility of 25–200 × 10^3 cm^2 V^{-1} s^{-1} has been attributed to carrier confinement and coherence [216–218]. In addition, graphene has shown a strong ambipolar electric field, i.e., charge carriers can be alternated between holes and electrons depending upon the nature of the gate voltage [183].

One important feature of graphene is its extremely low thickness. Thus, graphene light transmittance is extremely high with measured values of light transmittance of >97% for a single layer and >95% for a bi-layer of graphene (see Fig. 24) [197], making it potentially useful in optoelectronic applications as a highly transparent material [219]. It has been found that the optical and electronic properties are not affected if strongly bent or stretched [220, 221]. As a result, it is the most flexible and stretchable transparent conductor material known to date.

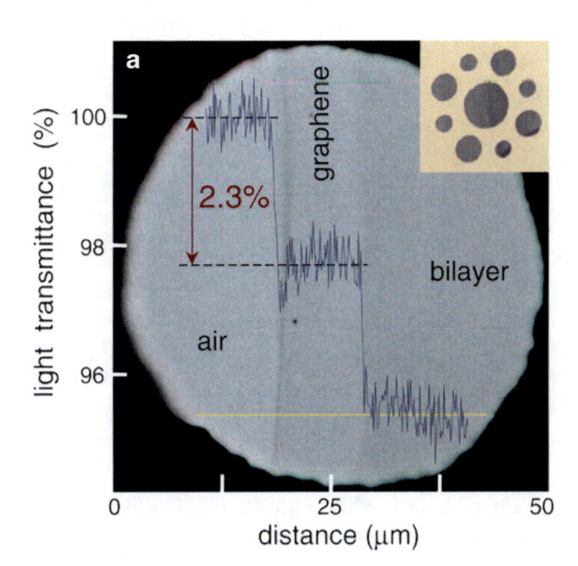

Fig. 24 Light transmittance of single- and bi-layer graphene sheets. (Reprinted with permission from [197])

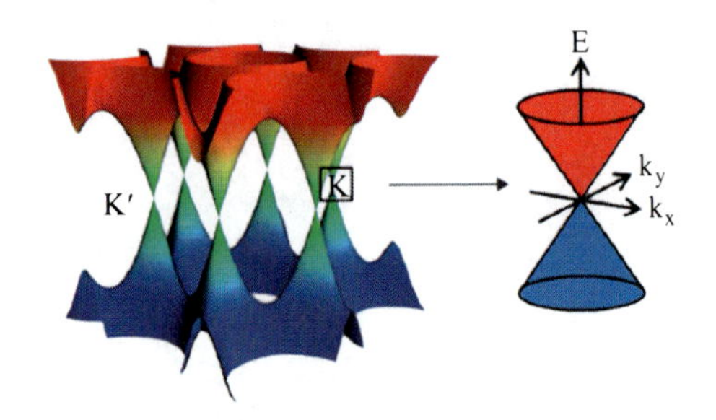

Fig. 25 Calculated band structure of single-layer graphene. (Reprinted with permission from [193])

The electronic band structure of single layer graphene has attracted many theoretical studies, in which (1 + 2)-dimensional Dirac equations seem to describe the electrons more accurately than the Schrödinger equation for a graphene layer [2, 192, 202, 204, 222, 223]. The calculated band structure shows an overlap in the Brillouin zone at two conically shaped points, K and K′ (see Fig. 25) [193]. The electrons are described as mass-less Dirac Fermions, i.e., electrons without rest mass [222]. In the bi-layer graphene, the two layers show a gapless state due to their near binding approximation, the interaction creating parabolic bands in the electronic structure at the overlapping point instead of conical bands [186, 193, 224]. Charge carriers have finite mass, and are thus called massive Dirac fermions [202, 204].

The electronic properties of graphene are highly dependent on the edges (e.g., zigzag, armchair), from which magnetic properties can arise due to the appearance of edge states [225–231]. The edge states can cause particular unconventional magnetism like ferromagnetism, spin glass behavior, and magnetic switching phenomena. Based on computational calculations, graphene is expected to show similar behavior to that of metals, semiconductors, and semi-metal materials [193, 202, 204, 232, 233]. It is predicted that by carefully controlling the shape of the edge on graphene, by specific directional cutting, or by chemical modification, its magnetic properties will become tunable [193, 232, 233]. Also, adsorption of molecules affects the electronic states at the edge and therefore the magnetic properties are also affected, depending on the nature of the interacting molecule [193, 233]. The substrates in which graphene is deposited affect its electronic behavior and its stacking sequence [234, 235]. Zhou et al. showed that even multi layer graphene (up to ten layers) can show similar electronic structure to single layer graphene due to graphene-substrate interactions and induced asymmetry in the layers [234].

Single layer graphene shows a half-integer QHE at room and low temperatures (see Fig. 26) [186–192]. Thus, graphene is exceptionally appropriate for molecular electronic applications. For bi-layer graphene, the QHE is also observed; nonetheless it keeps a metallic behavior at zero carrier concentration or neutrality [186]. Under the influence of a gate voltage, a semiconducting gap behavior is obtained, due to a change in the carrier concentration that introduces asymmetry between layers [237, 238]. It has also been found that substrate interactions and modification of lattice symmetry in graphene (single and multi layer) can cause a remarkable modification of its electronic band structure, leading to finite band gaps [234].

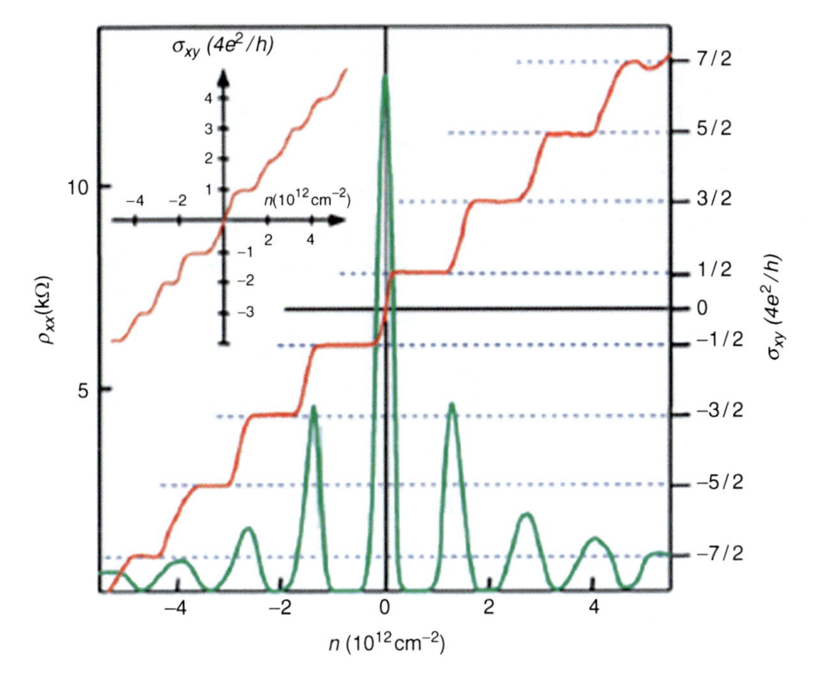

Fig. 26 Hall conductivity (σ_{xy}) and longitudinal resistivity (ρ_{xx}) of single-layer graphene as a function of carrier concentration. *Inset* shows Hall conductivity of bi-layer graphene. (Reprinted with permission from [236])

4.2.2 Multi Layer Graphene

In contrast with single and bi-layer graphene, multi layer graphene has no band gap in its electronic structure [239]. Thus, it shows a similar behavior to those of metallic materials. Use of composite electrodes with multi layer graphene for Li-ion battery applications has improved the performance of such devices due to its capacity to partake in the redox processes [193, 240, 241]. Studies in energy storage applications, e.g., batteries and supercapacitors, have shown promising results that suggest considerable future work [204, 242, 243] (see Fig. 27).

The appearance of magnetic properties in multi layer graphene has been studied [204, 227, 244–248]. Using graphene based activated carbon fibers it has been found that these materials show a Cuire–Weiss behavior, giving evidence of the presence of localized magnetic moments at its edges (see Fig. 28) [249].

4.3 Applications in Molecular Electronics

In this section, we will discuss fascinating results of graphene applications in molecular electronics, with the focus on optoelectronics and organic FETs.

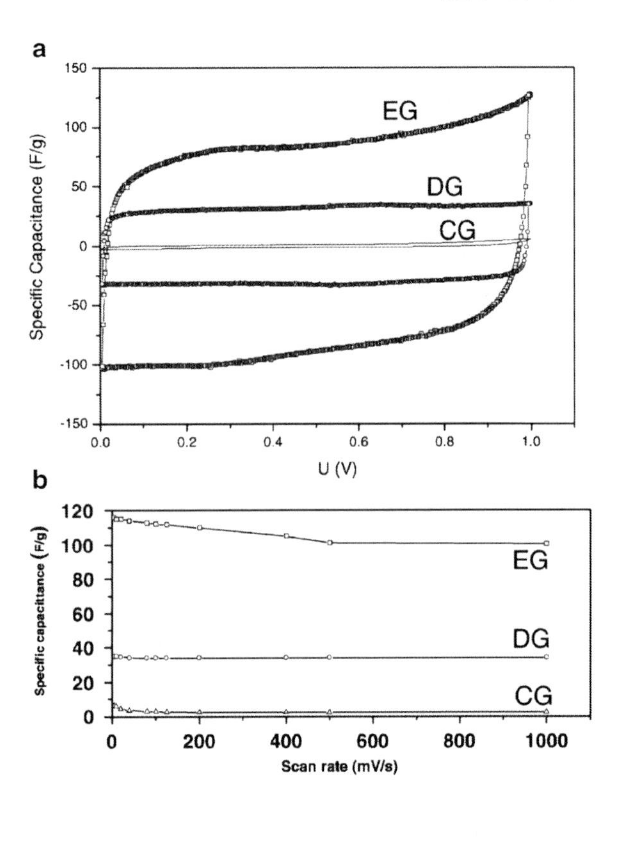

Fig. 27 Supercapacitor study using graphene electrodes (5 mg each). (**a**) Cyclic voltammogram of chemical vapor deposited graphene (CG); nanodiamond derived graphene (NG) and exfoliated graphene (EG). (**b**) Evolution of specific capacitance vs scan rate. (Reprinted with permission from [243])

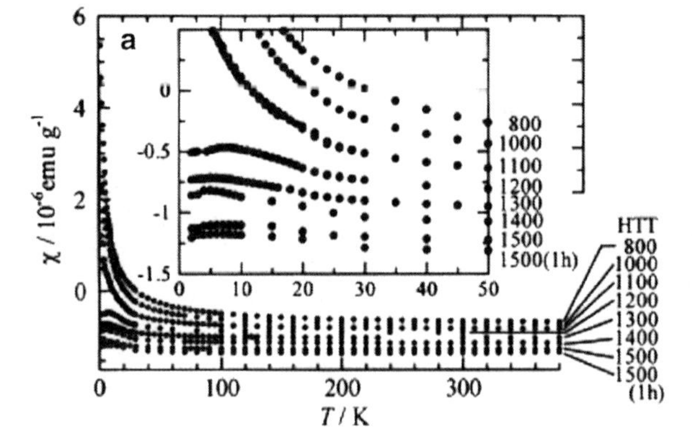

Fig. 28 Magnetic susceptibility (χ) dependence with temperature of activated carbon fibers heated at 1,500 °C. *Inset* shows behavior from 0 to 50 K. (Reprinted with permission from [249])

4.3.1 Organic Optoelectronic Applications

In order to consider a material suitable as an electrode for optoelectronic applications (e.g., organic light-emitting diodes, solar cells) it needs to fulfil two major requirements: (1) high light transmittance (>90%), i.e., light should not be absorbed and (2) electrical conductivity, i.e., regardless of the thickness, the layer should conduct electrical current efficiently [193, 250, 251]. However, nowadays these materials are also required to comply with new characteristics to improve the physicomechanical properties and efficiency of the commonly used materials (e.g., indium–tin oxide) [252–254]. Among these properties, the materials should be flexible, low-cost, thermally stable, and suitable for large-scale production procedures [252]. Thus, graphene is potentially an excellent candidate. Extensive research has been done in order to prove its usefulness as a transparent electrode [221, 250, 255–258].

The controlled synthesis and homogeneous coating of substrates with graphene are some of the major challenges to overcome. However, devices have been made that showed good results, although not comparable to those of indium–tin oxide based devices, but showing that careful design and further research will almost certainly lead to replacement of metal oxide based electrodes [2, 193, 252]. In this section, major results and device designs using graphene as anode/cathode electrodes in organic light-emitting devices (OLEDs) and solar cells will be described and discussed.

Organic Light-Emitting Devices

Since C. W. Tang reported in 1987 [259] the first light-emitting device based on organic polymers (OLEDs), these devices have attracted researchers due to the low-cost of manufacture and easy variability of the polymer physical properties [254]. LEDs are basically a light-emitting polymer trapped between two electrodes externally connected to a power supply (see Fig. 29) [253]. The cathode provides electrons to the LUMO of the polymer and the anode injects holes in the HOMO of the polymer. Once the electrons and holes recombine into the bulk of the polymer an exciton is produced which later releases its energy in the form of electromagnetic radiation [253].

In order for the radiation to be observed, one of the electrodes has to be transparent. This is made possible using ITO electrodes but several disadvantages

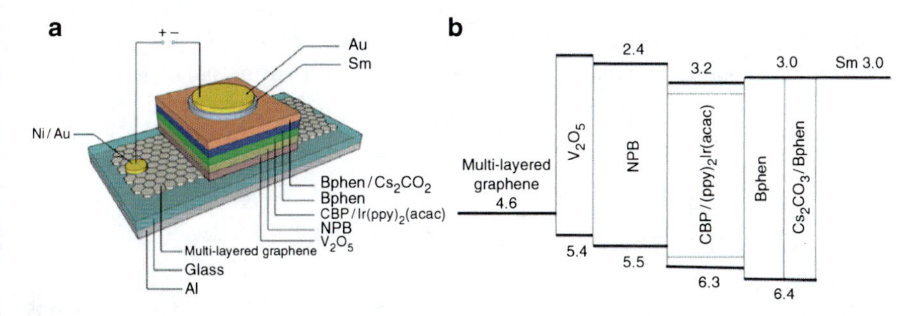

Fig. 29 (**a**) Architecture of graphene based OLED. (**b**) Energy band diagram of the device. (Reprinted with permission from [260])

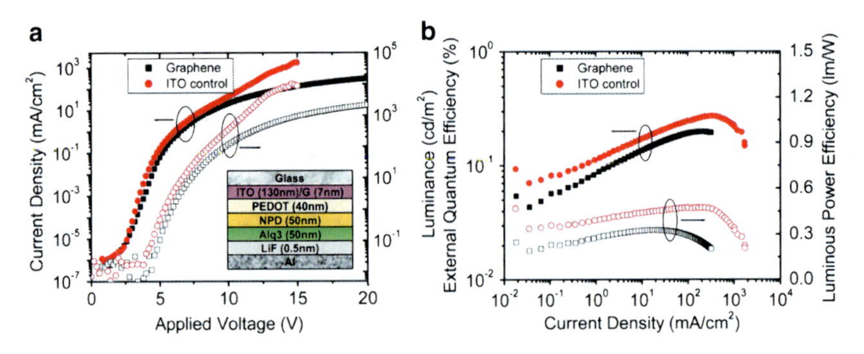

Fig. 30 ITO vs graphene based OLED. *Open circles* refer to plot on right axis. (Reprinted with permission from [254])

with this material [254] have forced researchers to look for alternative ones, including carbon nanotubes, metal nanowires and gratings, conductive oxides, and metal films. Recently, graphene based electrodes have been studied and the results are promising [253, 254, 260, 261]. P. Peumans and collaborators showed that graphene can be used as transparent electrodes in OLEDs with an enormous reduction on the film thickness (7 nm) compared to the commonly used ITO (150 nm) [254]. The manufactured devices showed results comparable to those of ITO-based devices (see Fig. 30). Qin et al. further proved the usefulness of graphene as anode electrodes in OLED [260]. A maximum luminance efficiency of 0.75 cd/A at 7.2 V and 0.38 lm/W at 5.8 V was obtained.

A different approach to incorporate graphene into light-emitting devices was reported by N. D. Robinson and collaborators [253]. A light-emitting electrochemical cell (LEC) device was constructed using graphene as the cathode and PEDOT-PSS as the transparent electrode. Here the light-emitting polymer is blended with an electrolyte. When a potential is applied to the cell, rearrangement of electrolyte ions produces high charge-density layers on the surface of the electrodes, allowing the injection of electrons and holes into the active polymer. The calculated power conversion efficiency and quantum efficiency for the cell at 4 V were 5 lm/W and 9 cd/A, respectively. These results are close to reported values for highly optimized LEC devices.

Graphene films are promising candidates as electrodes in OLED devices as transparent anodes and cathodes. However, optimization of graphene film characteristics (thickness and quality) is necessary in order to obtain device results comparable to ITO based devices [254, 260].

Dye-Sensitized Solar Cell Devices

Since Grätzel et al. introduced the dye-sensitized solar cells (DSSC) in 1991 using TiO_2 films as anode electrodes [262], these cells have become the focus of intense investigation. Its low cost and relatively high efficiencies for sunlight conversion

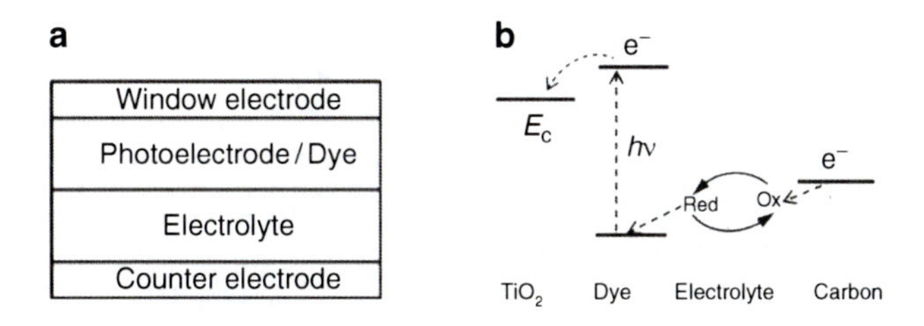

Fig. 31 (**a**) Schematic design of a dye-sensitized solar cell. (**b**) Energy band diagram of DSSC. (Reprinted with permission from [265])

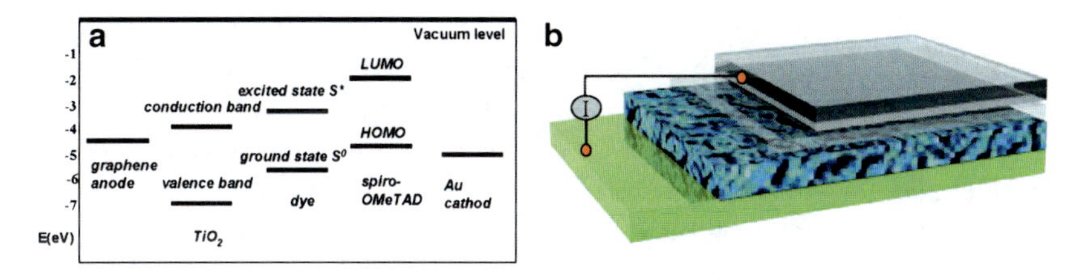

Fig. 32 (**a**) Energy level diagram, (**b**) Four layers from *bottom* to *top* are Au, dye-sensitized heterojunction, compact TiO_2, and graphene film. (Reprinted with permission from [267])

into electric current are the main reasons for the interest [263]. Conversion efficiencies of up to 11% have been reported, but further improvements and optimization are constantly being reported [264]. These devices operate through light energy absorption by the dye material which supplies the current into the system. The window electrode, a semiconductor (e.g., TiO_2), separates the charge together with the counter electrode (e.g., carbon) and the electrolyte, in which a redox process occurs promoted by the counter electrode (see Fig. 31) [265, 266].

In 2007, Müllen and collaborators reported the use of multi layer graphene as a transparent electrode in a DSSC [267]. The graphene layer functioned as the anode electrode (see Fig. 32) in the cell. Although the cell overall conversion efficiency was 0.26%, feasibility was proven. This result started a new area of applications for graphene in optoelectronics.

In the last few years, several studies have reported the design and use of different experimental methods to incorporate graphene into DSSCs with interesting results. In 2009, Kim et al. reported an improvement of photoconversion efficiency from 4.89 to 5.26% by incorporating graphene into the TiO_2 interfacial layer between a fluorine doped tin oxide (FTO) layer and a pristine TiO_2 film [263]. In this study, graphene oxide (GO) and TiO_2 nanocomposites were photocatalytically reduced successfully by UV-irradiation. In this case, graphene layers came to improve the connection between the photoactive layer and the electrode, reducing the back-transfer reaction of electrons into the FTO and pristine TiO_2 film interface.

In 2010 Yang, Zhai et al. reported a DSSC with a total conversion efficiency of 6.97% [268]. The introduction of graphene as 2D bridges into the electrodes increased the conversion efficiency by 39% compared with a nanocrystalline titanium dioxide photoanode. Also, the short-circuit current density was increased by 45% without affecting the open-circuit voltage of the cell. In this case, the composites were prepared by mixing GO with the TiO_2 material and then it was thermally reduced to graphene. In the same year, L. Gao and collaborators reported an overall energy conversion efficiency of 4.28% by incorporating 0.5 wt% graphene in the TiO_2 photoanode [269]. This represents an enhancement of 59% against the DSSC without graphene. It was concluded that the presence of graphene layers increased the dye adsorption efficiency and extended the exciton lifetime; the short-circuit photocurrent density was also increased by 66% to 8.38 mA/cm^2.

Electrophoretic deposition of graphene-TiO_2 composites on ITO have been studied by S. Lee and used in DSSC devices [270]. An improvement of up to five times in the power conversion efficiency was obtained by incorporating graphene into the device, with a maximum value of 1.68%. This was attributed to a reduction in the charge transfer resistance in the composite film which at the end facilitates the transport of electrons and lowers the probability of electron–hole recombination. The effect of the amount of graphene incorporated on the DSSC photovoltaic performance was also studied. It was found that the overall energy conversion efficiency increases and then decreases as the amount of graphene is increased (see Fig. 33). It was concluded that as the graphene amount increases the transmittance of the film decreases, affecting the overall cell performance.

In DSSCs, graphene incorporation into electrode composites increases the overall energy conversion efficiency by up to five times. Nonetheless, further research is in order to optimize the construction and characteristics of graphene based devices (e.g., graphene film thickness and amount, device fabrication conditions, and configuration) [267, 270].

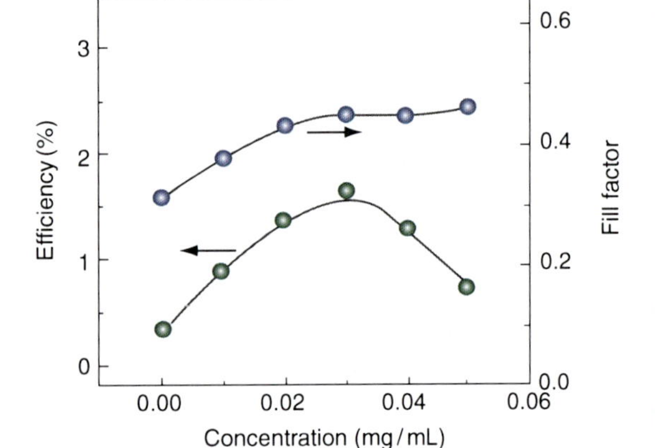

Fig. 33 Power conversion efficiency and fill factor of varying amounts of graphene concentration on DSSC devices. (Reprinted with permission from [270])

Bulk Heterojunction Solar Cell Devices

Bulk heterojunction (BHJ) solar cell devices have been the focus of many studies in the last two decades. The incorporation of organic based polymers and dyes, functionalized polymers, and composites made of single organic molecules have taken these devices into higher and higher energy conversion efficiencies [271]. New materials are designed and studied continuously in order to achieve higher conversion efficiencies and lower cost of production for these devices. Graphene has come to play an important role in the design of novel BHJ devices, due to its high electron mobility and transmittance. Many studies have employed it as transparent counter electrode, as the photoactive composite of the cell, and as a hole transporting and electron blocking layer [221, 258, 272–282].

Müllen et al. in 2008 reported the successful use of graphene films (TGF) as transparent electrodes (anodes) in BHJ solar cells composed of P3HT and PCBM as the active layer (see Fig. 34) [258]. The cell conversion efficiency under low intensity monochromatic light showed the same values as the ITO electrode (1.5%) and under simulated solar light the values were lower (0.29%) compared to the obtained values for ITO (1.17%) under the same conditions. Although the study under simulated solar light was not satisfactory, the BHJ solar cell shows promise for the use of graphene in this type of devices after further optimization of the cell.

Further research has taken the BHJ cells into more complex designs and new photoactive layers with values of energy conversion efficiencies from 0.13 [273] to 1.4% [282]. Graphene provides these cells with new characteristics impossible to match by ITO based cells, like functionality under extreme bending angles up to 138° under which ITO cells crack and fail after bending by 60° [221].

Graphene has also been studied as the acceptor within the active layer of the cell (see Fig. 35). In these studies hydrophilic GO and solution-processable functionalized graphene (SPFGraphene) act as efficient acceptor moieties [272, 280]. In these cells it has been determined that the energy conversion efficiency depends on the graphene content and the annealing temperature of the cell when constructed (see Fig. 36). Conversion efficiencies vary from 0.034 to 1.4% for SPFGraphene [280] and from 2.1 to 3.8% for GO based solar devices [277].

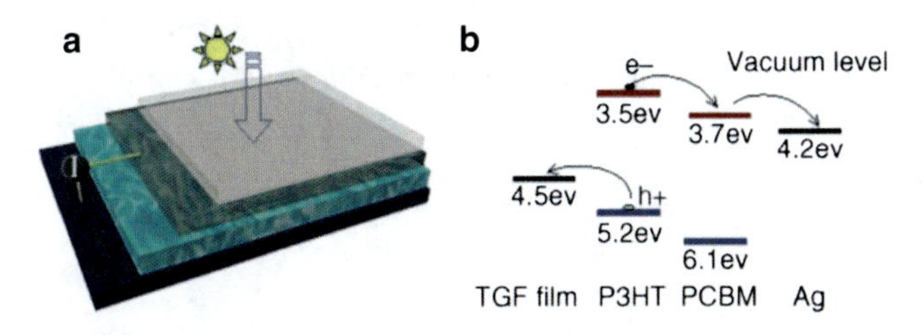

Fig. 34 Illustration of the BHJ solar cell. (**a**) The four layers from *bottom* to *top* are Ag, blend of P3HT and PCBM, graphene and quartz. (**b**) Energy level diagram of the cell. (Reprinted with permission from [258])

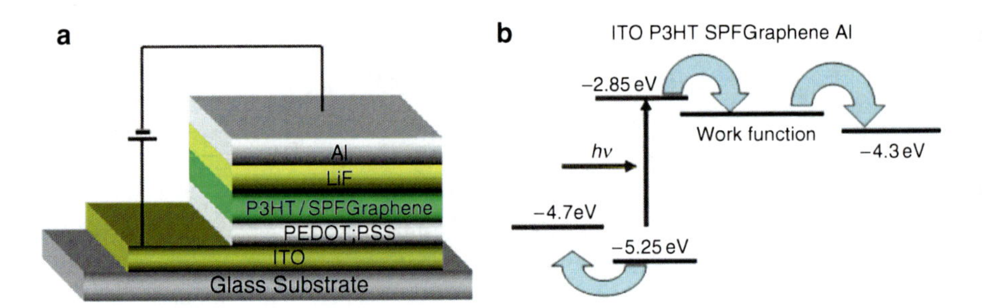

Fig. 35 (**a**) Schematic structure of BHJ cell with P3HT/SPFGraphene as active layer. (**b**) Energy band diagram. (Reprinted with permission from [278])

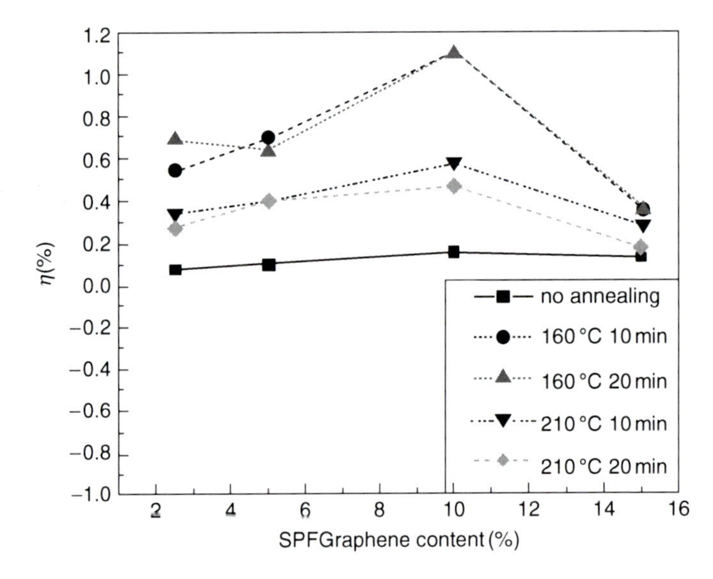

Fig. 36 Dependence of conversion efficiency (η) with annealing temperature and graphene content in BHJ devices. (Reprinted with permission from [272])

Optimization of these devices will lead to continuously higher conversion efficiencies, and possibly to the unequivocally replacement of ITO and metal oxide based electrodes from these devices to reduce cost and improve the physical–mechanical properties of the cells (e.g., bending, thermal stability) [221].

Field-Effect Transistor Applications

The graphene based FET is of great interest due to its intrinsic electro-mechanical properties [185]. Its physical–mechanical resistance makes it suitable for device manufacture. Graphene FET devices (see Fig. 37) have been made by standardized

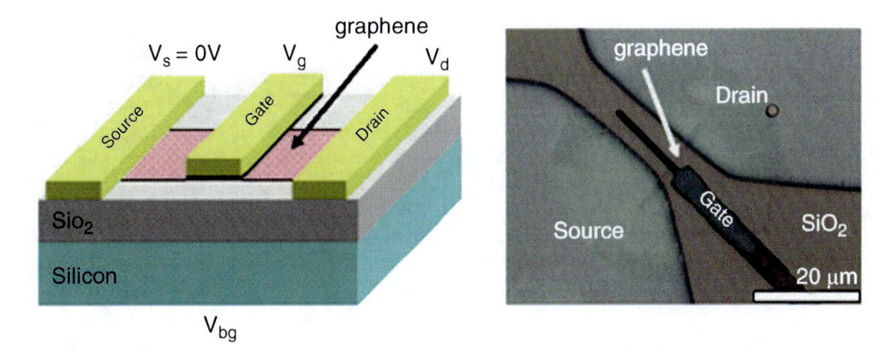

Fig. 37 *Left*: scheme of graphene FET device. *Right*: top-view micrograph of graphene FET. (Reprinted with permission from [185])

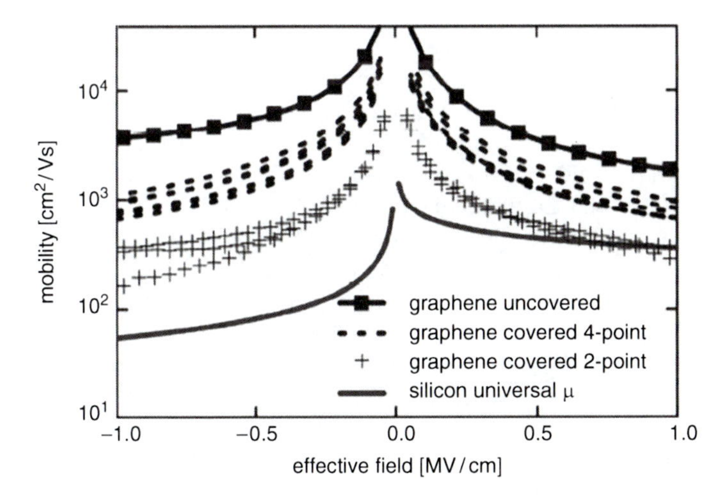

Fig. 38 Dependence of carrier mobility on electric field in graphene FETs. (Reprinted with permission from [185])

lithography methods also applied to other nanomaterials [201, 283–285]. These devices allowed measurement of carrier mobility on suspended graphene and top gated devices. It was observed that the substrate and gate potential affected greatly the carrier mobility on the graphene sheet [185, 193, 201, 286, 287]. Carrier mobility values on nonoptimized top gated devices have exceeded those of silicon based FETs (see Fig. 38) [185, 283, 285].

Due to its zero-gap electronic structure, large graphene sheets are not suitable for FET applications [185, 193]. It has been shown that graphene transistors conduct current even at the point of expected isolator behavior (i.e., Dirac point) [288]. Therefore, current modulation cannot be achieved using “macroscopic” graphene sheets [185]. In order to have a band gap on graphene, the use of narrow ribbons of

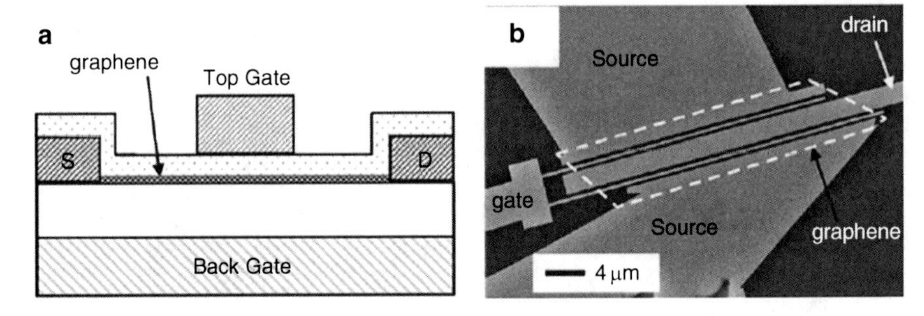

Fig. 39 (**a**) Device schematic of the dual-gate graphene transistor. (**b**) SEM image of a double-channel graphene transistor. The channel width is 27 μm, and the gate length is 350 nm for each channel. (Reprinted with permission from [298])

graphene (GNRs) have been considered [185, 193, 286, 287, 289, 290]. Computational calculations predict semiconductor to semi-metallic behaviors, based on the width and type of edge (armchair or zig-zag) [193, 291–295].

Experimental results confirmed the presence of band gaps on devices made of 20 nm width GNRs at low temperatures (1.7–4 K) [185, 289, 296]. In 2008, Ponomarenko et al. measured a band gap of 500 meV in a 1 nm width GNR FET device [297]. This device did not show conductance at room temperature, which is necessary for good performance of manually modulated transistors. Further research is needed in order to design devices carefully with smaller width graphene ribbons, as well as to control the type of edge obtained.

The design of dual gate FET devices (see Fig. 39) has also been explored, using single layer and bi-layer graphene, with remarkable results. Wide gate-tunable band gaps up to 250 meV in bi-layer devices [298], carrier mobilities of 8,000 cm^2/V at room temperature [299], and cutoff frequencies of 50 GHz [300] have been achieved, providing more evidence for the potential use of graphene in this type of molecular electronic devices.

4.4 Conclusions and Future Directions

Since 2004 [183], graphene research has evolved from a heavily theoretical and fundamental field into a variety of research areas [301]. Its electrical, magnetic, physical–mechanical, and chemical properties position it as the most promising material for molecular electronic and optoelectronic applications, possibly replacing the currently used silicon and metal oxide based devices. Nonetheless, further research is essential in order to control easily such properties and construct devices with specific and novel architectures to explore in depth all of these exciting properties, as well as to achieve the synthesis of large-scale, size- and layer-count controlled graphene.

References

1. Green JR, Korkin A, Labanowski J (2003) Nano and giga: challenges in microelectronics. Elsevier, Amsterdam
2. Avouris P, Chen Z, Perebeinos V (2007) Carbon-based electronics. Nat Nanotechnol 2:605–615
3. Theis TN, Solomon PM (2010) It's time to reinvent the transistor! Science 327:1600–1601
4. Lu W, Lieber CM (2007) Nanoelectronics from the bottom up. Nat Mater 6:841–850
5. Xu B, Tao NJ (2003) Measurement of single-molecule resistance by repeated formation of molecular junctions. Science 301:1221–1223
6. Xu BQ, Xiao XY, Yang X et al (2005) Large gate modulation in the current of a room temperature single molecule transistor. J Am Chem Soc 127:2386–2387
7. Akasaka T, Fred W, Nagase S (2010) Chemistry of nanocarbons. Wiley, Hoboken, NJ
8. Sablon K (2008) Nanoelectrodes for molecular devices: a controllable fabrication. Nanoscale Res Lett 3:268–270
9. Venkataraman L, Klare JE, Nuckolls C et al (2006) Dependence of single-molecule junction conductance on molecular conformation. Nature 442:904–907
10. Kroto HW, Heath JR, O'Brien SC et al (1985) C_{60}: buckminsterfullerene. Nature 318:162–163
11. Iijima S (1991) Helical microtubules of graphitic carbon. Nature (London) 354:56–58
12. Iijima S, Ichihashi T (1993) Single-shell carbon nanotubes of 1-nm diameter. Nature 363:603–605
13. Bethune DS, Kiang CH, de Vries MS et al (1993) Cobalt-catalyzed growth of carbon nanotubes with single-atomic-layer walls. Nature (London) 363:605–607
14. Iijima S (1980) Direct observation of the tetrahedral bonding in graphitized carbon black by high resolution electron microscopy. J Cryst Growth 50:675–683
15. Iijima S (1987) The 60-carbon cluster has been revealed. J Phys Chem 91:3466–3467
16. Ugarte D (1992) Curling and closure of graphitic networks under electron-beam irradiation. Nature 359:707–709
17. Kuznetsov VL, Chuvilin AL, Butenko YV et al (1994) Onion-like carbon from ultra-disperse diamond. Chem Phys Lett 222:343–348
18. Sano N, Wang H, Chhowalla M et al (2001) Synthesis of carbon 'onions' in water. Nature 414:506–507
19. Liu W, Meng QS (2009) An effective method of increasing production rate of onion-like fullerenes. J Phys Conf Ser 188:012035
20. Alexandrou I, Wang H, Sano N et al (2004) Structure of carbon onions and nanotubes formed by arc in liquids. J Chem Phys 120:1055–1058
21. Iijima S, Yudasaka M, Yamada R et al (1999) Nano-aggregates of single-walled graphitic carbon nano-horns. Chem Phys Lett 309:165–170
22. Heath JR, O'Brien SC, Zhang Q et al (1985) Lanthanum complexes of spheroidal carbon shells. J Am Chem Soc 107:7779–7780
23. Chai Y, Guo T, Jin C et al (1991) Fullerenes with metals inside. J Phys Chem 95:7564–7568
24. Dunsch L, Yang S (2007) Metal nitride cluster fullerenes: their current state and future prospects. Small 3:1298–1320
25. Chaur MN, Melin F, Ortiz AL et al (2009) Chemical, electrochemical, and structural properties of endohedral metallofullerenes. Angew Chem Int Ed 48:7514–7538
26. Stevenson S, Mackey MA, Stuart MA et al (2008) A distorted tetrahedral metal oxide cluster inside an icosahedral carbon cage. synthesis, isolation, and structural characterization of $Sc_4(\mu_3\text{-}O)_2@Ih\text{-}C_{80}$. J Am Chem Soc 130:11844–11845
27. Wang T, Chen N, Xiang J et al (2009) Russian-doll-type metal carbide endofullerene: synthesis, isolation, and characterization of $Sc_4C_2@C_{80}$. J Am Chem Soc 131:16646–16647
28. Chen N, Chaur MN, Moore C et al (2010) Synthesis of a new endohedral fullerene family, $Sc_2S@C_{2n}$ (n = 40–50) by the introduction of SO_2. Chem Commun 46:4818–4820

29. Dunsch L, Yang S, Zhang L et al (2010) Metal sulfide in a C(82) fullerene cage: a new form of endohedral clusterfullerenes. J Am Chem Soc 132:5413–5421
30. Kratschmer W, Lamb LD, Fostiropoulos K et al (1990) Solid C_{60}: a new form of carbon. Nature 347:354–358
31. Krätschmer W, Fostiropoulos K, Huffman DR (1990) The infrared and ultraviolet absorption spectra of laboratory-produced carbon dust: evidence for the presence of the C_{60} molecule. Chem Phys Lett 170:167–170
32. Haufler RE, Conceicao J, Chibante LPF et al (1990) Efficient production of C_{60} (buckminsterfullerene), $C_{60}H_{36}$, and the solvated buckide ion. J Phys Chem 94:8634–8636
33. Taylor R, Hare JP, Abdul-Sada AK et al (1990) Isolation, separation and characterization of the fullerenes C_{60} and C_{70}: the third form of carbon. J Chem Soc Chem Commun 20:1423–1425
34. Howard JB, McKinnon JT, Makarovsky Y et al (1991) Fullerenes C_{60} and C_{70} in flames. Nature (London) 352:139–141
35. McKinnon JT, Bell WL, Barkley RM (1992) Combustion synthesis of fullerenes. Combust Flame 88:102–112
36. Goel A, Hebgen P, Vander Sande JB et al (2002) Combustion synthesis of fullerenes and fullerenic nanostructures. Carbon 40:177–182
37. Takehara H, Fujiwara M, Arikawa M et al (2005) Experimental study of industrial scale fullerene production by combustion synthesis. Carbon 43:311–319
38. Alford JM, Bernal C, Cates M et al (2008) Fullerene production in sooting flames from 1,2,3,4-tetrahydronaphthalene. Carbon 46:1623–1625
39. Murayama H, Tomonoh S, Alford JM et al (2004) Fullerene production in tons and more: from science to industry. Fullerenes, Nanotubes, Carbon Nanostruct 12:1–9
40. Fulcheri L, Schwob Y, Fabry F et al (2000) Fullerene production in a 3-phase AC plasma process. Carbon 38:797–803
41. Song X, Liu Y, Zhu J (2006) The effect of furnace temperature on fullerene yield by a temperature controlled arc discharge. Carbon 44:1584–1586
42. Ahmad B, Riaz M, Ahmad M et al (2008) Synthesis and characterization of higher fullerene (C84) in dc arc discharge using Cu as a catalyst. Mater Lett 62:3367–3369
43. Huczko A, Lange H, Byszewski P et al (1997) Fullerene formation in carbon arc: electrode gap dependence and plasma spectroscopy. J Phys Chem A 101:1267–1269
44. Gonzalez-Aguilar J, Moreno M, Fulcheri L (2007) Carbon nanostructures production by gas-phase plasma processes at atmospheric pressure. J Phys D: Appl Phys 40:2361–2374
45. Kareev IE, Bubnov VP, Fedutin DN (2009) Electric-arc high-capacity reactor for the synthesis of carbon soot with a high content of endohedral metallofullerenes. Tech Phys 54:1695–1698
46. Yamada M, Akasaka T, Nagase S (2010) Endohedral metal atoms in pristine and functionalized fullerene cages Acc Chem Res 43:92–102
47. Tsuchiya T, Akasaka T, Nagase S (2010) New vistas in fullerene endohedrals: functionalization with compounds from main group elements. Pure Appl Chem 82:505–521
48. Dunsch L, Georgi P, Krause M et al (2003) New clusters in endohedral fullerenes: the metalnitrides. Synth Met 135–136:761–762
49. Dunsch L, Krause M, Noack J et al (2004) Endohedral nitride cluster fullerenes. Formation and spectroscopic analysis of $L_{3-x}M_xN@C_{2n}$ ($0 \leq x \leq 3$; N = 39,40). J Phys Chem Solids 65:309–315
50. Krause M, Ziegs F, Popov AA et al (2007) Entrapped bonded hydrogen in a fullerene: the five-atom cluster Sc_3CH in C_{80}. ChemPhysChem 8:537–540
51. Stevenson S, Rice G, Glass T et al (1999) Small-bandgap endohedral metallofullerenes in high yield and purity. Nature 401:55–57
52. Chen N, Klod S, Rapta P et al (2010) Direct Arc-discharge assisted synthesis of $C_{60}H_2(C_3H_5N)$: a cis-1-pyrrolino C_{60} fullerene hydride with unusual redox properties. Chem Mater 22:2608–2615

53. Haddon RC, Brus LE, Raghavachari K (1986) Electronic structure and bonding in icosahedral carbon cluster (C_{60}). Chem Phys Lett 125:459–464
54. Xie Q, Perez-Cordero E, Echegoyen L (1992) Electrochemical detection of $C_{60}{}^{6-}$ and $C_{70}{}^{6-}$: enhanced stability of fullerides in solution. J Am Chem Soc 114:3978–3980
55. Xie Q, Arias F, Echegoyen L (1993) Electrochemically-reversible, single-electron oxidation of C_{60} and C_{70}. J Am Chem Soc 115:9818–9819
56. Echegoyen L, Echegoyen LE (1998) Electrochemistry of fullerenes and their derivatives. Acc Chem Res 31:593–601
57. Anderson MR, Dorn HC, Stevenson SA et al (1998) The voltammetry of C_{84} isomers. J Electroanal Chem 444:151–154
58. Chaur MN, Athans AJ, Echegoyen L (2008) Metallic nitride endohedral fullerenes: synthesis and electrochemical properties. Tetrahedron 64:11387–11393
59. Lu X, Slanina Z, Akasaka T et al (2010) Yb@$C_{(2n)}$ ($n = 40, 41, 42$): new fullerene allotropes with unexplored electrochemical properties. J Am Chem Soc 132:5896–5905
60. Zhao J, Miao B, Zhao L et al (2004) Electronic transport properties of single C_{60} molecules and device applications. Int J Nanotechnol 1:157–169
61. Joachim C, Gimzewski JK (1995) Analysis of low-voltage I(V) characteristics of a single C_{60} molecule. Europhys Lett 30:409–414
62. Joachim C, Gimzewski J, Schlittler R et al (1995) Electronic transparence of a single C_{60} molecule. Phys Rev Lett 74:2102–2105
63. Néel N, Kröger J, Limot L et al (2007) Controlled contact to a C_{60} molecule. Phys Rev Lett 98:065502
64. Saffarzadeh A (2008) Electronic transport through a C_{60} molecular bridge: the role of single and multiple contacts. J Appl Phys 103:083705–083706
65. Mishra S (2005) Quantum transport through a C_{60}-X molecular bridge with the extra atom at the center. Phys Rev B 72:075421
66. Porath D, Levi Y, Tarabiah M et al (1997) Tunneling spectroscopy of isolated C_{60} molecules in the presence of charging effects. Phys Rev B 56:9829–9833
67. Porath D, Millo O (1997) Single electron tunneling and level spectroscopy of isolated C_{60} molecules. J Appl Phys 81:2241
68. Amman M, Wilkins R, Ben-Jacob E et al (1991) Analytic solution for the current-voltage characteristic of two mesoscopic tunnel junctions coupled in series. Phys Rev B 43:1146–1149
69. Allemand PM, Koch A, Wudl F et al (1991) Two different fullerenes have the same cyclic voltammetry. J Am Chem Soc 113:1050–1051
70. Imahori H, Tkachenko NV, Vehmanen V et al (2001) An extremely small reorganization energy of electron transfer in porphyrin–fullerene dyad. J Phys Chem A 105:1750–1756
71. Marcus RA (1956) The theory of oxidation-reduction reactions involving electron transfer. I J Chem Phys 24:966–978
72. Marcus RA, Sutin N (1985) Electron transfers in chemistry and biology. Biochim Biophys Acta Rev Bioenerg 811:265–322
73. Marcus RA (1993) Electron transfer reactions in chemistry: theory and experiment (Nobel lecture). Angew Chem Int Ed 32:1111–1121
74. Guldi DM, Illescas BM, Atienza CM et al (2009) Fullerene for organic electronics. Chem Soc Rev 38:1587–1597
75. Imahori H, Yamada H, Guldi DM et al (2002) Comparison of reorganization energies for intra- and intermolecular electron transfer. Angew Chem Int Ed 41:2344–2347
76. Schuster DI, Li K, Guldi DM et al (2007) Azobenzene-linked porphyrin-fullerene dyads. J Am Chem Soc 129:15973–15982
77. Imahori H, Guldi DM, Tamaki K et al (2001) Charge separation in a novel artificial photosynthetic reaction center lives 380 ms. J Am Chem Soc 123:6617–6628
78. Guldi DM, Imahori H, Tamaki K et al (2004) A molecular tetrad allowing efficient energy storage for 1.6 s at 163 K. J Phys Chem A 108:541–548

79. Ito O, Yamanaka K (2009) Roles of molecular wires between fullerenes and electron donors in photoinduced electron transfer. Bull Chem Soc Jpn 82:316–332
80. De la Torre G, Giacalone F, Segura JL et al (2005) Electronic communication through pi-conjugated wires in covalently linked porphyrin/C_{60} ensembles. Chem Eur J 11: 1267–1280
81. Ikemoto J, Takimiya K, Aso Y et al (2002) Porphyrin–oligothiophene–fullerene triads as an efficient intramolecular electron-transfer system. Org Lett 4:309–311
82. Guldi DM, Giacalone F, de la Torre G et al (2005) Topological effects of a rigid chiral spacer on the electronic interactions in donor-acceptor ensembles. Chem Eur J 11:7199–7210
83. Oike T, Kurata T, Takimiya K et al (2005) Polyether-bridged sexithiophene as a complexation-gated molecular wire for intramolecular photoinduced electron transfer. J Am Chem Soc 127:15372–15373
84. D'Souza F, Maligaspe E, Ohkubo K et al (2009) Photosynthetic reaction center mimicry: low reorganization energy driven charge stabilization in self-assembled cofacial zinc phthalocyanine dimer-fullerene conjugate. J Am Chem Soc 131:8787–8797
85. Megiatto JD, Schuster DI, Abwandner S et al (2010) [2]Catenanes decorated with porphyrin and [60]fullerene groups: design, convergent synthesis, and photoinduced processes. J Am Chem Soc 132:3847–3861
86. Takai A, Chkounda M, Eggenspiller A et al (2010) Efficient photoinduced electron transfer in a porphyrin tripod-fullerene supramolecular complex via pi-pi interactions in nonpolar media. J Am Chem Soc 132:4477–4489
87. De la Escosura A, Martinez-Diaz MV, Guldi DM et al (2006) Stabilization of charge-separated states in phthalocyanine-fullerene ensembles through supramolecular donor-acceptor interactions. J Am Chem Soc 128:4112–4118
88. Metzger RM (2006) Unimolecular rectifiers and what lies ahead. Colloids Surf A 284–285:2–10
89. Metzger RM (2006) Unimolecular rectifiers: methods and challenges. Anal Chim Acta 568:146–155
90. Aviram A, Ratner MA (1974) Molecular rectifiers. Chem Phys Lett 29:277–283
91. Viani L, dos Santos MC (2006) Comparative study of lower fullerenes doped with boron and nitrogen. Solid State Commun 138:498–501
92. Xie R, Bryant GW, Zhao J et al (2003) Tailorable acceptor C_{60}-nBn and donor C_{60}-mNm pairs for molecular electronics. Phys Rev Lett 90:206602/1–206602/4
93. Metzger RM (2003) One-molecule-thick devices: rectification of electrical current by three Langmuir-Blodgett monolayers. Synth Met 137:1499–1501
94. Metzger RM, Baldwin JW, Shumate WJ et al (2003) Electrical rectification in a Langmuir-Blodgett monolayer of dimethyanilinoazafullerene sandwiched between gold electrodes. J Phys Chem B 107:1021–1027
95. Wang B, Zhou Y, Ding X et al (2006) Conduction mechanism of Aviram-Ratner rectifiers with single pyridine-s-C_{60} oligomers. J Phys Chem B 110:24505–24512
96. Gayathri SS, Patnaik A (2006) Electrical rectification from a fullerene[60]-dyad based metal-organic-metal junction. Chem Commun (Cambridge, UK) 1977–1979
97. Matino F, Arima V, Piacenza M et al (2009) Rectification in supramolecular zinc porphyrin/fulleropyrrolidine dyads self-organized on gold(111). Chemphyschem 10:2633–2641
98. Acharya S, Song H, Lee J et al (2009) An amphiphilic C_{60} penta-addition derivative as a new U-type molecular rectifier. Org Electron 10:85–94
99. Koiry SP, Jha P, Aswal DK et al (2010) Diodes based on bilayers comprising of tetraphenyl porphyrin derivative and fullerene for hybrid nanoelectronics. Chem Phys Lett 485:137–141
100. Joachim C, Gimzewski JK (1997) An electromechanical amplifier using a single molecule. Chem Phys Lett 265:353–357
101. Joachim C, Gimzewski JK, Tang H (1998) Physical principles of the single-C_{60} transistor effect. Phys Rev B: Condens Matter Mater Phys 58:16407–16417

102. Park H, Park J, Lim AKL et al (2000) Nanomechanical oscillations in a single-C_{60} transistor. Nature (London) 407:58–60
103. Park H, Lim AKL, Alivisatos AP et al (1999) Fabrication of metallic electrodes with nanometer separation by electromigration. Appl Phys Lett 75:301
104. Winkelmann CB, Roch N, Wernsdorfer W et al (2009) Superconductivity in a single-C_{60} transistor. Nat Phys 5:876–879
105. Roch N, Winkelmann CB, Florens S et al (2008) Kondo effect in a C_{60} single-molecule transistor. Phys Status Solid B 245:1994–1997
106. Mentovich ED, Belgorodsky B, Kalifa I et al (2010) 1-Nanometer-sized active-channel molecular quantum-dot transistor. Adv Mater 22:2182–2186
107. Morita T, Lindsay S (2008) Reduction-induced switching of single-molecule conductance of fullerene derivatives. J Phys Chem B 112:10563–10572
108. Ortiz AL, Rivera DM, Athans AJ et al (2009) Regioselective addition of N-(4-thiocyanatophenyl)pyrrolidine addends to fullerenes. Eur J Org Chem 3396–3403:S3396/1–S3396/25
109. Ortiz AL, Echegoyen L (2010) Unexpected and selective formation of an (e, e, e, e)-tetrakis-[60]fullerene derivative via electrolytic retro-cyclopropanation of a D2h-hexakis-[60]fullerene adduct. J Mater Chem 21:1362–1364
110. Zhang S, Lukoyanova O, Echegoyen L (2006) Synthesis of fullerene adducts with terpyridyl- or pyridylpyrrolidine groups in trans-1 positions. Chem Eur J 12:2846–2853
111. Yu M, Lourie O, Dyer MJ et al (2000) Strength and breaking mechanism of multiwalled carbon nanotubes under tensile load. Science 287:637–640
112. Hamada N, Sawada S, Oshiyama A (1992) New one-dimensional conductors: graphitic microtubules. Phys Rev Lett 68:1579–1581
113. Pillai SK, Ray SS, Moodley M (2007) Purification of single-walled carbon nanotubes. J Nanosci Nanotechnol 7:3011–3047
114. Journet C, Maser WK, Bernier P et al (1997) Large-scale production of single-walled carbon nanotubes by the electric-arc technique. Nature 388:756–758
115. Guo T, Nikolaev P, Thess A et al (1995) Catalytic growth of single-walled manotubes by laser vaporization. Chem Phys Lett 243:49–54
116. Bonard J, Croci M, Conus F et al (2002) Watching carbon nanotubes grow. Appl Phys Lett 81:2836
117. Marchand M, Journet C, Guillot D et al (2009) Growing a carbon nanotube atom by atom: "and yet it does turn". Nano Lett 9:2961–2966
118. Meyyappan M (2009) A review of plasma enhanced chemical vapour deposition of carbon nanotubes. J Phys D 42:213001
119. Hou S, Chung D, Lin T (2009) Flame synthesis of carbon nanotubes in a rotating counter-flow. J Nanosci Nanotechnol 9:4826–4833
120. Sun BM, Liu YC, Ding ZY (2009) Carbon nanotubes preparation using carbon monoxide from the pyrolysis flame. Adv Mater Res 87–88:104–109
121. Zhang L, Zaric S, Tu X et al (2008) Assessment of chemically separated carbon nanotubes for nanoelectronics. J Am Chem Soc 130:2686–2691
122. Pillai SK, Ray SS, Moodley M (2008) Purification of multi-walled carbon nanotubes. J Nanosci Nanotechnol 8:6187–6207
123. Matlhoko L, Pillai SK, Ray SS et al (2008) Purification of laser synthesized SWCNTs by different methods: a comparative study. J Nanosci Nanotechnol 8:6023–6030
124. Matlhoko L, Pillai SK, Moodley M et al (2009) A comparison of purification procedures for multi-walled carbon nanotubes produced by chemical vapor deposition. J Nanosci Nanotechnol 9:5431–5435
125. Hersam MC (2008) Progress towards monodisperse single-walled carbon nanotubes. Nat Nanotechnol 3:387–394
126. Collins PG, Arnold MS, Avouris P (2001) Engineering carbon nanotubes and nanotube circuits using electrical breakdown. Science 292:706–709

127. Krupke R, Hennrich F, Lohneysen H et al (2003) Separation of metallic from semiconducting single-walled carbon nanotubes. Science 301:344–347
128. Arnold MS, Green AA, Hulvat JF et al (2006) Sorting carbon nanotubes by electronic structure using density differentiation. Nat Nanotechnol 1:60–65
129. Wei L, Lee CW, Li L et al (2008) Assessment of (n, m) selectively enriched small diameter single-walled carbon nanotubes by density differentiation from cobalt-incorporated MCM-41 for macroelectronics. Chem Mater 20:7417–7424
130. Zheng M, Jagota A, Strano MS et al (2003) Structure-based carbon nanotube sorting by sequence-dependent DNA assembly. Science 302:1545–1548
131. Zheng M, Jagota A, Semke ED et al (2003) DNA-assisted dispersion and separation of carbon nanotubes. Nat Mater 2:338–342
132. Huang X, Mclean RS, Zheng M (2005) High-resolution length sorting and purification of DNA-wrapped carbon nanotubes by size-exclusion chromatography. Anal Chem 77:6225–6228
133. Zheng M, Semke ED (2007) Enrichment of single chirality carbon nanotubes. J Am Chem Soc 129:6084–6085
134. Tu X, Zheng M (2008) A DNA-based approach to the carbon nanotube sorting problem. Nano Res 1:185–194
135. Zhang L, Tu X, Welsher K et al (2009) Optical characterizations and electronic devices of nearly pure (10,5) single-walled carbon nanotubes. J Am Chem Soc 131:2454–2455
136. Paolucci D, Franco MM, Iurlo M et al (2008) Singling out the electrochemistry of individual single-walled carbon nanotubes in solution. J Am Chem Soc 130:7393–7399
137. Kavan L, Rapta P, Dunsch L (2000) In situ Raman and Vis-NIR spectroelectrochemistry at single-walled carbon nanotubes. Chem Phys Lett 328:363–368
138. Kavan L, Rapta P, Dunsch L et al (2001) Electrochemical tuning of electronic structure of single-walled carbon nanotubes: in-situ Raman and Vis-NIR study. J Phys Chem B 105:10764–10771
139. Melle-Franco M, Marcaccio M, Paolucci D et al (2004) Cyclic voltammetry and bulk electronic properties of soluble carbon nanotubes. J Am Chem Soc 126:1646–1647
140. Guldi DM, Marcaccio M, Paolucci D et al (2003) Single-wall carbon nanotube-ferrocene nanohybrids: observing intramolecular electron transfer in functionalized SWNTs. Angew Chem Int Ed 42:4206–4209
141. Zheng M, Diner BA (2004) Solution redox chemistry of carbon nanotubes. J Am Chem Soc 126:15490–15494
142. Pénicaud A, Poulin P, Derré A et al (2005) Spontaneous dissolution of a single-wall carbon nanotube salt. J Am Chem Soc 127:8–9
143. Wang Z, Pedrosa H, Krauss T et al (2007) Reply. Phys Rev Lett 98:019702
144. Dukovic G, Wang F, Song D et al (2005) Structural dependence of excitonic optical transitions and band-gap energies in carbon nanotubes. Nano Lett 5:2314–2318
145. Saito R, Fujita M, Dresselhaus G et al (1992) Electronic structure of chiral graphene tubules. Appl Phys Lett 60:2204
146. Perello DJ, Chulim S, Chae SJ et al (2010) Anomalous Schottky barriers and contact band-to-band tunneling in carbon nanotube transistors. ACS Nano 4:3103–3108
147. Tans SJ, Devoret MH, Dai H et al (1997) Individual single-wall carbon nanotubes as quantum wires. Nature 386:474–477
148. Zhong Z, Gabor NM, Sharping JE et al (2008) Terahertz time-domain measurement of ballistic electron resonance in a single-walled carbon nanotube. Nat Nanotechnol 3: 201–205
149. Yao Z, Kane C, Dekker C (2000) High-field electrical transport in single-wall carbon nanotubes. Phys Rev Lett 84:2941–2944
150. Frank S, Poncharal P, Wang ZL et al (1998) Carbon nanotube quantum resistors. Science 280:1744–1746

151. Bachtold A, Hadley P, Nakanishi T et al (2001) Logic circuits with carbon nanotube transistors. Science 294:1317–1320
152. Javey A, Guo J, Paulsson M et al (2004) High-field quasiballistic transport in short carbon nanotubes. Phys Rev Lett 92:106804
153. Guo X, Nuckolls C (2009) Functional single-molecule devices based on SWNTs as point contacts. J Mater Chem 19:5470–5473
154. Bruque NA, Ashraf MK, Beran GJO et al (2009) Conductance of a conjugated molecule with carbon nanotube contacts. Phys Rev B Condens Matter Mater Phys 80:155455/1–155455/13
155. Shen X, Sun L, Benassi E et al (2010) Spin filter effect of manganese phthalocyanine contacted with single-walled carbon nanotube electrodes. J Chem Phys 132:054703/1–054703/6
156. Feldman AK, Steigerwald ML, Guo X et al (2008) Molecular electronic devices based on single-walled carbon nanotube electrodes. Acc Chem Res 41:1731–1741
157. Guo X, Small JP, Klare JE et al (2006) Covalently bridging gaps in single-walled carbon nanotubes with conducting molecules. Science 311:356–359
158. Whalley AC, Steigerwald ML, Guo X et al (2007) Reversible switching in molecular electronic devices. J Am Chem Soc 129:12590–12591
159. Wilson NR, Macpherson JV (2009) Carbon nanotube tips for atomic force microscopy. Nat Nanotechnol 4:483–491
160. Tans SJ, Verschueren ARM, Dekker C (1998) Room-temperature transistor based on a single carbon nanotube. Nature 393:49–52
161. Derycke V, Martel R, Appenzeller J et al (2001) Carbon nanotube inter- and intramolecular logic gates. Nano Lett 1:453–456
162. Javey A, Tu R, Farmer DB et al (2005) High performance n-type carbon nanotube field-effect transistors with chemically doped contacts. Nano Lett 5:345–348
163. Kim SM, Jang JH, Kim KK et al (2009) Reduction-controlled viologen in bisolvent as an environmentally stable n-type dopant for carbon nanotubes. J Am Chem Soc 131:327–331
164. Klinke C, Chen J, Afzali A et al (2005) Charge transfer induced polarity switching in carbon nanotube transistors. Nano Lett 5:555–558
165. Zhang Z, Liang X, Wang S et al (2007) Doping-free fabrication of carbon nanotube based ballistic CMOS devices and circuits. Nano Lett 7:3603–3607
166. Ding L, Wang S, Zhang Z et al (2009) Y-contacted high-performance n-type single-walled carbon nanotube field-effect transistors: scaling and comparison with Sc-contacted devices. Nano Lett 9:4209–4214
167. Martel R, Derycke V, Lavoie C et al (2001) Ambipolar electrical transport in semiconducting single-wall carbon nanotubes. Phys Rev Lett 87:256805
168. Xu G, Liu F, Han S et al (2008) Low-frequency noise in top-gated ambipolar carbon nanotube field effect transistors. Appl Phys Lett 92:223114
169. Yu WJ, Kim UJ, Kang BR et al (2009) Adaptive logic circuits with doping-free ambipolar carbon nanotube transistors. Nano Lett 9:1401–1405
170. Bandaru PR, Daraio C, Jin S et al (2005) Novel electrical switching behaviour and logic in carbon nanotube Y-junctions. Nat Mater 4:663–666
171. Kim D, Huang J, Rao BK et al (2006) Pseudo Y-junction single-walled carbon nanotube based ambipolar transistor operating at room temperature. IEEE Trans Nanotechnol 5: 731–736
172. Rosenblatt S, Yaish Y, Park J et al (2002) High performance electrolyte gated carbon nanotube transistors. Nano Lett 2:869–872
173. Allen B, Kichambare P, Star A (2007) Carbon nanotube field-effect-transistor-based biosensors. Adv Mater 19:1439–1451
174. Katsura T, Yamamoto Y, Maehashi K et al (2008) High-performance carbon nanotube field-effect transistors with local electrolyte gates. Jpn J Appl Phys 47:2060–2063
175. Liu S, Shen Q, Cao Y et al (2010) Chemical functionalization of single-walled carbon nanotube field-effect transistors as switches and sensors. Coord Chem Rev 254:1101–1116

176. Zhao Y, Hu L, Grüner G et al (2008) A tunable photosensor. J Am Chem Soc 130:16996–17003
177. Huang SC, Artyukhin AB, Misra N et al (2010) Carbon nanotube transistor controlled by a biological ion pump gate. Nano Lett 10:1812–1816
178. Javey A, Guo J, Wang Q et al (2003) Ballistic carbon nanotube field-effect transistors. Nature 424:654–657
179. Li S, Yu Z, Yen S et al (2004) Carbon nanotube transistor operation at 2.6 GHz. Nano Lett 4:753–756
180. Dürkop T, Getty SA, Cobas E et al (2004) Extraordinary mobility in semiconducting carbon nanotubes. Nano Lett 4:35–39
181. Martin-Fernandez I, Sansa M, Esplandiu MJ et al (2010) Massive manufacture and characterization of single-walled carbon nanotube field effect transistors. Microelectron Eng 87:1554–1556
182. Li D, Müller MB, Gilje S et al (2008) Processable aqueous dispersions of graphene nanosheets. Nat Nanotechnol 3:101–105
183. Novoselov KS, Geim AK, Morozov SV et al (2004) Electric field effect in atomically thin carbon films. Science 306:666–669
184. Chen JH, Jang C, Xiao S et al (2008) Intrinsic and extrinsic performance limits of graphene devices on SiO_2. Nat Nanotechnol 3:206–209
185. Lemme MC (2009) Current status of graphene transistors. Solid State Phenomena 156–158: 499–509
186. Novoselov KS, McCann E, Morozov SV et al (2006) Unconventional quantum Hall effect and Berry's phase of 2pi in bilayer graphene. Nat Phys 2:177–180
187. Ozyilmaz B, Jarillo-Herrero P, Efetov D et al (2007) Electronic transport and quantum hall effect in bipolar graphene p-n-p junctions. Phys Rev Lett 99:166804
188. Dubois SM, Zanolli Z, Declerck X et al (2009) Electronic properties and quantum transport in Graphene-based nanostructures. Eur Phys J B 72:1–24
189. Shibata N, Nomura K (2009) Fractional quantum Hall effects of graphene and its bilayer. J Phys Soc Jpn 78:104708/1–104708/7
190. Darancet P, Wipf N, Berger C et al (2008) Quenching of the quantum Hall effect in multilayered epitaxial graphene: the role of undoped planes. Phys Rev Lett 101:116806
191. Abanin DA, Novoselov KS, Zeitler U et al (2007) Dissipative quantum hall effect in graphene near the Dirac point. Phys Rev Lett 98:196806
192. Novoselov KS, Jiang Z, Zhang Y et al (2007) Room-temperature quantum hall effect in graphene. Science 315:1379
193. Kang YS, Seelaboyina R, Lahiri I et al (2010) Synthesis of graphene and its applications: a review. Crit Rev Solid State Mater Sci 35:52–71
194. Mohiuddin TMG, Zhukov AA, Elias DC et al (2009) Transverse spin transport in graphene. Int J Mod Phys B 23:2641–2646
195. Ponomarenko LA, Yang R, Mohiuddin TM et al (2009) Effect of a high-k environment on charge carrier mobility in graphene. Phys Rev Lett 102:206603/1–206603/4
196. Bolotin KI, Sikes KJ, Hone J et al (2008) Temperature-dependent transport in suspended graphene. Phys Rev Lett 101:096802/1–096802/4
197. Nair RR, Blake P, Grigorenko AN et al (2008) Fine structure constant defines visual transparency of graphene. Science 320:1308
198. Blake P (2008) Graphene-based liquid crystal device. Nano Lett 8:1704–1708
199. Booth TJ, Blake P, Nair RR et al (2008) Macroscopic graphene membranes and their extraordinary stiffness. Nano Lett 8:2442–2446
200. Wu J, Pisula W, Mullen K (2007) Graphenes as potential material for electronics. Chem Rev 107:718–747
201. Allen MJ, Tung VC, Kaner RB (2010) Honeycomb carbon: a review of graphene. Chem Rev 110:132–145

202. Castro Neto AH, Guinea F, Peres NMR et al (2009) The electronic properties of graphene. Rev Mod Phys 81:109–162
203. Geim AK (2009) Graphene: status and prospects. Science 324:1530–1534
204. Rao CNR, Biswas K, Subrahmanyam KS et al (2009) Graphene, the new nanocarbon. J Mater Chem 19:2457
205. Eda G, Lin Y, Miller S et al (2008) Transparent and conducting electrodes for organic electronics from reduced graphene oxide. Appl Phys Lett 92:233305/1–233305/3
206. Novoselov KS, Jiang D, Schedin F et al (2005) Two-dimensional atomic crystals. Proc Natl Acad Sci USA 102:10451–10453
207. Gass MH, Bangert U, Bleloch AL et al (2008) Free-standing graphene at atomic resolution. Nat Nanotechnol 3:676–681
208. Huc V, Bendiab N, Rosman N et al (2008) Large and flat graphene flakes produced by epoxy bonding and reverse exfoliation of highly oriented pyrolytic graphite. Nanotechnology 19:455601
209. Shukla A, Kumar R, Mazher J et al (2009) Graphene made easy: high quality, large-area samples. Solid State Commun 149:718–721
210. Mermin N (1968) Crystalline order in two dimensions. Phys Rev 176:250–254
211. Meyer JC, Geim AK, Katsnelson MI et al (2007) The structure of suspended graphene sheets. Nature 446:60–63
212. Stolyarova E, Rim KT, Ryu S et al (2007) High-resolution scanning tunneling microscopy imaging of mesoscopic graphene sheets on an insulating surface. Proc Natl Acad Sci USA 104:9209–9212
213. Sakhaeepour A (2009) Elastic properties of single-layered graphene sheet. Solid State Commun 149:91–95
214. Lee C, Wei X, Kysar JW et al (2008) Measurement of the elastic properties and intrinsic strength of monolayer graphene. Science 321:385–388
215. Balandin AA, Ghosh S, Bao W et al (2008) Superior thermal conductivity of single-layer graphene. Nano Lett 8:902–907
216. Borysiuk J, Bozek R, Strupinski W et al (2010) Graphene growth on C and Si-face of 4 H-SiC – TEM and AFM studies. Mater Sci Forum 645–648:577–580
217. Berger C, Song Z, Li X et al (2006) Electronic confinement and coherence in patterned epitaxial graphene. Science 312:1191–1196
218. Morozov SV, Novoselov KS, Katsnelson MI et al (2008) Giant intrinsic carrier mobilities in graphene and its bilayer. Phys Rev Lett 100:016602/1–016602/4
219. Yong V, Tour JM (2010) Theoretical efficiency of nanostructured graphene-based photovoltaics. Small 6:313–318
220. Kim KS, Zhao Y, Jang H et al (2009) Large-scale pattern growth of graphene films for stretchable transparent electrodes. Nature 457:706–710
221. Gomez De Arco L, Zhang Y, Schlenker CW et al (2010) Continuous, highly flexible, and transparent graphene films by chemical vapor deposition for organic photovoltaics. ACS Nano 4:2865–2873
222. Geim AK, Novoselov KS (2007) The rise of graphene. Nat Mater 6:183–191
223. Katsnelson MI, Novoselov KS, Geim AK (2006) Chiral tunnelling and the Klein paradox in graphene. Nat Phys 2:620–625
224. Partoens B, Peeters F (2006) From graphene to graphite: electronic structure around the K point. Phys Rev B 74:075404
225. Girit CO, Meyer JC, Erni R et al (2009) Graphene at the edge: stability and dynamics. Science 323:1705–1708
226. Joseph Joly VL, Kiguchi M, Hao Si-Jia et al (2010) Observation of magnetic edge state in graphene nanoribbons. Phys Rev B 81:245428
227. Matte HSSR, Subrahmanyam KS, Rao CNR (2009) Novel magnetic properties of graphene: presence of both ferromagnetic and antiferromagnetic features and other aspects. J Phys Chem C 113:9982–9985

228. Ugeda MM, Brihuega I, Guinea F et al (2010) Missing atom as a source of carbon magnetism. Phys Rev Lett 104:96804
229. Castro EV, Peres NMR, Lopes dos Santos JMB (2008) Localized states at zigzag edges of multilayer graphene and graphite steps. Europhys Lett 84:17001
230. Takai K, Suzuki T, Enoki T et al (2010) Structure and magnetic properties of curved graphene networks and the effects of bromine and potassium adsorption. Phys Rev B 81:205420
231. Enoki T, Takai K (2008) Unconventional electronic and magnetic functions of nanographene-based host-guest systems. Dalton Trans 29:3773–3781
232. Kim WY, Kim KS (2010) Tuning molecular orbitals in molecular electronics and spintronics. Acc Chem Res 43:111–120
233. Enoki T, Kobayashi Y, Fukui K (2007) Electronic structures of graphene edges and nanographene. Int Rev Phys Chem 26:609–645
234. Zhou SY, Gweon GH, Fedorov AV et al (2007) Substrate-induced bandgap opening in epitaxial graphene. Nat Mater 6:770–775
235. Hass J, Varchon F, Millán-Otoya J et al (2008) Why multilayer graphene on 4 H-SiC($000\bar{1}$) behaves like a single sheet of graphene. Phys Rev Lett 100:125504
236. Novoselov KS, Geim AK, Morozov SV et al (2005) Two-dimensional gas of massless Dirac fermions in graphene. Nature 438:197–200
237. McCann E (2006) Asymmetry gap in the electronic band structure of bilayer graphene. Phys Rev B 74:161403
238. Castro EV, Novoselov KS, Morozov SV et al (2007) Biased bilayer graphene: semiconductor with a gap tunable by the electric field effect. Phys Rev Lett 99:216802/1–216802/4
239. Morozov SV, Novoselov KS, Schedin F et al (2005) Two-dimensional electron and hole gases at the surface of graphite. Phys Rev B Condens Matter Mater Phys 72:201401/1–201401/4
240. Wang D, Choi D, Li J et al (2009) Self-assembled TiO_2-graphene hybrid nanostructures for enhanced Li-ion insertion. ACS Nano 3:907–914
241. Paek SM, Yoo E, Honma I (2009) Enhanced cyclic performance and lithium storage capacity of SnO_2/graphene nanoporous electrodes with three-dimensionally delaminated flexible structure. Nano Lett 9:72–75
242. Yu A, Roes I, Davies A et al (2010) Ultrathin, transparent, and flexible graphene films for supercapacitor application. Appl Phys Lett 96:253105
243. Vivekchand SRC, Rout CS, Subrahmanyam KS et al (2008) Graphene-based electrochemical supercapacitors. J Chem Sci 120:9–13
244. Wakabayashi K, Harigaya K (2003) Magnetic structure of nano-graphite Möbius ribbon. J Phys Soc Jpn 72:998–1001
245. Harigaya K, Yamashiro A, Shimoi Y et al (2004) Theoretical study on novel electronic properties in nanographite materials. J Phys Chem Solids 65:123–126
246. Harigaya K, Enoki T (2002) Theory on the mechanisms of novel magnetism in stacked nanographite. Mol Cryst Liq Cryst 386:205–209
247. Harigaya K, Kobayashi Y, Kawatsu N et al (2004) Tuning magnetism and novel electronic wave interference patterns in nanographite materials. Physica E Low Dimens Syst Nanostruct 22:708–711
248. Makarova TL (2004) Magnetic properties of carbon structures. Semiconductors 38:615–638
249. Enoki T, Kawatsu N, Shibayama Y et al (2001) Magnetism of nano-graphite and its assembly. Polyhedron 20:1311–1315
250. Wu J, Becerril HA, Bao Z et al (2008) Organic solar cells with solution-processed graphene transparent electrodes. Appl Phys Lett 92:263302/1–263302/3
251. Wang Y, Chen X, Zhong Y et al (2009) Large area, continuous, few-layered graphene as anodes in organic photovoltaic devices. Appl Phys Lett 95:063302/1–063302/3
252. Kumar A, Zhou C (2010) The race to replace tin-doped indium oxide: which material will win? ACS Nano 4:11–14

253. Matyba P, Yamaguchi H, Eda G et al (2010) Graphene and mobile ions: the key to all-plastic, solution-processed light-emitting devices. ACS Nano 4:637–642
254. Wu J, Agrawal M, Becerril HA et al (2010) Organic light-emitting diodes on solution-processed graphene transparent electrodes. ACS Nano 4:43–48
255. Tung VC, Chen L, Allen MJ et al (2009) Low-temperature solution processing of graphene-carbon nanotube hybrid materials for high-performance transparent conductors. Nano Lett 9:1949–1955
256. Eda G, Chhowalla M (2010) Chemically derived graphene oxide: towards large-area thin-film electronics and optoelectronics. Adv Mater 22:2392–2415
257. Li X, Li C, Zhu H et al (2010) Hybrid thin films of graphene nanowhiskers and amorphous carbon as transparent conductors. Chem Commun 46:3502–3504
258. Wang X, Zhi L, Tsao N et al (2008) Transparent carbon films as electrodes in organic solar cells. Angew Chem Int Ed 47:2990–2992
259. Tang CW, VanSlyke SA (1987) Organic electroluminescent diodes. Appl Phys Lett 51:913
260. Sun T, Wang ZL, Shi ZJ et al (2010) Multilayered graphene used as anode of organic light emitting devices. Appl Phys Lett 96:133301
261. Tongay S, Schumann T, Hebard AF (2009) Graphite based Schottky diodes formed on Si, GaAs, and 4H-SiC substrates. Appl Phys Lett 95:222103
262. O'Regan B, Grätzel M (1991) A low-cost, high-efficiency solar cell based on dye-sensitized colloidal TiO_2 films. Nature 353:737–740
263. Kim SR, Parvez MK, Chhowalla M (2009) UV-reduction of graphene oxide and its application as an interfacial layer to reduce the back-transport reactions in dye-sensitized solar cells. Chem Phys Lett 483:124–127
264. Péchy P, Renouard T, Zakeeruddin SM et al (2001) Engineering of efficient panchromatic sensitizers for nanocrystalline TiO_2-based solar cells. J Am Chem Soc 123:1613–1624
265. Zhu H, Wei J, Wang K et al (2009) Applications of carbon materials in photovoltaic solar cells. Solar Energy Mater Solar Cells 93:1461–1470
266. Kay A (1996) Low cost photovoltaic modules based on dye sensitized nanocrystalline titanium dioxide and carbon powder. Solar Energy Mater Solar Cells 44:99–117
267. Wang X, Zhi L, Muellen K (2008) Transparent, conductive graphene electrodes for dye-sensitized solar cells. Nano Lett 8:323–327
268. Yang N, Zhai J, Wang D et al (2010) Two-dimensional graphene bridges enhanced photoinduced charge transport in dye-sensitized solar cells. ACS Nano 4:887–894
269. Sun S, Gao L, Liu Y (2010) Enhanced dye-sensitized solar cell using graphene-TiO_2 photoanode prepared by heterogeneous coagulation. Appl Phys Lett 96:083113
270. Tang YB, Lee CS, Xu J et al (2010) Incorporation of graphenes in nanostructured TiO_2 films via molecular grafting for dye-sensitized solar cell application. ACS Nano 4:3482–3488
271. Thompson BC, Frechet JMJ (2008) Polymer-fullerene composite solar cells. Angew Chem Int Ed 47:58–77
272. Liu Q, Liu Z, Zhang X et al (2009) Polymer photovoltaic cells based on solution-processable graphene and P3HT. Adv Funct Mater 19:894–904
273. Xu Y, Long G, Huang L et al (2010) Polymer photovoltaic devices with transparent graphene electrodes produced by spin-casting. Carbon 48:3308–3311
274. Yin Z, Wu S, Zhou X et al (2010) Electrochemical deposition of ZnO nanorods on transparent reduced graphene oxide electrodes for hybrid solar cells. Small 6:307–312
275. Guo CX, Yang HB, Sheng ZM et al (2010) Layered graphene/quantum dots for photovoltaic devices. Angew Chem Int Ed 49:3014–3017
276. Liang M, Luo B, Zhi L (2009) Application of graphene and graphene-based materials in clean energy-related devices. Int J Energy Res 33:1161–1170
277. Yin B, Liu Q, Yang L et al (2010) Buffer layer of PEDOT: PSS/graphene composite for polymer solar cells. J Nanosci Nanotechnol 10:1934–1938

278. Liu Z, He D, Wang Y et al (2010) Solution-processable functionalized graphene in donor/acceptor-type organic photovoltaic cells. Solar Energy Mater Solar Cells 94:1196–1200
279. Li SS, Tu KH, Lin CC et al (2010) Solution-processable graphene oxide as an efficient hole transport layer in polymer solar cells. ACS Nano 4:3169–3174
280. Liu Z, Liu Q, Huang Y et al (2008) Organic photovoltaic devices based on a novel acceptor material: graphene. Adv Mater 20:3924–3930
281. Liu Q, Liu Z, Zhang X et al (2008) Organic photovoltaic cells based on an acceptor of soluble graphene. Appl Phys Lett 92:223303
282. Liu Z, He D, Wang Y et al (2010) Graphene doping of P3HT:PCBM photovoltaic devices. Synth Met 160:1036–1039
283. Echtermeyer TJ, Lemme MC, Bolten J et al (2007) Graphene field-effect devices. Eur Phys J Spec Top 148:19–26
284. Williams JR, Dicarlo L, Marcus CM (2007) Quantum Hall effect in a gate-controlled p-n junction of graphene. Science 317:638–641
285. Lemme MC, Echtermeyer TJ, Baus M et al (2007) A graphene field-effect device. IEEE Electron Device Lett 28:282–284
286. Burghard M, Klauk H, Kern K (2009) Carbon-based field-effect transistors for nanoelectronics. Adv Mater 21:2586–2600
287. Cao Y, Steigerwald ML, Nuckolls C et al (2010) Current trends in shrinking the channel length of organic transistors down to the nanoscale. Adv Mater 22:20–32
288. Martin J, Akerman N, Ulbricht G et al (2008) Observation of electron–hole puddles in graphene using a scanning single-electron transistor. Nat Phys 4:144–148
289. Han M, Özyilmaz B, Zhang Y et al (2007) Energy band-gap engineering of graphene nanoribbons. Phys Rev Lett 98:206805
290. Li X, Wang X, Zhang L et al (2008) Chemically derived, ultrasmooth graphene nanoribbon semiconductors. Science 319:1229–1232
291. Zhao P, Chauhan J, Guo J (2009) Computational study of tunneling transistor based on graphene nanoribbon. Nano Lett 9:684–688
292. Zhang Q, Fang T, Xing H et al (2008) Graphene nanoribbon tunnel transistors. IEEE Electron Device Lett 29:1344–1346
293. Muñoz-Rojas F, Fernández-Rossier J, Brey L et al (2008) Performance limits of graphene-ribbon field-effect transistors. Phys Rev B 77:045301
294. Ryzhii V, Ryzhii M, Satou A et al (2008) Current-voltage characteristics of a graphene-nanoribbon field effect transistor. J Appl Phys 103:094510
295. Ryzhii V, Ryzhii M, Satou A et al (2009) Device model for graphene bilayer field-effect transistor. J Appl Phys 105:104510
296. Chen Z, Lin Y, Rooks M et al (2007) Graphene nano-ribbon electronics. Physica E Low Dimens Syst Nanostruct 40:228–232
297. Ponomarenko LA, Schedin F, Katsnelson MI et al (2008) Chaotic dirac billiard in graphene quantum dots. Science 320:356–358
298. Zhang Y, Tang TT, Girit C et al (2009) Direct observation of a widely tunable bandgap in bilayer graphene. Nature 459:820–823
299. Kim S, Nah J, Jo I et al (2009) Realization of a high mobility dual-gated graphene field-effect transistor with Al_2O_3 dielectric. Appl Phys Lett 94:062107
300. Lin Y, Chiu H, Jenkins KA et al (2010) Dual-gate graphene FETs with f_{T} of 50 GHz. IEEE Electron Device Lett 31:68–70
301. Anonymous (2008) Graphene 2.0. Nat Nanotechnol 3:517

Top Curr Chem (2012) 312: 175–212
DOI: 10.1007/128_2011_219

Published online: 12 August 2011

Current Challenges in Organic Photovoltaic Solar Energy Conversion

Cody W. Schlenker and Mark E. Thompson

Abstract Over the last 10 years, significant interest in utilizing conjugated organic molecules for solid-state solar to electric conversion has produced rapid improvement in device efficiencies. Organic photovoltaic (OPV) devices are attractive for their compatibility with low-cost processing techniques and thin-film applicability to flexible and conformal applications. However, many of the processes that lead to power losses in these systems still remain poorly understood, posing a significant challenge for the future efficiency improvements required to make these devices an attractive solar technology. While semiconductor band models have been employed to describe OPV operation, a more appropriate molecular picture of the pertinent processes is beginning to emerge. This chapter presents mechanisms of OPV device operation, based on the bound molecular nature of the involved transient species. With the intention to underscore the importance of considering both thermodynamic *and* kinetic factors, recent progress in elucidating molecular characteristics that dictate photovoltage losses in heterojunction organic photovoltaics is also discussed.

Keywords Charge-transfer state · Organic electronics · Organic solar cells · Photovoltaics · Solar energy

Contents

C.W. Schlenker and M.E. Thompson (✉)
Department of Chemistry and the Center for Energy Nanoscience, University of Southern California, Los Angeles, CA 90089, USA
e-mail: mthompso@usc.edu

1 Introduction to Photovoltaics

1.1 The Global Energy Landscape

Developing a sustainable platform to meet the global demand for clean energy is arguably one of the greatest technological challenges of the twenty-first century [1] and one that will require significant innovation in the chemical sciences. As of 2007, anthropogenic emissions of CO_2 into the atmosphere due to energy consumption had risen to 29.9 billion metric tons per year, with no prior year showing a reduction in emission relative to the preceding year [2]. This release of CO_2, considered to be a radiative climate-forcing factor [3], corresponds to a world primary energy consumption of 502×10^{18} J and an average energy consumption rate of 15.9 TW (1 TW $= 10^{12}$ W). Both China and the United States, each emitting roughly 6 billion metric tons of CO_2 due to energy production, outpace all other countries with regard to energy-related emissions, as shown in Fig. 1a. As depicted in Fig. 1b, fossil fuel sources currently deliver the majority of the world's energy production. With comparable constitution of coal (~21%), petroleum (~37%), and natural gas (~25%) as primary energy sources, in the United States for example, only a small fraction of total energy production is derived from carbon-neutral renewable sources, such as solar photovoltaic (PV) energy conversion. This is illustrated graphically in Fig. 1c. As a result, realizing even the most modest stabilization levels in greenhouse gas emissions proposed by the Intergovernmental Panel on Climate Change (IPCC) in the next 5–15 years requires significant scientific and technological advances in clean energy production and utilization [4]. A major limitation to widespread deployment of solar PV technology to generate clean electricity is the high cost of solar panels. One potentially attractive avenue to bring down the cost of solar PV is to use thin-film organic materials that can be produced and processed cheaply to fabricate lightweight and flexible panels at minimal cost. Such organic solar converters have the advantage of potentially being 100% recyclable and deployable in a range of applications that are intractable for current technologies, due to weight and conformational restrictions. This chapter deals with recent advances and current challenges in solar energy conversion using organic photovoltaics.

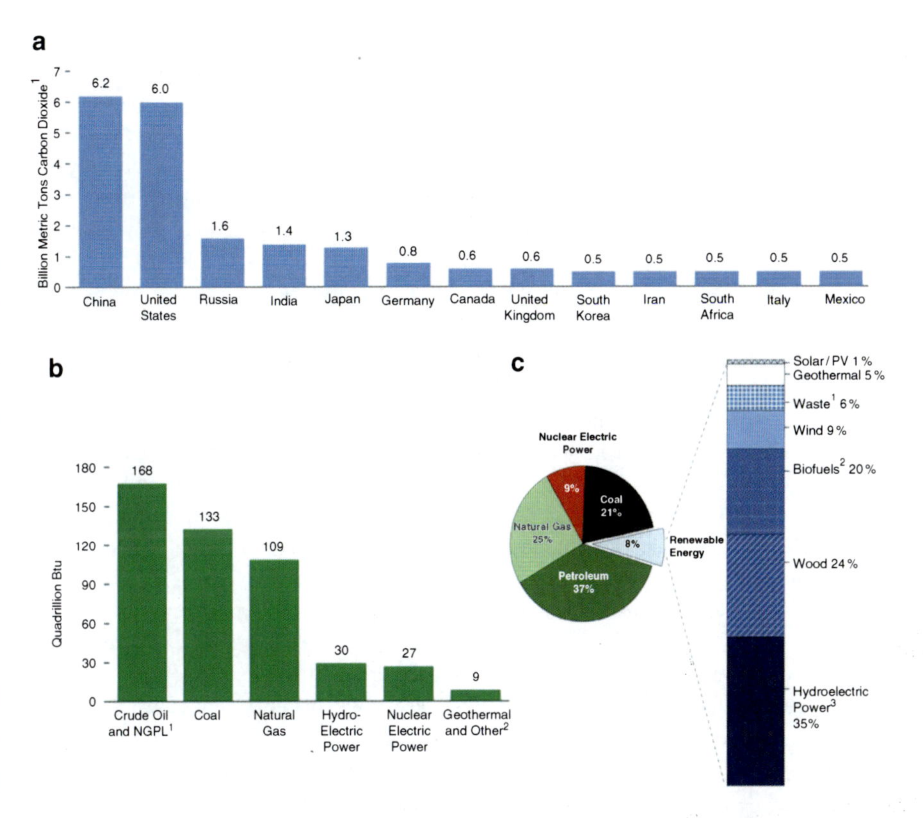

Fig. 1 (**a**) Global 2007 CO_2 emissions by country from energy consumption. (**b**) Global energy production by source in Btu. (**c**) US energy production by energy source. Source: U.S. Energy Information Administration (2009)

1.2 *The Photovoltaic Effect*

The photovoltaic (PV) effect, known since the mid-nineteenth century [5], generally describes the onset of, or change in, an electrical potential occurring upon illumination when two electrodes are separated by a suitable material. The material separating the two electrodes in a PV cell, which is any device exhibiting the photovoltaic effect, may be either a solid or a liquid [6]. Depending on the nature of this photoactive material and the nature of the electrical contacts, one of several different mechanisms [6–11] may be responsible for the separation of opposite polarity charges that constitutes the PV effect. In all cases, the photovoltaic generation of charge arises due to some form of asymmetry within the device, either static (existing both in the presence and absence of illumination) or photoinduced (occurring only under illumination). As a result, PV cells exhibit definite polarity with respect to their electrical terminals, a feature that distinguishes the photovoltaic effect from photoconductivity [12]. When charge separation in the PV cell results, due to electromagnetic radiation within the spectral irradiance of the sun, the device is referred to as a solar cell. Utilizing the voltage supplied by a solar cell, or a modular array of solar cells, to power common electrical appliances is the primary impetus behind the study of photovoltaic solar energy conversion. The capture and storage of solar energy through the selective cleavage and formation of chemical bonds, referred to as artificial photosynthesis [13] or solar fuel production [14], is a closely related topic, with many similar processes that may occur in PV cells.

Differences between various solar cell technologies are characterized most broadly by the nature of the material in the photoactive region. In solid-state solar cells, such a photoactive region may consist of a junction formed in a single material by a process known as doping. Doping entails the controlled (both in concentration and identity) introduction of impurities on opposite sides of the junction. Generally, electron-accepting materials are introduced to form a region of concentrated positive (p) charge carriers on one side of the junction, and electron-donating materials are introduced to form a region of concentrated negative (n) charge carriers on the opposite side of the junction. This type of junction is referred to as a *pn* junction.

Alternatively, a heterojunction may be formed between two different materials, each being either doped or undoped. Further distinctions can be drawn regarding the nature of the bonding between lattice sites, as well as the dielectric properties of the material in the photoactive region. Here two primary classes of solar cell device structures may be defined. The first class incorporates conventional non-carbon-based semiconductor materials, such as crystalline silicon, that possess strong covalent interactions between lattice sites and exhibit relatively high dielectric constants ($\varepsilon_m = 10$–15). The second class incorporates carbon-based materials, such as dye molecules and conjugated polymers, which may be processed at low temperature or from solution. Introducing the current understanding of the mechanism of action for this second class of organic PV devices is the primary focus in this chapter. The information is presented with the assumption that the reader has a basic understanding of chemistry and solid-state physics, insofar as to recognize the existence of molecular orbitals [15] and energy band structures [16].

1.3 The Solar Photovoltaic Industry

While silicon is not the ideal solar cell material, it currently dominates the solar PV market due to its prevalence in the microelectronics industry. Crystalline silicon (c-Si) is an inorganic semiconductor, in which the valence-band maximum and conduction-band minimum are not directly aligned in k-space, making c-Si an indirect bandgap material. The indirect nature of the bandgap in c-Si means that a considerable change in momentum is required for the promotion of an electron from the highest energy state in the valence band to the lowest energy state in the conduction band. As a result, the absorption coefficient for c-Si is relatively low in the wavelength region relevant to solar photon capture ($\lambda = 300$–1,300 nm) and cuts off completely near $\lambda = 1{,}100$ nm. To absorb a significant percentage (~90%) of the incident photons, a 100 μm thick c-Si layer is required, meaning that minority charge carriers must diffuse on the order of 200 μm, in order to be efficiently swept apart at the *pn* junction, and collected at the external contacts. Consequently, efficient c-Si devices require material of high purity and high crystal quality [17]. Despite these dubious characteristics, crystalline silicon in its various forms dominates the solar cell market with roughly 83% market share [11]. This is largely a result of the maturity of other silicon technologies, such as those of the transistor and microelectronics industries, that have developed infrastructure and processing techniques required for producing silicon solar cells with record power-conversion efficiency ($\eta_p = 25.0\%$). Several alternative high-performance inorganic technologies include III–V materials, such as crystalline GaAs ($\eta_p = 26.4\%$) and InP ($\eta_p = 22.1\%$), thin film chalcogenide systems, such as $CuInGaSe_2$ (known as CIGS), ($\eta_p = 19.4\%$) and CdTe ($\eta_p = 16.7\%$), and multi-junction devices, such as GaInP/GaAs/Ge ($\eta_p = 32.0\%$) and thin-film GaAs/CIS ($\eta_p = 25.8\%$) [18]. In total, alternative technologies such as these represent roughly 15% of the global PV market share.

1.4 Organic Solar Cells

While high-efficiency PV solar energy conversion is technologically feasible, the high cost of solar panels utilizing present systems is an impediment to the technology's wide deployment [11]. Economic viability is paramount in establishing a PV platform as a ubiquitous carbon-neutral route to meeting the ever-increasing global energy demand [19–21]. With predicted practical power conversion efficiencies of $\eta_p = 10$–15% [22, 23] and the potential for low-cost processing [24], organic photovoltaic (OPV) devices have recently garnered considerable attention [25–27] as an energy source that may eventually become cost-competitive with fossil fuels. Significant achievements have been seen over the last 20 years in the field of organic electronics, with the demonstrated success in commercializing organic light-emitting diodes [28] and performance advances in organic field-effect transistor devices [29–31] and sensors [32].

Interest in OPVs for solar energy conversion on the part of the scientific community began in earnest only within the last decade. This is illustrated in Fig. 2 by the low percentage (~5%) of reports appearing in peer-reviewed scientific journals pertaining to the topic of "organic solar photovoltaics," relative to all such "solar photovoltaics" reports through the year 1999. From the year 2000 to the present the relative percentage of organic solar photovoltaic reports has increased to greater than 35% of the total. Correspondingly, the OPV community has

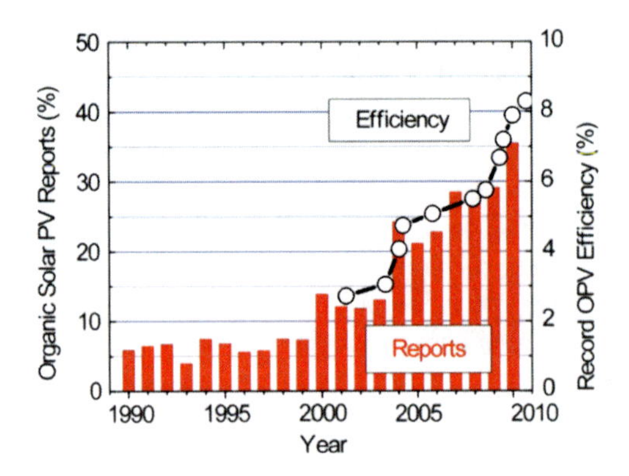

Fig. 2 Number of organic solar PV reports as a percentage of total solar PV reports in scholarly peer-reviewed journals between the years of 1990 and 2010, illustrating the redoubled research effort in organic solar cells beginning ca. 2000. Overlaid on the right axis are academic and industry reports of maximum OPV power conversion efficiencies between 2000 and 2010, as tabulated by the National Renewable Energy Laboratory (NREL; Golden, CO, USA) and the Fraunhofer Institute for Solar Energy Systems (ISE; Freiburg, Germany)

demonstrated several noteworthy milestones over the past decade. For a planar heterojunction architecture device, incorporating copper phthalocyanine (CuPc) and fullerene C_{60}, efficiencies of $\eta_p = 3.6\%$ at 1.5 suns (1 sun = 1 kW/m^2) [33] and $\eta_p = 4.2\%$ at high intensity ~4 suns [34] have been achieved. Incorporating these materials in a multilayer mixed heterojunction architecture [35] has lead to reports of improved efficiency with CuPc/C_{60} to η_p ~ 5% and tandem cells [36, 37] with efficiency of $\eta_p = 5.2$–5.7%. Conductivity doping has recently been identified as a promising route to enhanced efficiencies in both tandem and single junction cells, leading to efficiencies of η_p ~ 4% [38, 39] and higher. Notable recent reports for high-efficiency polymer-fullerene-composite solar cells [18, 40] are in the range of $\eta_p = 7.4$–7.9%. Currently, certified world records above 8% continue [41, 42] to creep toward the balance of systems threshold considered to be practical for many applications.

With these encouraging findings, rudimentary market analysis [43], device degradation studies [27, 44], and high throughput process characterization [24, 45, 46] have been performed for OPV solar cells and modules as well. Solar OPV components are also beginning to find their way into select niche applications, such as backpack-integrated solar charging modules for portable consumer electronics, and solar powered mass transit shelters [47]. While these demonstrations are very promising, OPV solar-energy conversion is presently a relatively immature technology, with device performance lagging behind that of its counterpart technologies, and falling short of the predictions for practical efficiencies. Thus, further development is required to identify promising new materials and suppress loss mechanisms for future high-performance organic solar cells. The following sections present current conceptions and perspectives on these latter topics.

2 Organic Materials and Mechanisms

2.1 Challenges for OPV

The present limit to improved organic solar cell efficiency may lie in balancing robust spectral coverage, leading to high current density, with controlled energy-level offsets, leading to enhanced open-circuit voltage ($V_{oc}, J = 0$) and short-circuit current density (J_{sc}, $V = 0$). There are a considerable number of photons in the near-infrared (NIR) and infrared (IR) regions of the solar spectrum that are currently uncollected in a typical OPV. The J_{sc} could be markedly improved by converting these photons to electrical charges. However, since obtaining a high maximum output power density (P_{max}), corresponding to high power conversion efficiency (η_p), also requires maintaining large V_{oc}, care must be taken when considering low-energy absorbing materials, so that improved photon collection does not come at the expense of electric potential. This chapter is intended to introduce a molecular perspective on the present state-of-the-art for OPV devices. First, characteristics of archetypal materials used in OPV devices are described from a photochemical standpoint, highlighting their excitonic character and electrical behavior. Recent demonstrations of suppressed voltage losses and strategies for controlling exciton dynamics are related to molecular properties. The scientific and technological implications of these findings are discussed. While this chapter is meant to emphasize the benefit that examining a wide array of materials for use in OPV devices can lend to a robust understanding of the involved processes, it is in no way intended as a comprehensive review of the number of materials and device architectures that have been explored. However, there are a number of both molecular and polymeric materials that have been reported for application to OPV. The reader is referred to several exceptional recent reviews for more detailed information regarding materials and OPVs [25, 26, 48–54].

2.2 Excitonic Materials

In conventional semiconductor (Si, GaAs, CdTe, etc.) PV devices, the site of photon absorption intimately shares its valence electrons with its neighbors in a strong covalent network. Therefore, the resulting excited state can be spatially delocalized over many lattice sites. The electronic excited state resulting from photon absorption by the condensed-phase material is termed an exciton, which one may consider as a quasi-particle capable of transporting electronic excitation with no net transport of mass or charge. The energy required to dissociate this exciton into a free hole and electron in inorganic semiconductors is on the order of kT [55–57], the thermal energy available at room temperature. Optical excitation of organic molecules does not directly generate free electron and hole pairs. This is because strong covalent bonding exists only intramolecularly, while the local intermolecular interactions in the condensed-phase organic material are comparatively weak. The excited state in such a system is spatially localized, generally on a single molecule. These excitons typically have energies 0.3–1.0 eV below that of the free electron and hole. As a result, OPV device operation often exhibits features reminiscent of molecular excitation- and charge-transfer reactions, while retaining relatively well-defined vibronic features associated with the isolated molecule.

Dielectric constants in organic semiconductors are commonly lower than their inorganic counterparts by a factor of four. Thus, the inchoate charge separation, induced during photon absorption, is relatively unscreened by the surrounding dielectric. The coulombic attraction between opposing partial charges of the electronically excited molecule results in an energetic stabilization of the exciton, compared to the free electron and radical anion (electron-polaron) or cation (hole-polaron) species. Both of these effects work in concert to produce energetically bound and spatially localized Frenkel-type excitons, with binding energies on the order of 0.5–1.5 eV in organic materials [58]. Thus, room- temperature optical excitation in conjugated organic materials leads to excitons, while generally leading to free charges in an inorganic semiconductor device.

A molecule in its electronically excited state can be a potent oxidizing agent, as well as a potent reducing agent. The efficient photoinduced charge-transfer required for converting solar photons to electrical or chemical energy may be realized at the interface between an excited-state electron donor (D) and a well chosen electron acceptor (A). This interface, termed the D/A heterojunction, is the contact point, where a charge-transfer event between two different materials can take place. The D/A heterojunction in its various configurations is the defining feature central to contemporary organic solar cell devices, as depicted in Fig. 3. The chemical potential energy gradient in organic D/A OPVs drives the photoconversion process [59]. Orbital energies, such as those implied by Fig. 3, are commonly used to estimate the driving force for charge-transfer and other important processes. However, as we will discuss in Sect. 3, the "LUMO" of the donor and the "HOMO" must be replaced by more relevant quantities, in order to understand fully donor/acceptor interactions in organic solar cells.

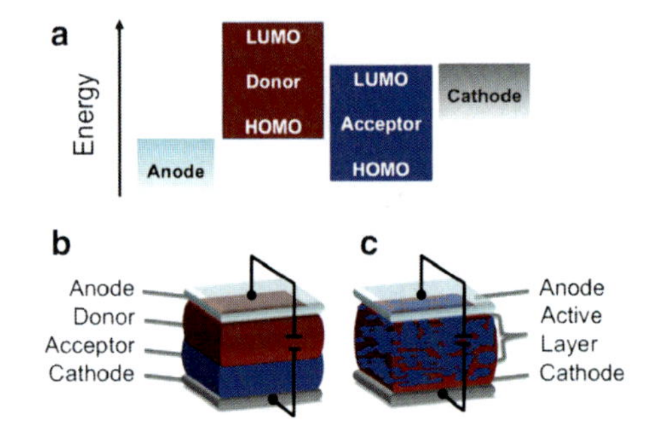

Fig. 3 Contemporary organic solar cell devices are based on donor/acceptor heterojunction device architectures. (**a**) Energy level diagram. (**b**) Planar heterojunction configuration. (**c**) Bulk heterojunction configuration

2.3 Photophysical Processes in Organic Solar Cells

The physical phenomenon of current generation in simple D/A systems can be thought of in terms of the six chemical steps depicted in Fig. 4. (1) The absorption of a photon leads to a localized exciton with energy E_{00} on either the donor ($^{1,3}D^*$) or acceptor ($^{1,3}A^*$), the superscripts denoting the spin multiplicity of the excited state. (2) This exciton diffuses to the donor/acceptor interface via an energy-transfer mechanism (i.e., no net transport of mass or charge occurs). (3) Charge-transfer quenching of the exciton at the D/A interface produces a charge- transfer (CT) state, in the form of a coulombically interacting donor/acceptor complex (D^+A^-). The nomenclature used to describe this species has been relatively imprecise, and has

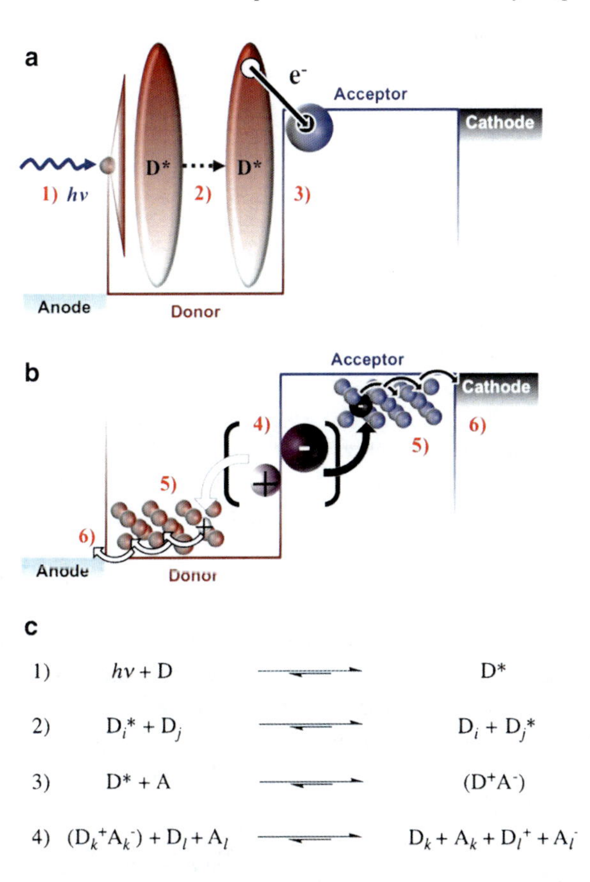

Fig. 4 Schematic illustration of the processes leading to photocurrent generation in organic solar cells. (**a**) Photon absorption in Step 1 leads to excitons that may diffuse in Step 2 to the donor/acceptor (D/A) interface. Quenching of the exciton at the D/A interface in Step 3 leads to formation of the charge-transfer (CT) state. Note that processes analogous to Steps 1–3 may also occur in the acceptor material. (**b**) Charge separation in Step 4 leads to free polarons that are transported through the organic layers and collected at the electrodes in Steps 5 and 6, respectively. (**c**) The equilibria involved in Steps 1–4 strongly influence device efficiency

encompassed terms such as geminate polaron pair, geminate hole-electron pair, charge-transfer exciton, and exciplex. In Fig. 4 the mixing between ionic radical and locally excited modes is expected to determine the degree of charge-transfer character [60–63] and we will use the general term "charge-transfer state" to refer to this species. In Fig. 4 the situation is illustrated for an optically excited donor and an acceptor in its ground state; however, it is important to note that a similar process can take place with an optically excited acceptor and a ground state donor. Utilizing photoexcitation of both the donor and acceptor materials is important to achieving the broadest possible coverage of the solar spectrum. (4) Subsequent spatial separation of the charges making up the CT state proceeds to produce fully ionized $^2D^+$ (hole) and $^2A^-$ (electron) polaron species. (5) Transport of opposite-polarity carriers in the donor and acceptor layers proceeds via self-exchange between localized hopping sites. (6) Electrical contacts facilitate charge collection in the external circuit, by regenerating the neutral ground-state molecular species. The thermodynamic and kinetic factors associated with each of these processes ultimately determine the power-conversion efficiency of the fabricated device. These processes will be explained in more detail in the following sections, as reversible chemical reactions with associated reactants, products, kinetic equilibria, and changes in free energy.

2.3.1 Photon Absorption

Molecules used in OPVs typically possess exceptionally high absorption coefficients (α), compared to conventional inorganic semiconductors. As a result, a relatively thin film of molecules absorbs a substantial fraction of the incident solar photon flux (Φ). Among the most widely studied are the metal phthalocyanines, reported for use in nonlinear optics [64], as sensitizers for photodynamic therapy [65, 66], and for organic thin-film transistors [30, 31]. In Fig. 5 the absorption for a typical organic D/A pair, copper phthalocyanine (CuPc) and fullerene C_{60}, is compared with the common direct and indirect bandgap inorganic semiconductors, gallium arsenide (GaAs) [67] and crystalline silicon (c-Si) [68]. Assuming a two-pass optical path in each case, and calculating the percentage of solar photons captured by 1,000 Å of each of these three active layers, we see that at 630 nm, near the peak of the solar spectral irradiance, a typical organic D/A pair captures roughly 90% of the incident solar photons, while GaAs and c-Si capture only 50% and 10% respectively. Note that for the organic D/A pair, this estimate is rather conservative, as there exist promising molecular materials with substantially higher α, compared with CuPc/C_{60} [69]. Nonetheless, from Fig. 5 one may conclude that OPV active-layer thicknesses on the nanometer length scale are sufficient for efficient photon capture, in contrast to active layer thicknesses of several microns required for c-Si. Thus, with significantly thinner films that absorb the same fraction of light, far less material is necessary for OPV devices, making the energy and charge conduction requirements less stringent.

Since thermal deactivation from high-lying electronically excited states occurs on the picosecond timescale [70, 71] for conjugated molecules, excitation energy

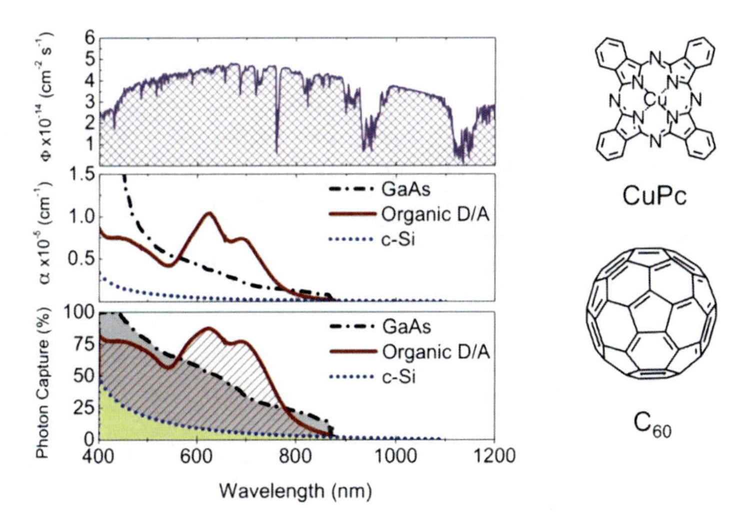

Fig. 5 Solar photon flux Φ (derived from ASTM G173-03 AM1.5 G spectral irradiance) is plotted as a function of wavelength in the *top panel*. The absorption coefficient α for the active region of an archetypal organic donor/acceptor pair, CuPc/C_{60} (*solid red trace*, Organic D/A), is compared in the intermediate panel with that of gallium arsenide (*dashed black trace*, GaAs), a direct bandgap semiconductor, and both are contrasted with crystalline silicon (*dotted blue trace*, c-Si), an indirect bandgap semiconductor. The *bottom panel* illustrates the percentage of solar photons captured by 1,000 Å active layers with α identical to those in the *middle panel*, assuming a two-pass optical path. Chemical structures for CuPc and C_{60} are depicted to the *right*

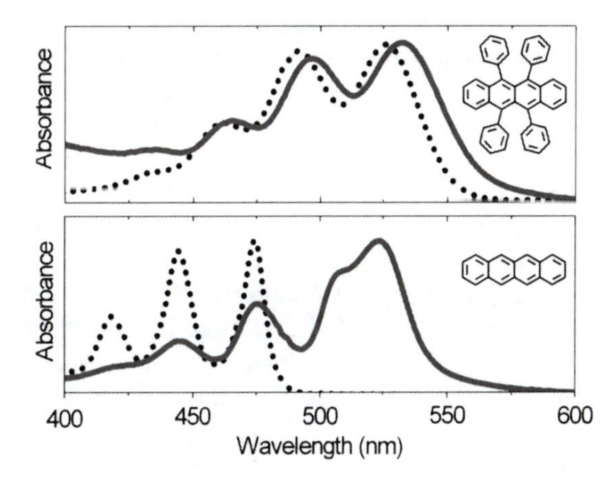

Fig. 6 Absorbance spectra for solution (*dotted trace*) and thin film (*solid trace*) samples of rubrene in the *upper panel* and tetracene in the *lower panel*. Molecular aggregation in the tetracene thin film gives rise to the splitting of its absorption bands. *Inset* are chemical structures for rubrene and tetracene

absorbed in excess of the lowest energy electronic state (E_{00}) is transferred as heat to the surrounding medium, as the site relaxes to the lowest-energy exciton. The spectral bands associated with this excitonic state tend to be broadened and bathochromically shifted, while retaining similar vibronic structure to the molecular species. For example, in Fig. 6, the thin-film absorption spectrum for the oligoacene rubrene is slightly broadened and red-shifted, compared with its solution spectrum.

In certain extreme cases, dimeric or excimeric species, arising from partial charge-transfer character between neighboring chromophores, may lead to distinct low-energy excitonic features that do not occur in the isolated species. Such is the case for the oligoacene tetracene shown in Fig. 6. Strong tetracene–tetracene interactions in the thin film give rise to splitting of the absorption bands, and a red-shift of ca. 100 nm relative to the solution spectrum. For simplicity, however, in the following discussion the excitonic species will be considered as a molecular excitation, with energy E_{00} equal to the energy of the intersection of the excitation and emission spectra of the thin-film sample.

2.3.2 Exciton Diffusion

In order to generate charge efficiently, when a molecule that is spatially remote from the D/A interface absorbs a photon, the energy of that excitation must diffuse to the D/A interface. This energy-transfer process is called exciton diffusion. The length over which excitation can propagate, prior to decay of the exciton population to $1/e$ (roughly 35%) of its initial value, is the exciton diffusion length (L_D). Ideally, the percentage of the solar photon flux absorbed within L_D of the D/A heterojunction will be 100%. In practice, however, this can be as low as 10%, due to short exciton diffusion length in many organic materials. This means that, in a film thick enough to absorb nearly all the incident photons, 90% of the resulting excitons are more than an exciton diffusion length from the D/A interface. In this case, a substantial fraction of the absorbed photons lead to excitons that do not undergo charge-transfer and subsequent charge separation. As a result, researchers have circumnavigated this problem by developing bulk-heterojunction architectures consisting of interdigitating domains of donor and acceptor molecules, with phase-segregation on a nanometer length scale. As illustrated in Fig. 3, such a structure is attractive for ensuring that excitons are formed within L_D of the D/A interface. The photocurrent generated by such bulk-heterojunction devices can be substantial; however, their performance is directly linked to local heterogeneity on the nanometer length-scale. The characterization and control of these features is currently a major limitation to bulk-heterojunction device performance [72, 73].

Understanding exciton diffusion is a current topic of scientific interest. Generally, excitons diffuse by either coulombic coupling, or by electron exchange. The former results from resonant interaction between transition dipole moments of the excited and ground-state molecules, and can occur on length scales considerably larger than the sum of their van der Waals radii. This form of energy migration is known as Förster resonant excitation transfer (FRET) and is the primary mechanism for singlet exciton diffusion in the OPVs. Conceptually, the coulombic interaction between the electric field of a photon and the π-system of a molecule can be directly extended to describe the FRET process. The elementary rate may be expressed as $k_{FRET} = (8.8 \times 10^{-28}/n^4\tau_o r^6)\, jK^2$ mol, where n is the index of refraction of the medium, τ_o is the radiative lifetime of the energy donor, r is the intermolecular separation, j is the spectral overlap integral, and K is an orientation factor. This dipolar interaction is the primary mechanism for singlet excitation transfer in OPV structures. In practice, predicting exciton dynamics using the above expression is complicated by conformational distortion, molecular disorder, and the presence of trap states in the structures used in OPV devices [74–76]. Consequently, there is significant interest in developing accurate methodologies for probing exciton dynamics [75, 77] in OPV structures [49, 78–85].

The other non-trivial mechanism for exciton diffusion arises from electron exchange between a chromophore in its electronically excited state and a ground-state molecule in close proximity. This process, known as Dexter excitation transfer (DET), occurs through direct wave-function overlap and is, therefore, limited to length scales on the order of the van der Waals radii of the two molecules. Triplet excited-state lifetimes are commonly in the microsecond to millisecond regime, and DET is the primary mechanism for triplet exciton diffusion. Although the formal rate dependence for DET is proportional to the spectral overlap integral and attenuates as $k_{\mathrm{DET}} \propto \exp(-2r)$, it contains terms that cannot easily be related to physically measurable quantities. Therefore, in general it is difficult to adequately address even the elementary rate of triplet exciton diffusion. The utility of triplet exciton diffusion in OPVs is a current area of interest [86–88].

2.3.3 Charge Transfer

Following exciton diffusion to the D/A interface, efficient charge-transfer quenching of the excited state must occur in order to initiate charge generation. The driving force for charge-transfer may be understood by considering the molecular orbital mixing of D* and A, as the two species begin to interact. According to perturbation theory, as D* encounters A, the electrophilic singly occupied orbital of D*, corresponding to the donor HOMO, will interact with the acceptor HOMO, while the nucleophilic singly-occupied orbital of D* corresponding to the donor LUMO will interact with the acceptor LUMO [89]. As shown in Fig. 7, the four resulting new MOs are split in energy, relative to the original MOs of either isolated species. One of the new HOMO-type levels is lower in energy than in either isolated species, and one of the new HOMO-type orbitals is higher in energy than in either of the isolated species. The new energy levels of the complex corresponding to the LUMO orbitals are similarly split above and below the isolated LUMOs.

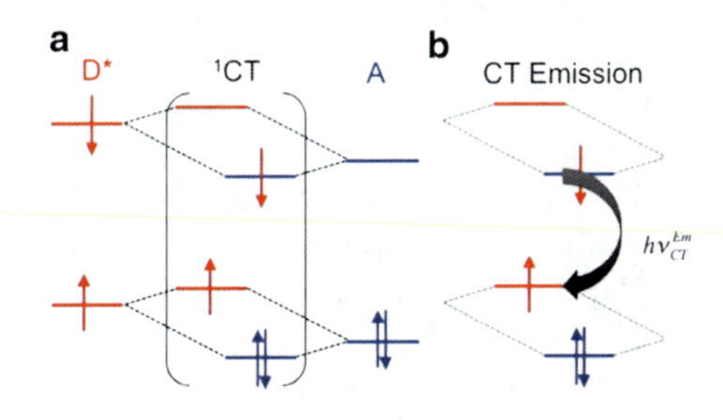

Fig. 7 (**a**) Molecular orbital (MO) description for the charge-transfer state formation in organic donor/acceptor systems. (**b**) Description for CT state emission energy $hv_{\mathrm{CT}}^{\mathrm{EM}}$ using exciplex MOs in (**a**)

As the electrons undergo redistribution from their initial non-interacting orbitals, three electrons are stabilized, while only one of the electrons is destabilized upon forming the charge-transfer state. This electronic stabilization is the driving force for populating the charge-transfer state relative to the initial non-interacting D* + A state.

In principle, the thermodynamic requirements for charge-transfer are straightforward. The energy of an exciton interacting with a charge acceptor must be greater than the CT-state energy. In practice, however, assigning enthalpic quantities associated with the lowest unoccupied molecular orbital (LUMO) in condensed phase organic materials is often complicated by sample degradation [90, 91]. Additionally, the physics of semiconductor devices has historically been discussed in terms of band models for covalent crystalline materials, where the free-electron approximation is appropriate, which is not the case for most organic systems. Terms like conduction band, LUMO, electron affinity, transport level, and optical LUMO are often inappropriately taken to be interchangeable. Significant attention [92–104] has been given to developing a physically relevant energy-level description for organic electronic materials, where it is more appropriate to discuss weakly interacting molecules in condensed-phase low-dielectric media.

Of particular importance to formation of the CT state is the energy of the excited state. When a molecule at the D/A interface becomes electronically excited by FRET, DET, or photon absorption, it may take part in a charge-transfer reaction with the adjacent charge-acceptor molecule. The thermodynamic requirements to form the CT state species (D^+A^-) at the D/A interface can be understood by considering the change in enthalpy for the photoinduced charge-transfer reaction $^{1,3}D^* + A \rightarrow (D^+A^-)$. The ionization energy of the excitonic state (E_i^*), sometimes referred to as the "optical LUMO," is an extremely useful quantity for assessing whether forward electron transfer will be exothermic. Succinctly, E_i^* is the oxidation potential of the excited state $[D^* \xrightarrow{E_i^*} D^+ + e^-]$, which can be used to determine which acceptors will engage in efficient charge-transfer with D*. An upper limit for E_i^* may be estimated from the combination of ultraviolet photoelectron spectroscopy (UPS) and UV-vis absorption/emission spectroscopy by numerically subtracting $-E_{00}$ from the ionization energy of the neutral ground state species, $E_i^* = E_i - (-E_{00})$, where E_{00} is given by the spectral intersection of the thin-film absorption and emission bands for fluorescent materials. For materials with high triplet-exciton yields, a value for E_{00} may be estimated from the high-energy onset of the phosphorescence band. Determining values for E_{00} in materials that relax to non-radiative states can be complicated, since information about the energy of the exciton cannot be obtained from their emission bands. In materials where emission is absent, an upper limit for E_{00} can be taken as the low-energy onset of the neat-film absorption.

If E_i^* is greater than the electron affinity (E_a) of the accepter, then the photoinduced forward electron-transfer reaction will be exothermic. This approach can be used to determine the E_i^* for both singlet and triplet excitons from the corresponding singlet and triplet state E_{00} values. A similar calculation of an acceptor exciton's electron affinity (E_a^*) may be made, to assess whether acceptor exciton-dissociation

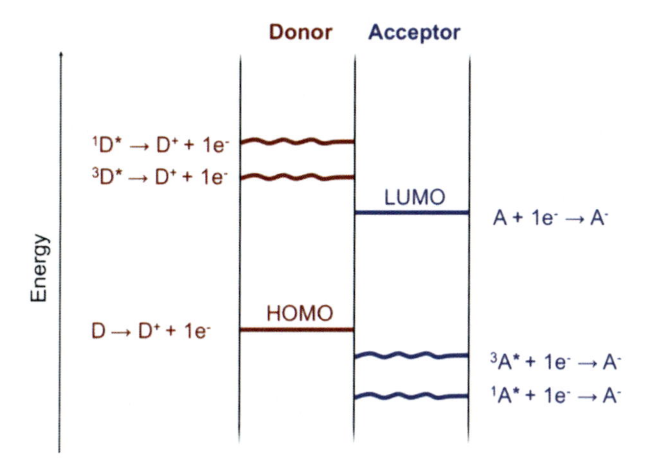

Fig. 8 Schematic illustration of donor/acceptor energies relevant for charge-transfer in organic solar cells. *Straight lines* represent ground state binding energies, while *wavy lines* represent excited state binding energies

will be exothermic. Thus, one can also define an "optical HOMO" for OPV acceptors from the electron affinity and E_{00}, which can be used to assess which donors will be appropriate for dissociation of an exciton in a given acceptor (Fig. 8).

Although a cogent kinetic description for the overall charge-generation process is still developing, during exciton dissociation some generalized charge-transfer (CT) state [92, 105] is thought to form (see above). Very little screening of charge takes place, since the dielectric constant, $\varepsilon_m \approx 3$, of the surrounding medium is relatively low, and the attractive electrostatic potential energy $E_c = q^2/4\pi\varepsilon_0\varepsilon_m r$ between (D^+A^-) pairs is on the order of 25 kJ mol^{-1} for an average polaron spacing of $r = 20$ Å. This means that the coulombic stabilization is roughly ten times greater than kT at room temperature, and, if not efficiently surmounted, may severely limit the output power of OPV devices by markedly reducing the fill factor [106, 107]. Understanding the process of separating the charges generated during charge-transfer state formation is a current topic of OPV research and is discussed in the next section.

2.3.4 Charge Separation

Once formed, the charge-transfer state may decay via one of two general reversible pathways, excluding photochemical degradation. The first pathway results in the regeneration of some neutral species (excited-state or ground-state). This process, referred to as recombination, because it involves combining conjugate charges, can be a major loss mechanism. The second is the preferred pathway for CT state decay, in which a series of electron self-exchange reactions occur between the partial

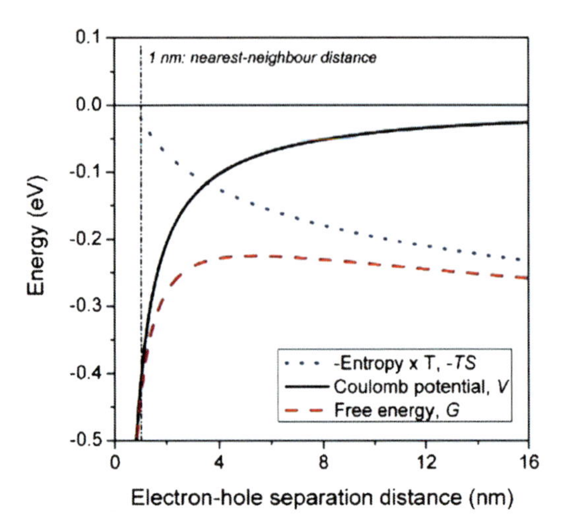

Fig. 9 Illustration highlighting the possible importance of entropy in the charge separation process. Positive and negative polarons experience a strong coulombic attraction that may be offset by entropy to achieve charge separation. Reproduced with permission from [54]. Copyright 2010 American Chemical Society

positive (negative) charge of the CT complex and the surrounding neutral ground state donor (acceptor) molecules. A comprehensive picture for the charge separation process (as well as the exciton dissociation process) has yet to materialize. However, it has been noted by Clarke and Durrant [54], as shown in Fig. 9, that entropy may play an important role, since the change in free energy ($\Delta G = \Delta H - T\Delta S$) must of course be $\Delta G < 0$ for charge separation to occur spontaneously. In the case of the relaxed CT_1 state, the change in enthalpy (ΔH) for charge separation will be positive. Therefore, at a given temperature (T) the change in entropy is necessarily $\Delta S > 0$. In this case, charge separation is driven by an entropic effect. In general, the relative importance of ΔS will depend on the relaxation rate for $CT_n \rightarrow CT_1$, relative to any other competing process.

A major experimental challenge in this area is the development of ultrafast two-dimensional optical spectroscopies with interfacial specificity to characterize reliably the electronic coupling and CT state dynamics for a broad range of materials and architectures [108–111]. Recently, there has been substantial interest in more precisely probing the time scale of charge separation for donor/acceptor OPV materials [112–117]. Of specific concern has been the role of higher-lying (hot) CT_n states [118] that may enable sub-picosecond free-polaron generation with no resolvable CT state formation, compared with bound CT state relaxation, followed by charge separation as a two-step process [112]. The primary tool in such studies is transient absorption spectroscopy [113]; however, there are notable difficulties in uniquely distinguishing the low-energy optical signatures, as they are generally broad and unstructured, with significant spectral overlap between polaron species [119]. In many respects the work being taken up in this area parallels the observed dependence on the dielectric constant of the solvent [120] whether solvent-separated or contact ion-pairs are formed in donor/acceptor systems in solution [61, 121] as developed by Weller [122].

2.3.5 Charge Transport

Following charge separation, weak electronic coupling in OPV materials leads to localized radicals surrounded by polarized neutral molecules. Transport of this polaron species generally proceeds via localized hopping. Ideally, the hopping rate may be expressed [123, 124] in terms of Marcus theory, to account for electronic coupling and for the internal reorganization energy required for the neutral and charged species to reach the same geometry. Inhomogeneity often leads to carrier trapping in amorphous OPV materials, decreasing the charge-transport rates below those given by Marcus theory. Thus, the use of crystalline materials in OPVs is being explored, which would eliminate carrier traps of this type, and improve OPV performance [101, 123, 125, 126].

2.3.6 Charge Collection

The nature of the electrical contacts employed in device preparation is largely determined by interface states between the electrode and the organic materials, and can strongly impact device performance. Properties of this organic/electrode interface are an area of substantial interest, since the density of states profile can be quite complex, due to interface chemical and polarization effects [93–95, 127]. A general theoretical treatment for the surface recombination velocity at the interface of an amorphous organic semiconductor and a metal electrode has been developed by Scott and Malliaras [128]. This treatment implies that one may expect a finite surface recombination velocity for the extraction of photogenerated charges. While indium tin oxide (ITO) is currently the dominant transparent electrode technology for D/A OPV devices, its ultimate utility is questionable. Several promising alternatives to ITO are currently being investigated, such as nanowire or nanotube arrays fabricated from various materials [129–134]. Graphene based transparent electrodes for optoelectronic applications are also noteworthy [135–138].

3 Electrical Response

The preceding sections described molecular interactions important in organic solar cells. This section discusses the impact of those interactions on the overall device behavior. Simulated electrical behavior for a typical solar cell is illustrated in Fig. 10. Under forward bias voltages $0 < V < V_{oc}$, a typical photovoltaic device under illumination supplies power ($P = J \times V$) to the external circuit (cf. lower panel of Fig. 10, dashed trace in first quadrant). The formalism used here implies that, under reverse bias, the organic material is reduced at the anode and oxidized at the cathode, while, under forward bias, the organic material is oxidized at the anode and reduced at the cathode. The short circuit current, J_{sc}, is approximately equal to

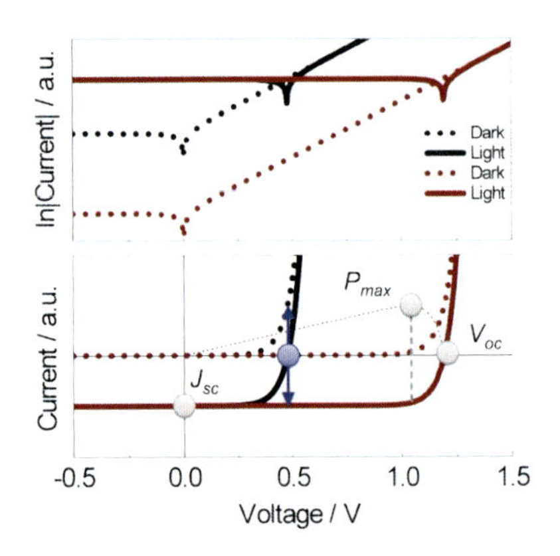

Fig. 10 Simulated solar cell electrical behavior in the *dark* (*dotted traces*) and under illumination (*solid traces*) comparing the effect of the saturation current parameter J_s on V_{oc}. The *black traces* represent a device with $J_s \times 10^6$ that of the device represented by the *red traces*. The sharp inflection points in the semilog plots (*upper panel*) are the points where the current switches from positive to negative. Also illustrated in the linear representation (*lower panel*) are the short circuit current density, J_{sc}, and the maximum output power, P_{max}, given by the product of current and voltage. The *blue arrows* represent the point at which the dark current and the current under illumination are equal in magnitude. The corresponding potential marked in *blue* on the voltage axis is V_{oc} for the *black trace*

the photocurrent and open circuit voltage, V_{oc}, is the point at which a sufficient bias has been applied to "shut off" the OPV. It is easy to see in this plot that the V_{oc} occurs at the point where the magnitude of the dark current in forward bias matches the photocurrent (see the blue arrows in Fig. 10). At this point the rate of injection of charge is equal to the rate of its photogeneration, and the device is under a steady-state condition, with $J = 0$. The point of maximum output power density is denoted by P_{max} in Fig. 10, corresponding to the maximum product of current and voltage. A useful quantity for assessing P_{max} relative to the photocurrent and photovoltage is the fill factor, given by $FF = P_{max}/J_{sc} \times V_{oc}$. Typical FF values for OPVs range from 0.3 to 0.7. Finally, the power conversion efficiency (η_P) is calculated as the numerical quotient of P_{max} and the total integrated spectral irradiance (P_o), giving $\eta_P = P_{max}/P_o$. Note that precise standards of measurement and calibration have been developed, including spectral mismatch correction, for accurately reporting η_P. The active researcher is admonished to adhere to the standards outlined in the literature [139–142] whenever reporting OPV performance metrics.

3.1 Photocurrent

Photocurrent generation, involving the CT state, has often been treated [106, 107, 143–145] according to the Onsager–Braun model [146, 147], to describe the field dependence for charge separation and charge recombination from the CT state. The original model proposed by Onsager predicts that the probability for a charged particle undergoing Brownian motion, to escape the Coulomb potential of its conjugate charge, will depend on the initial charge separation and on the applied electric field. This model was later extended by Braun to account for the finite lifetime of the charge-transfer state. As such, the probability for charge separation from the CT_1 state is taken as $P(E) = k_d(E)/[k_d(E) + k_f(E)] = k_d(E)\tau(E)$, where $k_d(E)$ is the rate constant for CT_1 separation, k_f is the rate constant for charge recombination, and τ is the field-dependent charge-transfer state lifetime. A salient feature of the Onsager–Braun formulation is that the dissociation rate $k_d(E)$ is a linear function of the (spatially averaged) charge-carrier mobility. Several modifications to the Onsager–Braun model have been proposed to improve its validity in describing the field dependent photocurrent in organic solar cells. For example the impact of carrier diffusion was implemented by Mihailetchi et al. [148] for PPV: PCBM based on the Sokel–Hughes [149] model and this has been extended to the P3HT:PCBM system by Limpinsel et al. [150], as illustrated in Fig. 11. Several other authors have suggested improvements to the Onsager–Braun description, such as to account for its breakdown in dealing with species of high charge mobility [151, 152]. The model has also been modified to include a more appropriate description of charge separation and recombination kinetics [153, 154].

In practice, poor charge mobility, energetic disorder, carrier trapping, and physical aberrations complicate device characterization. The effects of these non-idealities are often modeled according to an equivalent circuit shown in Fig. 12. Incorporating all specific series resistive elements as R_s, and all specific parallel resistances as R_p, one obtains the expression

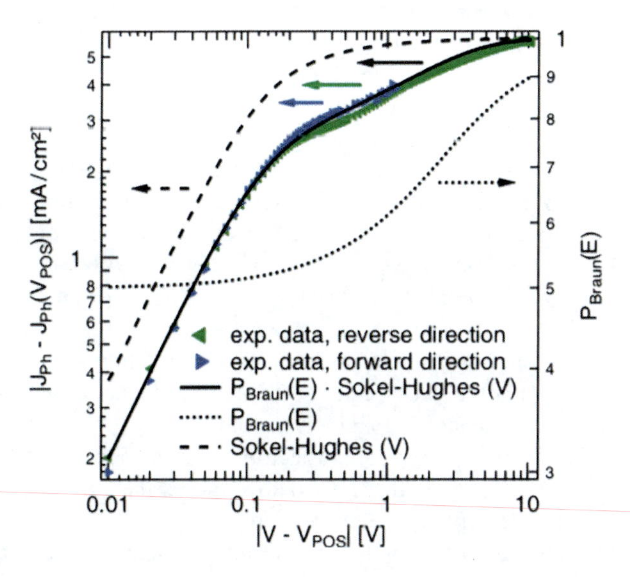

Fig. 11 Adapted Onsager–Braun model for photocurrent analysis applied to P3HT:PCBM. Reprinted figure with permission from [150]. Copyright 2010 by the American Physical Society

Fig. 12 Single diode equivalent circuit model commonly employed in estimating solar cell losses

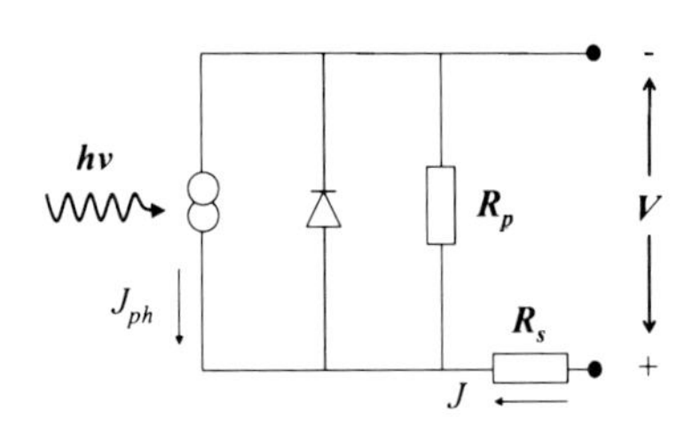

$$J = J_\mathrm{s}\left\{\exp\left(q\frac{V - JR_\mathrm{s}}{nkT}\right) - 1\right\} + \frac{V - JR_\mathrm{s}}{R_\mathrm{p}} - J_\mathrm{ph}, \tag{1}$$

where q, k, T, and J_ph are elementary charge, Boltzman's constant, temperature, and photocurrent density, respectively [106, 107, 155–159]. Ideality factors of $n \approx 2$ are common for organic devices and are associated with current–voltage relationships dominated by recombination [16]. Developing a physically relevant interpretation for the commonly observed phenomenological J–V behavior [34, 49, 160], represented by (1), is an active area of OPV device research [106, 107, 155, 156, 161–163]. Giebink et al. [106, 107] proposed that the current–voltage characteristics in donor/fullerene small-molecule OPVs can be described based on the kinetics of CT state separation and trap-limited recombination at the D/A interface. This description more accurately reproduces low-temperature and low-bias characteristics, where the fidelity of (1) is often poor.

In the dark, the device behaves as a diode, exhibiting saturation current density (J_s), flowing with limited voltage dependence under reverse bias and an exponential rise (rectification) in current density in forward bias. In a hypothetical device with no resistive losses, the current in the dark follows $J = J_\mathrm{s}\exp[(qV/nkT) - 1]$. While an oversimplification, this expression highlights that J_s relates the current density under reverse bias to the current density under forward bias. For the latter, Fig. 13 represents the processes that determine the magnitude of J_s in the recombination controlled regime. In general, the magnitude of J_s will depend on the rate (k_inj) of carrier injection from the electrodes, the charge mobility of the materials, and the rate of net charge recombination (k_rec) at the donor/acceptor interface, i.e., $(\mathrm{D^+A^-}) \rightarrow (\mathrm{D + A})$ or $(\mathrm{D^* + A})$ or $(\mathrm{D + A^*})$. As we will see, the saturation current can be a useful concept for discussing the magnitude of the open circuit voltage.

Under illumination, the charge-transfer state concentration at the D/A interface is increased, due to photoinduced electron transfer. A reducing potential, applied to the anode, facilitates charge separation and hole collection by regenerating $\mathrm{D^+}$ polarons at the anode/donor interface. Electrons are analogously collected at the cathode. Switching polarity at the electrodes opposes the charge-separation and collection processes. In Fig. 10, the expression $J = J_\mathrm{s}\exp[(qV/nkT) - 1] - J_\mathrm{ph}$ is plotted to illustrate this behavior for two different values of J_s. For the simple expression in Fig. 10, with no applied potential, $J = J_\mathrm{sc} = -J_\mathrm{ph}$. At open circuit no current flows through the device, and the photovoltage is $V_\mathrm{oc} = q^{-1}\,nkT$

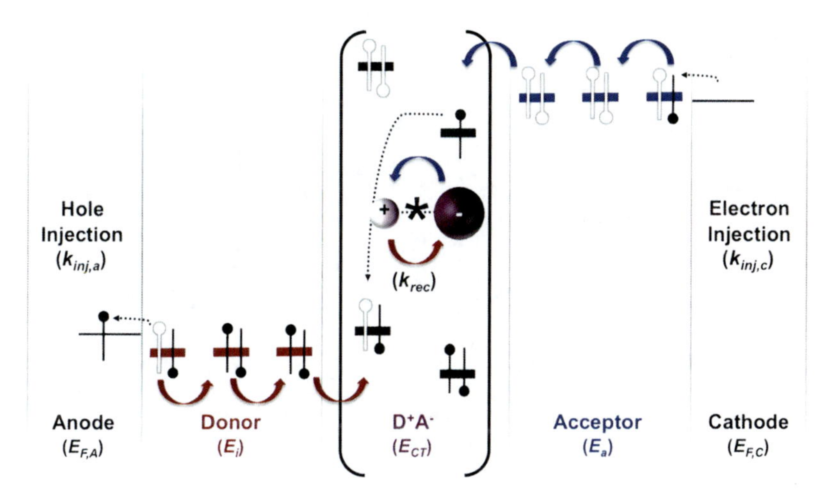

Fig. 13 Recombination losses occurring under forward bias in a typical OPV device. Holes injected from the anode Fermi level ($E_{F,A}$) into the HOMO level (E_i) of the donor and electrons injected from the cathode Fermi level ($E_{F,C}$) into the LUMO level (E_a) of the acceptor are transported to the D/A interface. Coulombic attraction between holes and electrons yields the (D^+A^-) CT state with energy E_{CT}. Charge recombination reaction (D^+A^-) → D + A occurs with rate constant k_{rec}

ln (J_{sc}/J_s + 1). For $n \approx 2$, the magnitude of the photovoltage depends on a ratio of rates, where J_{sc} represents how frequently photogenerated charges are collected, and J_s represents how frequently charges recombine. As shown in Fig. 10, when all other parameters are invariant, small J_s (i.e., slow recombination) leads to large V_{oc}. Since $P_{max} = J_{sc} \times V_{oc}/FF$, the saturation current directly influences device performance. Thus, a key to attaining high V_{oc} is achieving a low J_s, which requires that the recombination rate be very low.

3.2 *Photovoltage*

The photovoltage for inorganic photodiodes is directly limited by the "built-in" electrical potential (V_{bi}) across the *pn* junction, as illustrated in Fig. 14a [164–166]. However, the most general description for the open-circuit voltage limit in any photovoltaic device has been addressed by Gregg [59]. According to the free energy of the system, the obtainable photovoltage will depend on both the electrical and chemical potential energy gradients in the device. Since the equilibrium charge-carrier concentrations and carrier mobilities in most heavily-doped inorganic PV devices are extremely high, compared with organic molecules, the effect of the chemical potential in classic silicon-based cells is negligible.

In organic solar cells, the chemical potential must be considered in addition to the electrical potential. For example, the magnitude and polarity of the photovoltage produced by the first modern donor/acceptor OPV device [9] was noted to

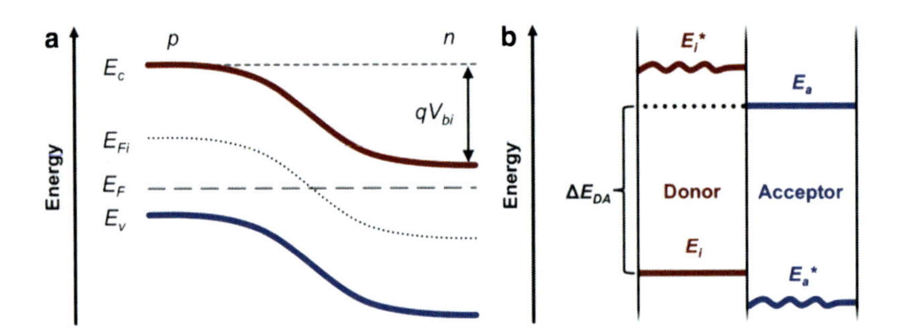

Fig. 14 (**a**) Equilibrium energy diagram for a *pn* junction in an inorganic semiconductor material with intrinsic Fermi energy E_{Fi}, conduction band energy E_{c}, valence band energy E_{v}. The quantity V_{bi} represents the total built-in electrical potential due to band bending. (**b**) Energy diagram illustrating ionization energies for an organic donor in both the ground (E_{i}) and excited $\left(E_{\mathrm{i}}^{*}\right)$ states, the acceptor electron affinities in the ground (E_{a}) and excited $\left(E_{\mathrm{a}}^{*}\right)$ states, and the energy level offset ΔE_{DA}

depend only weakly on the built-in potential across the device. Although estimates for V_{bi} have been made by measuring the open-circuit voltage produced from composites of various conjugated polymers [167] and derivatized fullerenes [168], the relationship between V_{bi} and V_{oc} can be complicated, even for organic single-layer devices [169]. In fact, the photovoltaic effect has been observed for organic molecular systems with no intrinsic electrical asymmetry [170–172]; hence $V_{\mathrm{bi}} = 0$. The offset (ΔE_{DA}), as depicted in Fig. 14, between the ionization energy (E_{i}) associated with the donor HOMO level and the electron affinity (E_{a}) associated with the acceptor LUMO level, appears to strongly influence the V_{oc} [22, 167, 168, 173–178]. In other words, the energy required to remove an electron from the HOMO of the donor and place it in the LUMO of the acceptor appears to be directly related to V_{oc}. However, since both electrical *and* chemical potential energy gradients affect OPVs, V_{oc} losses are also controlled by the interfacial kinetics [170] for depopulation of the CT state. In the following sections we discuss the population of the charge-transfer state and its relationship to the open-circuit voltage.

3.3 Charge-Transfer State Population

The MO description of the CT state illustrated in Fig. 7a not only describes the energetic driving force for charge-transfer to occur, but also helps to characterize important processes that this transient state may undergo. The charge-transfer complex illustrated in Fig. 7 should be regarded generally as possessing its own set of higher-lying electronic excited states and associated electronic transitions between states and their respective vibrational manifolds. For example, dissociation of the charge-transfer complex from the CT_1 electronic state via radiative decay, giving off a photon of energy $h\nu_{\mathrm{CT}}^{\mathrm{Em}}$, has been observed for a large number of D/A systems [60, 106, 179–186]. One may consider this process to be synonymous with exciplex emission [122, 187] for cases when the ground state is dissociative. Radiationless decay pathways from the CT state to the ground state are also possible, and have been implicated as a major efficiency-loss mechanism in some devices [188]. Such radiationless decay can occur through direct thermal relaxation or via singlet-triplet intersystem crossing [189]. It is also possible to imagine a D/A system in which the CT_1 state is close enough in energy to the D* + A state to allow appreciable dynamic equilibrium between population of the $^{1,3}D^* + A$ and that of (D^+A^-), allowing for the efficient formation of singlet and triplet excitons from the CT state. This is the operational mechanism for an organic light emitting diode (OLED).

To suppress open-circuit voltage losses in OPVs it is highly desirable to arrest all parasitic non-radiative CT deactivation pathways. That is, given two D/A systems exhibiting CT bands with similar oscillator strength and energy, the system with the higher rates of non-radiative CT decay will undergo faster charge recombination. Recombination via thermally activated non-radiative CT state decay is known to follow the standard Marcus treatment. Temperature-independent non-radiative components have been treated, in some cases, as radiationless quantum transitions for exciplex decay [190].

The general implication of the Onsager–Braun mechanism for OPV devices is that the efficiency of the charge separation process, and therefore the efficiency of the device, will be linked to the lifetime of the charge-transfer state. Processes that act to depopulate the CT_1 state more rapidly than charge separation proceeds will deleteriously impact device performance. In some cases the charge-transfer state properties can be probed using steady-state optical signatures. Donor/acceptor combinations exhibiting direct photon absorption to, or emission from, the charge-transfer state are of particular interest in defining the CT_1 energy and identifying potential losses. Low-energy absorption [191], emission [192], and electroluminescence [185], observed experimentally, may be visualized schematically, as illustrated by the simplified MO picture in Fig. 7b. An excited electron, populating the singly occupied orbital with largely acceptor LUMO character, relaxes to fill the singly occupied orbital with largely donor HOMO character, as excess optical energy is emitted by the charge-transfer complex. For example, several reports have suggested that low-energy bands observed in the luminescence spectra for blended polymer/fullerene composites, as shown in Fig. 15, may be ascribed to emission from the charge-transfer state formed between the polymer donor and the fullerene acceptor [179–184]. Similar effects have also been observed in small-molecule donor/acceptor systems [106, 185, 186].

Processes contributing to CT state deactivation are important since, in many cases, for efficient photocurrent generation, charge separation must kinetically compete with alternate deactivation pathways. As we have already mentioned, there are two general loss pathways, radiative and non-radiative CT deactivation. The probability for transient charge-transfer states in polymer:fullerene blends to undergo charge separation, radiative decay, or non-radiative deactivation has been analyzed by Vandewal et al. [188], according to the principle of detailed balance [164, 193]. The relative photovoltaic efficiency losses can be determined using a

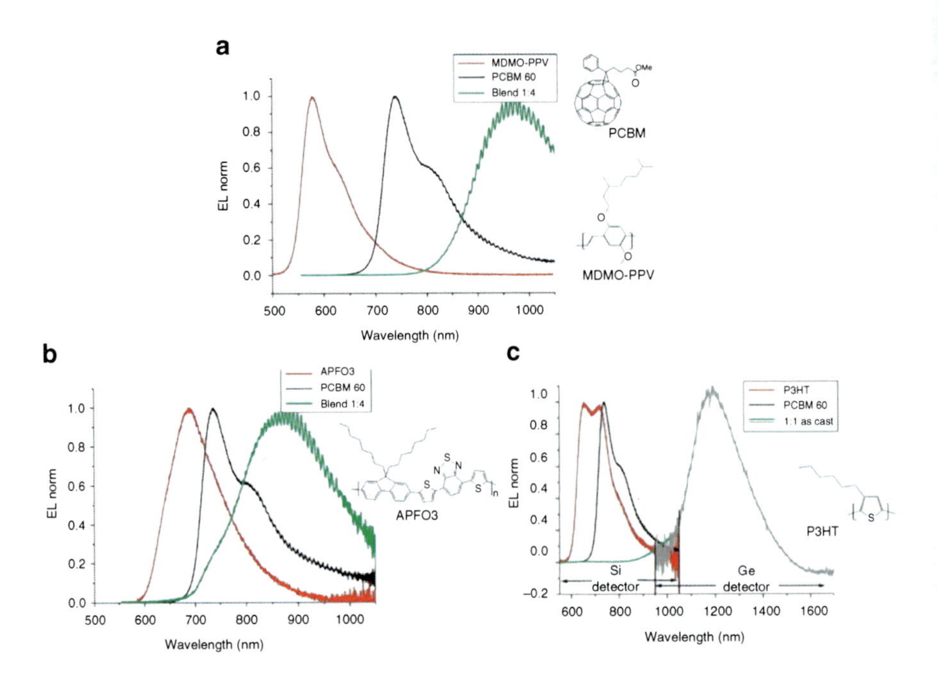

Fig. 15 Charge-transfer state electroluminescence (EL) for several polymer: fullerene blends used in donor/acceptor organic solar cells. Adapted with permission from [184]. Copyright 2009 American Chemical Society

combination of Fourier-transform photocurrent spectroscopy (FTPS) [194] to measure the photovoltaic external quantum-efficiency (EQE_{PV}) and electroluminescence (EL) measurements. This method affords an accurate estimate of the CT state energy, as illustrated by the intersection of the EQE_{PV} and EL curves for MDMO-PPV:PCBM blend in Fig. 16. This detailed-balance approach, based simply on photovoltaic and electroluminescence quantum efficiencies, has been used to accurately predict [195] the photovoltage produced by several polymer:fullerene blends, using a range of process conditions. The agreement between the calculated V_{oc} and the measured V_{oc} is illustrated in Fig. 17, and the structures for PCPDTBT and LBPP5 polymers are shown below.

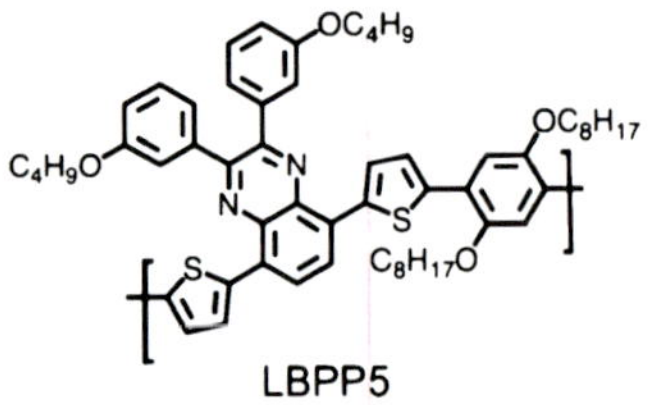

LBPP5 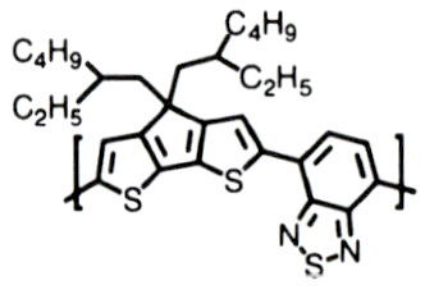 PCPDTBT

In accord with the work of Rau [193], the EQE_{PV} and EL data collected over a given range of temperatures and PV illumination intensities for PCBM blends with MDMO-PPV, P3HT, and APFO3 donor polymers were then used, for each blend,

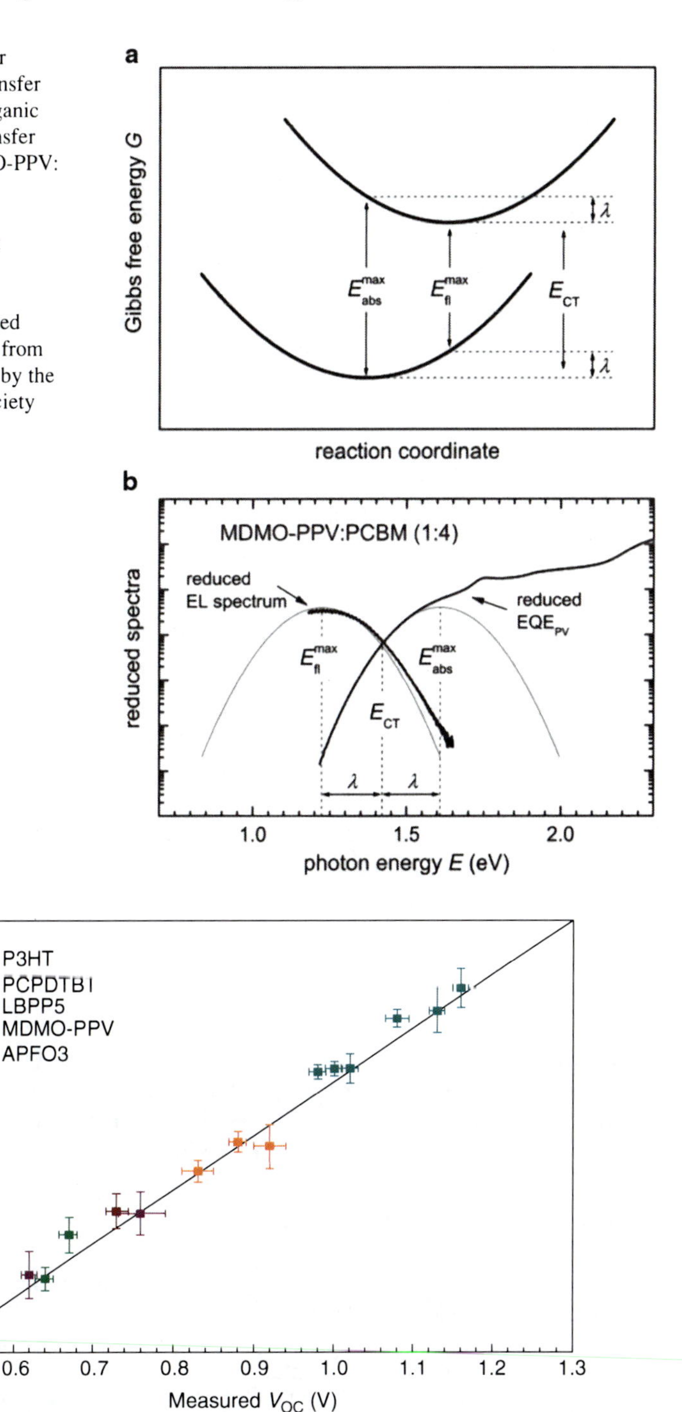

Fig. 16 Parameters for defining the charge-transfer state energy E_{CT} in organic solar cells. Charge-transfer state energy for MDMO-PPV: PCBM blend device determined by Fourier transform photocurrent spectroscopy and electroluminescence measurements. Reprinted figure with permission from [188]. Copyright 2010 by the American Physical Society

Fig. 17 Detailed balance approach for determining open circuit voltage based on the energy of the D/A charge-transfer state and its coupling to recombination loss modes. Reproduced with permission from [195]. Copyright 2009 Macmillan Publishers Limited

Table 1 Radiative and non-radiative voltage losses in polymer:fullerene solar cells

	Total loss	Radiative	Non-radiative
MDMO-PPV:PCBM (1:4)	0.58	0.24	0.34
P3HT:PCBM (1:1)	0.53	0.11	0.42
APFO3:PCBM (1:4)	0.59	0.24	0.35
APFO3:PCBM (1:1)	0.59	0.25	0.34

Voltage losses, ΔV, given in V

to calculate the total relative photovoltage loss and the contribution from both radiative losses and non-radiative losses, as summarized in Table 1. These data illustrate that the non-radiative loss component can be considerable. For example, in the case of P3HT, non-radiative CT state decay accounts for 80% of the total loss incurred for the device. Factors influencing the magnitude of this non-radiative loss component to charge separation are a current topic of OPV research. Strategies for mitigating these recombination losses are briefly discussed in the next section.

3.4 Charge Recombination and Electronic Coupling

The values in Table 1 illustrate that many organic solar cells incur V_{oc} losses related to the rate of both radiative and non-radiative recombination events. When an oxidizing potential is applied to the donor, this rate is directly related to the dark saturation current J_{s}, as shown in Fig. 13. Therefore, minimum J_{s} leads to maximum V_{oc}. Based on Marcus theory, when the CT state energy is high enough, thermally activated charge recombination occurs in the inverted regime. This means that, as the driving force for recombination increases, the rate decreases and, likewise, J_{s} decreases. By necessity, recombination rates must be slow enough for charge separation to out-compete charge recombination. The commonly observed trend of increasing V_{oc} with increasing ΔE_{DA} is a direct result of this inverted regime behavior, and the fact that the CT state energy tends to correlate with ΔE_{DA}.

In practice, suppressing charge-recombination rates by controlling ΔE_{DA} is undesirable, since it leads to absorption overlap between the donor and acceptor, thus limiting the photocurrent. Alternatively, consider Fig. 18, illustrating the electronic coupling element H_{ij} for recombination from the initial CT state, i, to the final neutral state, j. The rate constant for non-radiative recombination in the Marcus inverted regime is

$$k_{\mathrm{rec}}^{\mathrm{NR}} = \frac{4\pi^2}{h} H_{ij}^2 \frac{1}{\sqrt{4\pi\lambda kT}} \exp\left\{\frac{-(\Delta G^\circ + \lambda)^2}{4\lambda kT}\right\}, \qquad (2)$$

where h is Planck's constant, λ is the free energy for geometric reorganization, ΔG° is the total change in free energy, and kT is the available thermal energy. In Fig. 18, the activation energy, ΔG^*, for charge recombination exhibits a quadratic

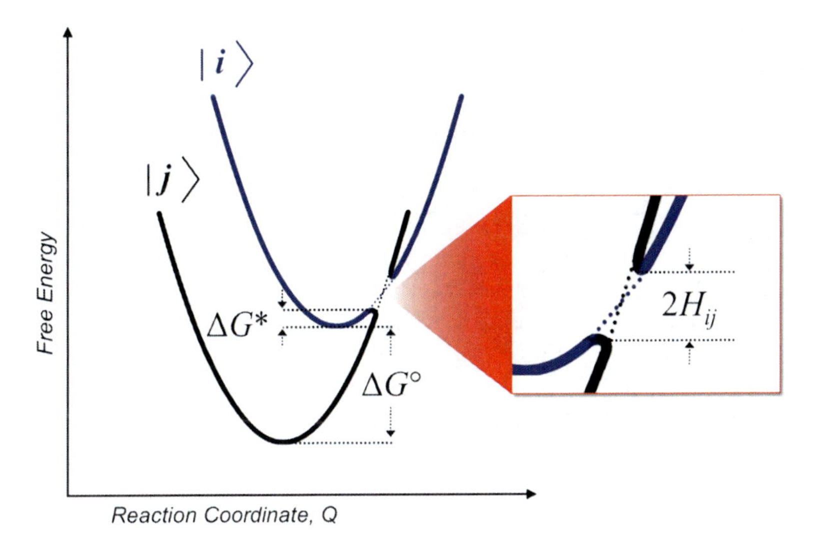

Fig. 18 Free energy surfaces illustrating the activation energy ΔG^* for charge recombination in the Marcus inverted region and the coupling term H_{ij} between the initial state, *i*, and the final state, *j*

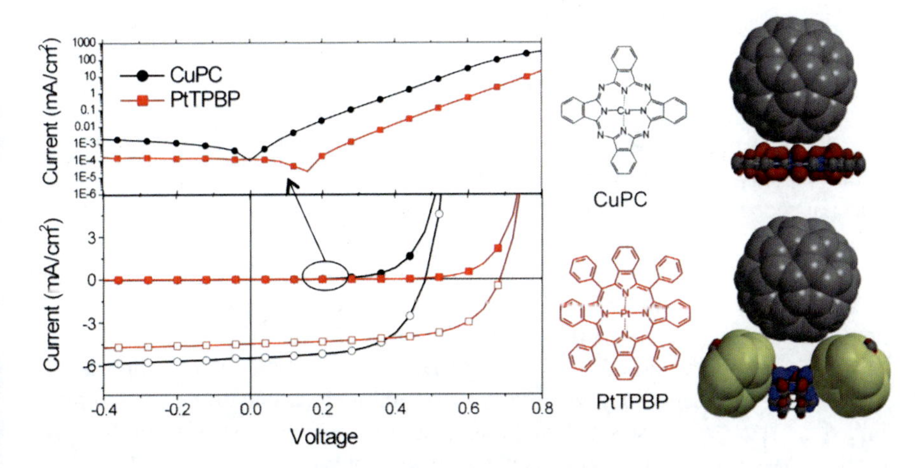

Fig. 19 Current–voltage characteristics for ITO/donor/C_{60}/BCP/Al based OPVs, donor = CuPC or PtTPBP. Images to the *right* show the structures of the donors and their HOMO and LUMO orbitals in *blue* and *red*, respectively. Steric interactions with the PtTPBP phenyl groups prevent close association of the donor and C_{60}

dependence on ΔG°. However, in two similar systems with equivalent thermodynamics, the system with weaker electronic coupling (smaller H_{ij}) will exhibit slower recombination and V_{oc} closer to the thermodynamic limit.

A practical approach to control H_{ij} is to control the distance between the π-systems of D^+ and A^- at the D/A interface. By tuning the steric interactions between bound polaron species, the length scale, over which charge recombination must occur, can be increased [196]. As a result, recombination rates diminish, while the charge-separation rate is either unaffected or increases. Several systems have been identified in which steric effects are thought to play a major role in determining V_{oc}. For example, crowding of the phenyl rings in platinum tetraphenylbenzoporphyrin (PtTPBP), shown in Fig. 19, causes saddling of the macrocycle, and the phenyls partially block access to the porphyrin π-system. While ΔE_{DA} is 1.4 eV for the D/A heterojunction formed between fullerene C_{60} and PtTPBP, ΔE_{DA} for CuPc/C_{60} is 1.7 eV. Based solely on these values, one might expect the photovoltage produced by the PtTPBP/C_{60} device to be significantly lower than that of the CuPc/C_{60} heterojunction. However, the PtTPBP device exhibits $V_{oc} = 0.69$ V, 40% higher than the CuPc device, with $V_{oc} = 0.48$ V. Similar effects have been observed between tetracene and rubrene [196], and related small molecules, as well as for polymer:fullerene blends [197].

4 Conclusions and Perspectives

Over the last 10 years, research in photovoltaic energy conversion has increasingly focused on organic solar cells, due to their potential to deliver power in low-cost, lightweight, flexible, and portable electronics. As a result, very significant improvements in device efficiencies have been achieved. New discoveries in this field have also revealed interesting prospects and challenges for future device improvements. In this chapter we highlight the role that transient charge-transfer states appear to have in determining photocurrent, photovoltage, and the resulting power-conversion efficiency.

Energy conversion in organic solar cells often involves several bound transient species. As a result, the relaxation pathways available to these intermediates can have a significant impact on the performance of the device. Charge-transfer states formed between the donor and acceptor appear to be particularly important in determining the open-circuit voltage. Understanding the origin of the V_{oc} is directly linked to energy losses, since photons near the solar peak maximum, ~2.0 eV (~600 nm), are efficiently absorbed by OPVs, yet qV_{oc} commonly ranges from only 0.5 to 1.0 eV. Gains in efficiency are further limited by V_{oc} losses, when one considers that a large fraction of incident solar photons are in the near-infrared (NIR). By eliminating voltage losses and harvesting NIR photons, future efficiencies two to three times that of present devices may be achievable.

Tuning frontier orbital energies has demonstrated that V_{oc} is related to ΔE_{DA} at low temperature. This effect has been shown, by detailed balance, to be a direct result of increasing the energy of the charge-transfer state and, therefore, increasing the Marcus reorganization energy for recombination. However, at room temperature or low light intensity, thermally activated recombination losses can lower V_{oc} from what is expected, based on the CT state energy.

Variations in molecular structure are known to substantially impact electron-transfer kinetics in chemical [13, 198, 199] and biological [200, 201] systems. The voltage produced by solution-based photoelectrochemical cells has also been linked to charge-transfer rates [202, 203]. Recent results have suggested that recombination losses can be controlled in thin film OPVs as well by tuning the electronic coupling at the D/A interface. Practical routes to controlled coupling have been explored for oligoacenes, porphyrinic materials, polymers, and perylene acceptors. These results suggest that V_{oc} losses can be recouped when the π-system of either the donor or the acceptor is sterically shielded from back electron transfer. However, a major challenge is that strong electronic coupling is desirable for efficient exciton diffusion, forward electron transfer, and charge transport. The large substituents used to frustrate electronic coupling also tend to diminish the absorption coefficient of these materials. Thus, significant effort will be needed to find new OPV materials systems that simultaneously give high J_{sc} and V_{oc}.

Acknowledgements The authors gratefully acknowledge support from the Center for Energy Nanoscience, an Energy Frontier Research Center funded by the U.S. Department of Energy, Office of Science, Office of Basic Energy Sciences under award number DE-SC0001013, from the Center for Advanced Molecular Photovoltaics (Award No KUS-C1-015-21), made by King Abdullah University of Science and Technology (KAUST), and from Global Photonic Energy Corporation.

References

1. Lewis NS (2007) Powering the planet. MRS Bull 32:808
2. U.S. Energy Information Administration (2009) In: DOE/EIA-0384(2009). U.S. Government Printing Office, Washington DC
3. Hansen J, Sato M, Ruedy R (1997) Radiative forcing and climate response. J Geophys Res-Atmos 102:6831
4. Metz B, Davidson OR, Bosch PR, Dave R, Meyer LA (2007) Climate Change 2007: Mitigation of Climate Change (Contribution of Working Group III to the Fourth Assessment Report of the Intergovernmental Panel on Climate Change, 2007). Cambridge University Press, Cambridge, UK
5. Becquerel E (1839) Mémoire sur les effets électriques produits sous l'influence des rayons solaires. C R Acad Sci 9:561
6. Copeland AW, Black OD, Garrett AB (1942) The photovoltaic effect. Chem Rev 31:177
7. Oregan B, Gratzel M (1991) A low-cost, high-efficiency solar-cell based on dye-sensitized colloidal TiO_2 films. Nature 353:737
8. Razeghi M, Rogalski A (1996) Semiconductor ultraviolet detectors. J Appl Phys 79:7433
9. Tang CW (1986) 2-Layer organic photovoltaic cell. Appl Phys Lett 48:183
10. Memming R (1988) Photoelectrochemical solar-energy conversion. Top Curr Chem 143:79
11. Goetzberger A, Hebling C, Schock HW (2003) Photovoltaic materials, history, status and outlook. Mater Sci Eng R 40:1
12. Petritz RL (1956) Theory of photoconductivity in semiconductor films. Phys Rev 104:1508
13. Wasielewski MR (1992) Photoinduced electron-transfer in supramolecular systems for artificial photosynthesis. Chem Rev 92:435
14. Gray HB (2009) Powering the planet with solar fuel. Nat Chem 1:7

15. McQuarrie DA, Simon JD (1997) Physical chemistry: a molecular approach. University Science Books, Sausalito, CA
16. Neamen DA (2003) Semiconductor physics and devices. McGraw-Hill, New York
17. Green MA (1993) Silicon solar-cells – evolution, high-efficiency design and efficiency enhancements. Semicond Sci Technol 8:1
18. Green MA, Emery K, Hishikawa Y, Warta W (2010) Solar cell efficiency tables (version 36). Prog Photovoltaics 18:346
19. Lewis NS, Nocera DG (2006) Powering the planet: chemical challenges in solar energy utilization. Proc Natl Acad Sci USA 103:15729
20. Hoffert MI, Caldeira K, Jain AK, Haites EF, Harvey LDD, Potter SD, Schlesinger ME, Schneider SH, Watts RG, Wigley TML, Wuebbles DJ (1998) Energy implications of future stabilization of atmospheric CO_2 content. Nature 395:881
21. Hoffert MI, Caldeira K, Benford G, Criswell DR, Green C, Herzog H, Jain AK, Kheshgi HS, Lackner KS, Lewis JS, Lightfoot HD, Manheimer W, Mankins JC, Mauel ME, Perkins LJ, Schlesinger ME, Volk T, Wigley TML (2002) Advanced technology paths to global climate stability: energy for a greenhouse planet. Science 298:981
22. Rand BP, Burk DP, Forrest SR (2007) Offset energies at organic semiconductor heterojunctions and their influence on the open-circuit voltage of thin-film solar cells. Phys Rev B 75:115327
23. Heremans P, Cheyns D, Rand BP (2009) Strategies for increasing the efficiency of heterojunction organic solar cells: material selection and device architecture. Acc Chem Res 42:1740
24. Helgesen M, Sondergaard R, Krebs FC (2010) Advanced materials and processes for polymer solar cell devices. J Mater Chem 20:36
25. Thompson BC, Frechet JMJ (2008) Organic photovoltaics – polymer-fullerene composite solar cells. Angew Chem Int Ed 47:58
26. Spanggaard H, Krebs FC (2004) A brief history of the development of organic and polymeric photovoltaics. Sol Energy Mater Sol Cells 83:125
27. Krebs FC, Gevorgyan SA, Gholamkhass B, Holdcroft S, Schlenker C, Thompson ME, Thompson BC, Olson D, Ginley DS, Shaheen SE, Alshareef HN, Murphy JW, Youngblood WJ, Heston NC, Reynolds JR, Jia SJ, Laird D, Tuladhar SM, Dane JGA, Atienzar P, Nelson J, Kroon JM, Wienk MM, Janssen RAJ, Tvingstedt K, Zhang FL, Andersson M, Inganas O, Lira-Cantu M, de Bettignies R, Guillerez S, Aernouts T, Cheyns D, Lutsen L, Zimmermann B, Wurfel U, Niggemann M, Schleiermacher HF, Liska P, Gratzel M, Lianos P, Katz EA, Lohwasser W, Jannon B (2009) A round robin study of flexible large-area roll-to-roll processed polymer solar cell modules. Sol Energy Mater Sol Cells 93:1968
28. Forrest SR (2004) The path to ubiquitous and low-cost organic electronic appliances on plastic. Nature 428:911
29. Bao Z, Dodabalapur A, Lovinger AJ (1996) Soluble and processable regioregular poly(3-hexylthiophene) for thin film field-effect transistor applications with high mobility. Appl Phys Lett 69:4108
30. Bao Z, Lovinger AJ, Dodabalapur A (1996) Organic field-effect transistors with high mobility based on copper phthalocyanine. Appl Phys Lett 69:3066
31. Dimitrakopoulos CD, Malenfant PRL (2002) Organic thin film transistors for large area electronics. Adv Mater 14:99
32. Roberts ME, Sokolov AN, Bao ZN (2009) Material and device considerations for organic thin-film transistor sensors. J Mater Chem 19:3351
33. Peumans P, Forrest SR (2001) Very-high-efficiency double-heterostructure copper phthalocyanine/C_{60} photovoltaic cells. Appl Phys Lett 79:126
34. Xue JG, Uchida S, Rand BP, Forrest SR (2004) 4.2% efficient organic photovoltaic cells with low series resistances. Appl Phys Lett 84:3013
35. Xue JG, Rand BP, Uchida S, Forrest SR (2005) A hybrid planar-mixed molecular heterojunction photovoltaic cell. Adv Mater 17:66

36. Xue JG, Uchida S, Rand BP, Forrest SR (2004) Asymmetric tandem organic photovoltaic cells with hybrid planar-mixed molecular heterojunctions. Appl Phys Lett 85:5757
37. Cheyns D, Rand BP, Heremans P (2010) Organic tandem solar cells with complementary absorbing layers and a high open-circuit voltage. Appl Phys Lett 97:3
38. Wynands D, Levichkova M, Riede M, Pfeiffer M, Baeuerle P, Rentenberger R, Denner P, Leo K (2010) Correlation between morphology and performance of low bandgap oligothiophene:C_{60} mixed heterojunctions in organic solar cells. J Appl Phys 107:6
39. Schueppel R, Timmreck R, Allinger N, Mueller T, Furno M, Uhrich C, Leo K, Riede M (2010) Controlled current matching in small molecule organic tandem solar cells using doped spacer layers. J Appl Phys 107:6
40. Liang YY, Xu Z, Xia JB, Tsai ST, Wu Y, Li G, Ray C, Yu LP (2010) For the bright future-bulk heterojunction polymer solar cells with power conversion efficiency of 7.4%. Adv Mater 22:E135
41. http://www.heliatek.com. Accessed Nov 2010
42. http://www.solarmer.com. Accessed Nov 2010
43. Brabec CJ (2004) Organic photovoltaics: technology and market. Sol Energy Mater Sol Cells 83:273
44. Jorgensen M, Norrman K, Krebs FC (2008) Stability/degradation of polymer solar cells. Sol Energy Mater Sol Cells 92:686
45. Krebs FC, Tromholt T, Jorgensen M (2010) Upscaling of polymer solar cell fabrication using full roll-to-roll processing. Nanoscale 2:873
46. Krebs FC (2009) Fabrication and processing of polymer solar cells: a review of printing and coating techniques. Sol Energy Mater Sol Cells 93:394
47. http://www.konarka.com/. Accessed Nov 2010
48. Kippelen B, Bredas JL (2009) Organic photovoltaics. Energy Environ Sci 2:251
49. Peumans P, Yakimov A, Forrest SR (2003) Small molecular weight organic thin-film photodetectors and solar cells. J Appl Phys 93:3693
50. Hoppe H, Sariciftci NS (2004) Organic solar cells: an overview. J Mater Res 19:1924
51. Steim R, Kogler FR, Brabec CJ (2010) Interface materials for organic solar cells. J Mater Chem 20:2499
52. Bredas JL, Norton JE, Cornil J, Coropceanu V (2009) Molecular understanding of organic solar cells: the challenges. Acc Chem Res 42:1691
53. Li C, Liu M, Pschirer NG, Baumgarten M, Müllen K (2010) Polyphenylene-based materials for organic photovoltaics. Chem Rev 110:6817
54. Clarke TM, Durrant JR (2010) Charge photogeneration in organic solar cells. Chem Rev 110:6736
55. Mathieu H, Lefebvre P, Christol P (1992) Simple analytical method for calculating exciton binding-energies in semiconductor quantum-wells. Phys Rev B 46:4092
56. Blossey DF (1970) Wannier exciton in an electric field. I. Optical absorption by bound and continuum states. Phys Rev B 2:3976
57. Blossey DF (1971) Wannier exciton in an electric field. 2. Electroabsorption in direct-band-gap solids. Phys Rev B 3:1382
58. Hill IG, Kahn A, Soos ZG, Pascal RA (2000) Charge-separation energy in films of pi-conjugated organic molecules. Chem Phys Lett 327:181
59. Gregg BA (2003) Excitonic solar cells. J Phys Chem B 107:4688
60. Jenekhe SA, Osaheni JA (1994) Excimers and exciplexes of conjugated polymers. Science 265:765
61. Gould IR, Young RH, Moody RE, Farid S (1991) Contact and solvent-separated geminate radical ion-pairs in electron-transfer photochemistry. J Phys Chem 95:2068
62. Masuhara H, Mataga N (1981) Ionic photo-dissociation of electron-donor–acceptor systems in solution. Acc Chem Res 14:312
63. Beens H, Knibbe H, Weller A (1967) Dipolar nature of molecular complexes formed in excited state. J Chem Phys 47:1183

64. de la Torre G, Vazquez P, Agullo-Lopez F, Torres T (1998) Phthalocyanines and related compounds: organic targets for nonlinear optical applications. J Mater Chem 8:1671
65. Henderson BW, Dougherty TJ (1992) How does photodynamic therapy work. Photochem Photobiol 55:145
66. Bonnett R (1995) Photosensitizers of the porphyrin and phthalocyanine series for photodynamic therapy. Chem Soc Rev 24:19
67. Casey HC, Sell DD, Wecht KW (1975) Concentration-dependence of absorption-coefficient for n-type and p-type GaAs between 1.3 and 1.6 ev. J Appl Phys 46:250
68. Sze SM (1981) Physics of semiconductor devices. Wiley, New York
69. Wang SY, Mayo EI, Perez MD, Griffe L, Wei GD, Djurovich PI, Forrest SR, Thompson ME (2009) High efficiency organic photovoltaic cells based on a vapor deposited squaraine donor. Appl Phys Lett 94:233304
70. Maier JP, Seilmeier A, Laubereau A, Kaiser W (1977) Ultrashort vibrational population lifetime of large polyatomic-molecules in vapor-phase. Chem Phys Lett 46:527
71. Shank CV, Ippen EP, Teschke O (1977) Sub-picosecond relaxation of large organic-molecules in solution. Chem Phys Lett 45:291
72. Giridharagopal R, Ginger DS (2010) Characterizing morphology in bulk heterojunction organic photovoltaic systems. J Phys Chem Lett 1:1160
73. Groves C, Reid OG, Ginger DS (2010) Heterogeneity in polymer solar cells: local morphology and performance in organic photovoltaics studied with scanning probe microscopy. Acc Chem Res 43:612
74. Saini S, Bagchi B (2010) Photophysics of conjugated polymers: interplay between Forster energy migration and defect concentration in shaping a photochemical funnel in PPV. PCCP 12:7427
75. Hennebicq E, Pourtois G, Scholes GD, Herz LM, Russell DM, Silva C, Setayesh S, Grimsdale AC, Mullen K, Bredas JL, Beljonne D (2005) Exciton migration in rigid-rod conjugated polymers: an improved Forster model. J Am Chem Soc 127:4744
76. Jackson B, Silbey R (1983) On the calculation of transfer rates between impurity states in solids. J Chem Phys 78:4193
77. Scholes GD, Jordanides XJ, Fleming GR (2001) Adapting the Forster theory of energy transfer for modeling dynamics in aggregated molecular assemblies. J Phys Chem B 105:1640
78. Fidder H, Knoester J, Wiersma DA (1991) Optical-properties of disordered molecular aggregates – a numerical study. J Chem Phys 95:7880
79. Nguyen TQ, Martini IB, Liu J, Schwartz BJ (2000) Controlling interchain interactions in conjugated polymers: the effects of chain morphology on exciton–exciton annihilation and aggregation in MEH-PPV films. J Phys Chem B 104:237
80. Stubinger T, Brutting W (2001) Exciton diffusion and optical interference in organic donor-acceptor photovoltaic cells. J Appl Phys 90:3632
81. Markov DE, Amsterdam E, Blom PWM, Sieval AB, Hummelen JC (2005) Accurate measurement of the exciton diffusion length in a conjugated polymer using a heterostructure with a side-chain cross-linked fullerene layer. J Phys Chem A 109:5266
82. Scully SR, McGehee MD (2006) Effects of optical interference and energy transfer on exciton diffusion length measurements in organic semiconductors. J Appl Phys 100:034907
83. Lunt RR, Benziger JB, Forrest SR (2010) Relationship between crystalline order and exciton diffusion length in molecular organic semiconductors. Adv Mater 22:1233
84. Lunt RR, Giebink NC, Belak AA, Benziger JB, Forrest SR (2009) Exciton diffusion lengths of organic semiconductor thin films measured by spectrally resolved photoluminescence quenching. J Appl Phys 105:053711
85. Gregg BA, Sprague J, Peterson MW (1997) Long-range singlet energy transfer in perylene bis(phenethylimide) films. J Phys Chem B 101:5362
86. Roberts ST, Schlenker CW, Barlier VS, McAnally RE, Zhang Y, Mastron JN, Thompson ME, Bradforth SE (2011) Observation of triplet exciton formation in a platinum sensitized organic photovoltaic device. J Phys Chem Lett 2:48

87. Rand BP, Schols S, Cheyns D, Gommans H, Girotto C, Genoe J, Heremans P, Poortmans J (2009) Organic solar cells with sensitized phosphorescent absorbing layers. Org Electron 10:1015
88. Luhman WA, Holmes RJ (2009) Enhanced exciton diffusion in an organic photovoltaic cell by energy transfer using a phosphorescent sensitizer. Appl Phys Lett 94:153304
89. Dewar MJS, Rogers H (1962) Pi-complexes. 2. Charge transfer spectra of pi-complexes formed by tetracyanoethylene with polycyclic aromatic hydrocarbons and with heteroaromatic boron compounds. J Am Chem Soc 84:395
90. Tsutsumi K, Yoshida H, Sato N (2002) Unoccupied electronic states in a hexatriacontane thin film studied by inverse photoemission spectroscopy. Chem Phys Lett 361:367
91. Kanai K, Akaike K, Koyasu K, Sakai K, Nishi T, Kamizuru Y, Ouchi Y, Seki K (2009) Determination of electron affinity of electron accepting molecules. Appl Phys A 95:309
92. Silinsh EA, Čápek V (1994) Organic molecular crystals. American Institute of Physics, New York
93. Ishii H, Sugiyama K, Ito E, Seki K (1999) Energy level alignment and interfacial electronic structures at organic metal and organic organic interfaces. Adv Mater 11:605
94. Hwang J, Wan A, Kahn A (2009) Energetics of metal-organic interfaces: new experiments and assessment of the field. Mater Sci Eng R 64:1
95. Kahn A, Koch N, Gao WY (2003) Electronic structure and electrical properties of interfaces between metals and pi-conjugated molecular films. J Polym Sci B Polym Phys 41:2529
96. Alloway DM, Armstrong NR (2009) Organic heterojunctions of layered perylene and phthalocyanine dyes: characterization with UV-photoelectron spectroscopy and luminescence quenching. Appl Phys A 95:209
97. Brumbach M, Placencia D, Armstrong NR (2008) Titanyl phthalocyanine/C_{60} heterojunctions: band-edge offsets and photovoltaic device performance. J Phys Chem C 112:3142
98. Duhm S, Heimel G, Salzmann I, Glowatzki H, Johnson RL, Vollmer A, Rabe JP, Koch N (2008) Orientation-dependent ionization energies and interface dipoles in ordered molecular assemblies. Nat Meter 7:326
99. Kera S, Yamane H, Fukagawa H, Hanatani T, Okudaira KK, Seki K, Ueno N (2007) Angle resolved UV photoelectron spectra of titanyl phthalocynine monolayer film on graphite. J Electron Spectrosc Relat Phenom 156:135
100. Kera S, Yamane H, Honda H, Fukagawa H, Okudaira KK, Ueno N (2004) Photoelectron fine structures of uppermost valence band for well-characterized ClAl-phthalocyanine ultrathin film: UPS and MAES study. Surf Sci 566:571
101. Placencia D, Wang WN, Shallcross RC, Nebesny KW, Brumbach M, Armstrong NR (2009) Organic photovoltaic cells based on solvent-annealed, textured titanyl phthalocyanine/C_{60} heterojunctions. Adv Funct Mater 19:1913
102. Sato N, Seki K, Inokuchi H (1981) Polarization energies of organic-solids determined by ultraviolet photoelectron-spectroscopy. J Chem Soc, Faraday Trans 77:1621
103. Yamane H, Honda H, Fukagawa H, Ohyama M, Hinuma Y, Kera S, Okudaira KK, Ueno N (2004) Homo-band fine structure of OTi- and Pb-phthalocyanine ultrathin films: effects of the electric dipole layer. J Electron Spectrosc Relat Phenom 137:223
104. Yamane H, Yabuuchi Y, Fukagawa H, Kera S, Okudaira KK, Ueno N (2006) Does the molecular orientation induce an electric dipole in Cu-phthalocyanine thin films? J Appl Phys 99:5
105. Deibel C, Strobel T, Dyakonov V (2010) Role of the charge transfer state in organic donor-acceptor solar cells. Adv Mater 22:4097
106. Giebink NC, Lassiter BE, Wiederrecht GP, Wasielewski MR, Forrest SR (2010) Ideal diode equation for organic heterojunctions. II. The role of polaron pair recombination. Phys Rev B 82:155306
107. Giebink NC, Wiederrecht GP, Wasielewski MR, Forrest SR (2010) Ideal diode equation for organic heterojunctions. I. Derivation and application. Phys Rev B 82:155305

108. Bredenbeck J, Ghosh A, Nienhuys HK, Bonn M (2009) Interface-specific ultrafast two-dimensional vibrational spectroscopy. Acc Chem Res 42:1332
109. Bredenbeck J, Ghosh A, Smits M, Bonn M (2008) Ultrafast two dimensional-infrared spectroscopy of a molecular monolayer. J Am Chem Soc 130:2152
110. Khalil M, Demirdoven N, Tokmakoff A (2003) Coherent 2D IR spectroscopy: molecular structure and dynamics in solution. J Phys Chem A 107:5258
111. Brixner T, Stenger J, Vaswani HM, Cho M, Blankenship RE, Fleming GR (2005) Two-dimensional spectroscopy of electronic couplings in photosynthesis. Nature 434:625
112. Lee J, Vandewal K, Yost SR, Bahlke ME, Goris L, Baldo MA, Manca JV, Van Voorhis T (2010) Charge transfer state versus hot exciton dissociation in polymer-fullerene blended solar cells. J Am Chem Soc 132:11878
113. Howard IA, Laquai F (2010) Optical probes of charge generation and recombination in bulk heterojunction organic solar cells. Macromol Chem Phys 211:2063
114. Morteani AC, Sreearunothai P, Herz LM, Friend RH, Silva C (2004) Exciton regeneration at polymeric semiconductor heterojunctions. Phys Rev Lett 92:247402
115. Ohkita H, Cook S, Astuti Y, Duffy W, Tierney S, Zhang W, Heeney M, McCulloch I, Nelson J, Bradley DDC, Durrant JR (2008) Charge carrier formation in polythiophene/fullerene blend films studied by transient absorption spectroscopy. J Am Chem Soc 130:3030
116. Howard IA, Hodgkiss JM, Zhang XP, Kirov KR, Bronstein HA, Williams CK, Friend RH, Westenhoff S, Greenham NC (2010) Charge recombination and exciton annihilation reactions in conjugated polymer blends. J Am Chem Soc 132:328
117. Drori T, Holt J, Vardeny ZV (2010) Optical studies of the charge transfer complex in polythiophene/fullerene blends for organic photovoltaic applications. Phys Rev B 82:075207
118. Howard IA, Mauer R, Meister M, Laquai F (2010) Effect of morphology on ultrafast free carrier generation in polythiophene: fullerene organic solar cells. J Am Chem Soc 132:14866
119. Muller JG, Lupton JM, Feldmann J, Lemmer U, Scharber MC, Sariciftci NS, Brabec CJ, Scherf U (2005) Ultrafast dynamics of charge carrier photogeneration and geminate recombination in conjugated polymer: fullerene solar cells. Phys Rev B 72:195208
120. Masuhara H, Hino T, Mataga N (1975) Ionic photodissociation of excited electron donor-acceptor systems. 1. Empirical equation on relationship between yield and solvent dielectric-constant. J Phys Chem 79:994
121. Mohammed OF, Adamczyk K, Banerji N, Dreyer J, Lang B, Nibbering ETJ, Vauthey E (2008) Direct femtosecond observation of tight and loose ion pairs upon photoinduced bimolecular electron transfer. Angew Chem Int Ed 47:9044
122. Weller A (1982) Exciplex and radical pairs in photochemical electron-transfer. Pure Appl Chem 54:1885
123. Coropceanu V, Cornil J, da Silva DA, Olivier Y, Silbey R, Bredas JL (2007) Charge transport in organic semiconductors. Chem Rev 107:926
124. Nelson J, Kwiatkowski JJ, Kirkpatrick J, Frost JM (2009) Modeling charge transport in organic photovoltaic materials. Acc Chem Res 42:1768
125. Servet B, Horowitz G, Ries S, Lagorsse O, Alnot P, Yassar A, Deloffre F, Srivastava P, Hajlaoui R, Lang P, Garnier F (1994) Polymorphism and charge-transport in vacuum-evaporated sexithiophene films. Chem Mater 6:1809
126. Ma WL, Yang CY, Gong X, Lee K, Heeger AJ (2005) Thermally stable, efficient polymer solar cells with nanoscale control of the interpenetrating network morphology. Adv Funct Mater 15:1617
127. Hirose Y, Kahn A, Aristov V, Soukiassian P, Bulovic V, Forrest SR (1996) Chemistry and electronic properties of metal-organic semiconductor interfaces: Al, Ti, In, Sn, Ag, and Au on PTCDA. Phys Rev B 54:13748
128. Scott JC, Malliaras GG (1999) Charge injection and recombination at the metal-organic interface. Chem Phys Lett 299:115
129. Hu LB, Kim HS, Lee JY, Peumans P, Cui Y (2010) Scalable coating and properties of transparent, flexible, silver nanowire electrodes. Acs Nano 4:2955

130. Lee JY, Connor ST, Cui Y, Peumans P (2010) Semitransparent organic photovoltaic cells with laminated top electrode. Nano Lett 10:1276
131. Lee JY, Connor ST, Cui Y, Peumans P (2008) Solution-processed metal nanowire mesh transparent electrodes. Nano Lett 8:689
132. Wu H, Hu LB, Rowell MW, Kong DS, Cha JJ, McDonough JR, Zhu J, Yang YA, McGehee MD, Cui Y (2010) Electrospun metal nanofiber webs as high-performance transparent electrode. Nano Lett 10:4242
133. Rowell MW, Topinka MA, McGehee MD, Prall HJ, Dennler G, Sariciftci NS, Hu LB, Gruner G (2006) Organic solar cells with carbon nanotube network electrodes. Appl Phys Lett 88:233506
134. Zhang DH, Ryu K, Liu XL, Polikarpov E, Ly J, Tompson ME, Zhou CW (2006) Transparent, conductive, and flexible carbon nanotube films and their application in organic light-emitting diodes. Nano Lett 6:1880
135. Wu JB, Becerril HA, Bao ZN, Liu ZF, Chen YS, Peumans P (2008) Organic solar cells with solution-processed graphene transparent electrodes. Appl Phys Lett 92:263302
136. De Arco LG, Zhang Y, Schlenker CW, Ryu K, Thompson ME, Zhou CW (2010) Continuous, highly flexible, and transparent graphene films by chemical vapor deposition for organic photovoltaics. Acs Nano 4:2865
137. Li XS, Zhu YW, Cai WW, Borysiak M, Han BY, Chen D, Piner RD, Colombo L, Ruoff RS (2009) Transfer of large-area graphene films for high-performance transparent conductive electrodes. Nano Lett 9:4359
138. Wang X, Zhi LJ, Mullen K (2008) Transparent, conductive graphene electrodes for dye-sensitized solar cells. Nano Lett 8:323
139. Emery K (2003) In: Luque A, Hegedus S (eds) Handbook of photovoltaic science and engineering. Wiley, Chichester, UK
140. Emery KA, Osterwald CR (1986) Solar-cell efficiency measurements. Sol Cells 17:253
141. Matson RJ, Emery KA, Bird RE (1984) Terrestrial solar spectra, solar simulation and solar-cell short-circuit current calibration – a review. Sol Cells 11:105
142. Shrotriya V, Li G, Yao Y, Moriarty T, Emery K, Yang Y (2006) Accurate measurement and characterization of organic solar cells. Adv Funct Mater 16:2016
143. Peumans P, Forrest SR (2004) Separation of geminate charge-pairs at donor-acceptor interfaces in disordered solids. Chem Phys Lett 398:27
144. Emelianova EV, van der Auweraer M, Bassler H (2008) Hopping approach towards exciton dissociation in conjugated polymers. J Chem Phys 128:224709
145. Blom PWM, Mihailetchi VD, Koster LJA, Markov DE (2007) Device physics of polymer: fullerene bulk heterojunction solar cells. Adv Mater 19:1551
146. Onsager L (1938) Initial recombination of ions. Phys Rev 54:554
147. Braun CL (1984) Electric-field assisted dissociation of charge-transfer states as a mechanism of photocarrier production. J Chem Phys 80:4157
148. Mihailetchi VD, Koster LJA, Hummelen JC, Blom PWM (2004) Photocurrent generation in polymer-fullerene bulk heterojunctions. Phys Rev Lett 93:216601
149. Sokel R, Hughes RC (1982) Numerical-analysis of transient photoconductivity in insulators. J Appl Phys 53:7414
150. Limpinsel M, Wagenpfahl A, Mingebach M, Deibel C, Dyakonov V (2010) Photocurrent in bulk heterojunction solar cells. Phys Rev B 81:085203
151. Tachiya M (1988) Breakdown of the Onsager theory of geminate ion recombination. J Chem Phys 89:6929
152. Pal SK, Kesti T, Maiti M, Zhang FL, Inganas O, Hellstrom S, Andersson MR, Oswald F, Langa F, Osterman T, Pascher T, Yartsev A, Sundstrom V (2010) Geminate charge recombination in polymer/fullerene bulk heterojunction films and implications for solar cell function. J Am Chem Soc 132:12440

153. Wojcik M, Tachiya M (2009) Accuracies of the empirical theories of the escape probability based on Eigen model and Braun model compared with the exact extension of Onsager theory. J Chem Phys 130:104107
154. Sano H, Tachiya M (1979) Partially diffusion-controlled recombination. J Chem Phys 71:1276
155. Potscavage WJ, Sharma A, Kippelen B (2009) Critical interfaces in organic solar cells and their influence on the open-circuit voltage. Acc Chem Res 42:1758
156. Moliton A, Nunzi JM (2006) How to model the behaviour of organic photovoltaic cells. Polym Int 55:583
157. Ortiz-Conde A, Sanchez FJG, Muci J (2000) Exact analytical solutions of the forward non-ideal diode equation with series and shunt parasitic resistances. Solid-State Electron 44:1861
158. Jain A, Kapoor A (2004) Exact analytical solutions of the parameters of real solar cells using Lambert W-function. Sol Energy Mater Sol Cells 81:269
159. Banwell TC, Jayakumar A (2000) Exact analytical solution for current flow through diode with series resistance. Electron Lett 36:291
160. Yoo S, Domercq B, Marder SR, Armstrong NR, Kippelen B (2004) Modeling of organic photovoltaic cells with large fill factor and high efficiency. Proc SPIE-Int Soc Opt Eng 5520:110
161. Yoo S, Domercq B, Kippelen B (2005) Intensity-dependent equivalent circuit parameters of organic solar cells based on pentacene and C_{60}. J Appl Phys 97:103706
162. Mazhari B (2006) An improved solar cell circuit model for organic solar cells. Sol Energy Mater Sol Cells 90:1021
163. Cuiffi J, Benanti T, Nam WJ, Fonash S (2010) Modeling of bulk and bilayer organic heterojunction solar cells. Appl Phys Lett 96:143307
164. Shockley W, Queisser HJ (1961) Detailed balance limit of efficiency of P-N junction solar cells. J Appl Phys 32:510
165. Guha S, Yang J, Nath P, Hack M (1986) Enhancement of open circuit voltage in high-efficiency amorphous-silicon alloy solar-cells. Appl Phys Lett 49:218
166. Hack M, Shur M (1985) Physics of amorphous-silicon alloy p-i-n solar-cells. J Appl Phys 58:997
167. Scharber MC, Wuhlbacher D, Koppe M, Denk P, Waldauf C, Heeger AJ, Brabec CL (2006) Design rules for donors in bulk-heterojunction solar cells – towards 10% energy-conversion efficiency. Adv Mater 18:789
168. Brabec CJ, Cravino A, Meissner D, Sariciftci NS, Fromherz T, Rispens MT, Sanchez L, Hummelen JC (2001) Origin of the open circuit voltage of plastic solar cells. Adv Funct Mater 11:374
169. Malliaras GG, Salem JR, Brock PJ, Scott JC (1998) Photovoltaic measurement of the built-in potential in organic light emitting diodes and photodiodes. J Appl Phys 84:1583
170. Gregg BA, Fox MA, Bard AJ (1990) Photovoltaic effect in symmetrical cells of a liquid-crystal porphyrin. J Phys Chem 94:1586
171. Kallmann H, Pope M (1959) Photovoltaic effect in organic crystals. J Chem Phys 30:585
172. Geacinto N, Pope M, Kallmann H (1966) Photogeneration of charge carriers in tetracene. J Chem Phys 45:2639
173. Mutolo KL, Mayo EI, Rand BP, Forrest SR, Thompson ME (2006) Enhanced open-circuit voltage in subphthalocyanine/C_{60} organic photovoltaic cells. J Am Chem Soc 128:8108
174. Ma BW, Woo CH, Miyamoto Y, Frechet JMJ (2009) Solution processing of a small molecule, subnaphthalocyanine, for efficient organic photovoltaic cells. Chem Mater 21:1413
175. Verreet B, Schols S, Cheyns D, Rand BP, Gommans H, Aernouts T, Heremans P, Genoe J (2009) The characterization of chloroboron (III) subnaphthalocyanine thin films and their application as a donor material for organic solar cells. J Mater Chem 19:5295
176. Kooistra FB, Knol J, Kastenberg F, Popescu LM, Verhees WJH, Kroon JM, Hummelen JC (2007) Increasing the open circuit voltage of bulk-heterojunction solar cells by raising the LUMO level of the acceptor. Org Lett 9:551

177. Lenes M, Wetzelaer G, Kooistra FB, Veenstra SC, Hummelen JC, Blom PWM (2008) Fullerene bisadducts for enhanced open-circuit voltages and efficiencies in polymer solar cells. Adv Mater 20:2116
178. Brabec CJ, Cravino A, Meissner D, Sariciftci NS, Rispens MT, Sanchez L, Hummelen JC, Fromherz T (2002) The influence of materials work function on the open circuit voltage of plastic solar cells. Thin Solid Films 403:368
179. Loi MA, Toffanin S, Muccini M, Forster M, Scherf U, Scharber M (2007) Charge transfer excitons in bulk heterojunctions of a polyfluorene copolymer and a fullerene derivative. Adv Funct Mater 17:2111
180. Kim H, Kim JY, Park SH, Lee K, Jin Y, Kim J, Suh H (2005) Electroluminescence in polymer-fullerene photovoltaic cells. Appl Phys Lett 86:183502
181. Hallermann M, Haneder S, Da Como E (2008) Charge-transfer states in conjugated polymer/fullerene blends: below-gap weakly bound excitons for polymer photovoltaics. Appl Phys Lett 93:053307
182. Veldman D, Ipek O, Meskers SCJ, Sweelssen J, Koetse MM, Veenstra SC, Kroon JM, van Bavel SS, Loos J, Janssen RAJ (2008) Compositional and electric field dependence of the dissociation of charge transfer excitons in alternating polyfluorene copolymer/fullerene blends. J Am Chem Soc 130:7721
183. Zhou Y, Tvingstedt K, Zhang FL, Du CX, Ni WX, Andersson MR, Inganas O (2009) Observation of a charge transfer state in low-bandgap polymer/fullerene blend systems by photoluminescence and electroluminescence studies. Adv Funct Mater 19:3293
184. Tvingstedt K, Vandewal K, Gadisa A, Zhang FL, Manca J, Inganas O (2009) Electroluminescence from charge transfer states in polymer solar cells. J Am Chem Soc 131:11819
185. Ng AMC, Djurisic AB, Chan WK, Nunzi JM (2009) Near infrared emission in rubrene: fullerene heterojunction devices. Chem Phys Lett 474:141
186. Pandey AK, Nunzi JM (2007) Rubrene/fullerene heterostructures with a half-gap electroluminescence threshold and large photovoltage. Adv Mater 19:3613
187. Kavarnos GJ, Turro NJ (1986) Photosensitization by reversible electron-transfer – theories, experimental-evidence, and examples. Chem Rev 86:401
188. Vandewal K, Tvingstedt K, Gadisa A, Inganas O, Manca JV (2010) Relating the open-circuit voltage to interface molecular properties of donor: acceptor bulk heterojunction solar cells. Phys Rev B 81:125204
189. Turro NJ (1991) Modern molecular photochemistry. University Science Books, Sausalito, CA
190. Kuzmin MG, Soboleva IV, Dolotova EV (2007) The behavior of exciplex decay processes and interplay of radiationless transition and preliminary reorganization mechanisms of electron transfer in loose and tight pairs of reactants. J Phys Chem A 111:206
191. Panda P, Veldman D, Sweelssen J, Bastiaansen J, Langeveld-Voss BMW, Meskers SCJ (2007) Charge transfer absorption for pi-conjugated polymers and oligomers mixed with electron acceptors. J Phys Chem B 111:5076
192. Haarer D (1980) Optical and photoelectric properties of organic charge transfer crystals. Festkor-Adv Solid St 20:341
193. Rau U (2007) Reciprocity relation between photovoltaic quantum efficiency and electroluminescent emission of solar cells. Phys Rev B 76:085303
194. Vanecek M, Poruba A (2002) Fourier-transform photocurrent spectroscopy of microcrystalline silicon for solar cells. Appl Phys Lett 80:719
195. Vandewal K, Tvingstedt K, Gadisa A, Inganas O, Manca JV (2009) On the origin of the open-circuit voltage of polymer-fullerene solar cells. Nat Meter 8:904
196. Perez MD, Borek C, Forrest SR, Thompson ME (2009) Molecular and morphological influences on the open circuit voltages of organic photovoltaic devices. J Am Chem Soc 131:9281
197. Yang LQ, Zhou HX, You W (2010) Quantitatively analyzing the influence of side chains on photovoltaic properties of polymer-fullerene solar cells. J Phys Chem C 114:16793

198. Marcus RA (1993) Electron-transfer reactions in chemistry – theory and experiment. Rev Mod Phys 65:599
199. Barbara PF, Meyer TJ, Ratner MA (1996) Contemporary issues in electron transfer research. J Phys Chem 100:13148
200. Gray HB, Winkler JR (1996) Electron transfer in proteins. Annu Rev Biochem 65:537
201. Fedurco M (2000) Redox reactions of heme-containing metalloproteins: dynamic effects of self-assembled monolayers on thermodynamics and kinetics of cytochrome c electron-transfer reactions. Coord Chem Rev 209:263
202. Pichot F, Gregg BA (2000) The photovoltage-determining mechanism in dye-sensitized solar cells. J Phys Chem B 104:6
203. Gregg BA, Pichot F, Ferrere S, Fields CL (2001) Interfacial recombination processes in dye-sensitized solar cells and methods to passivate the interfaces. J Phys Chem B 105:1422

Top Curr Chem (2012) 312: 213–238
DOI: 10.1007/128_2011_220

Published online: 2 August 2011

Molecular Monolayers as Semiconducting Channels in Field Effect Transistors

Cherie R. Kagan

Abstract This chapter describes the fundamental study of charge transport through single layers of π-conjugated molecules organized to form the semiconducting channels of field-effect transistors (FETs). Physical and chemical methods of evaporation, Langmuir-Blodgett assembly and transfer, and self-assembly have been used by the community to realize single molecular monolayers on the gate or gate dielectric surface of FETs. Advancements in molecular design and chemical modification of FET interfaces continue to improve measured charge transport properties in FETs. These monolayer FETs have been integrated in electronic circuitry and demonstrated as chemical sensors, where they promise the ultimate in performance as the entire molecular monolayer is modulated by the applied gate field and is accessed by analytes, respectively.

Keywords Circuits · Field-effect transistor · Langmuir-Blodgett · Monolayer · Self-assembly · Sensors

Contents

C.R. Kagan
Department of Electrical and Systems Engineering, University of Pennsylvania, Philadelphia, PA 19104 USA

Department of Materials Science and Engineering, University of Pennsylvania, Philadelphia, PA 19104 USA

Department of Chemistry, University of Pennsylvania, Philadelphia, PA 19104 USA
e-mail: kagan@seas.upenn.edu

Molecular electronics is the study of the motion of charge that depends on the chemical structure, conformation or organization, and environment of molecular species. Assembling circuitry through molecular design was first theorized by Aviram and Ratner [1]. This chapter describes fundamental studies of the behavior of charge in molecular monolayers through solid-state measurements of charge transport in monolayers organized at surfaces to form field-effect transistors (FETs), and the application of these transistors in electronics and as chemical sensors. Figure 1 is a general schematic of a molecular monolayer FET, where the molecules are organized into an ordered, close packed single sheet, providing intermolecular π–π coupling, so charges are transported in the monolayer plane. Synthetic chemistry allows molecules to be designed considering both the

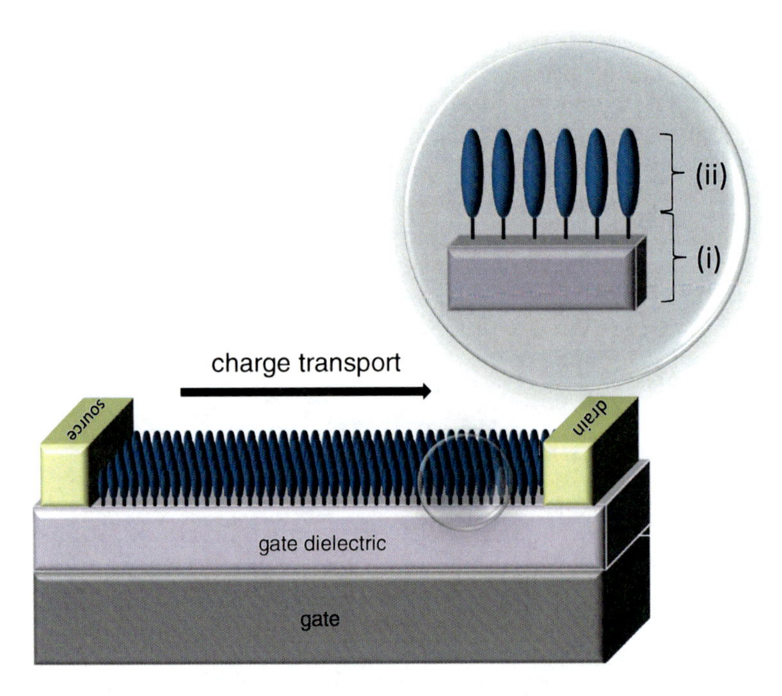

Fig. 1 Schematic of a molecular monolayer field-effect transistor. The bubble highlights (*i*) the substrate and/or the monolayer may comprise the electrically insulating, gate dielectric layer of the FET, and (*ii*) contains a π-conjugated core for charge transport from source-to-drain, modulated by the voltage applied to the gate electrode. Regions (*i*) and (*ii*) of the molecule may also contain binding groups for self-assembly or aliphatic chemistry for Langmuir–Blodgett assembly and transfer from the air–water interface

electronic behavior of the FET and the methods, such as chemically directed self-assembly and Langmuir–Blodgett transfer, used to organize the monolayers at device surfaces. For example, segment (1) of the molecule may be designed to incorporate part, or all, of the gate dielectric layer, and may have functional groups to allow chemical self-assembly on device surfaces. Segment (2) may incorporate π-conjugated charge-transporting groups, polymerizable chemistries, and/or aliphatic chemistry to facilitate Langmuir–Blodgett assembly on the water surface and transfer to device surfaces. The architecture of the FET provides a junction to probe charge transport in the molecular solid state, allowing electron and hole transport behavior to be separated and characterized. The junction geometry also introduces metal–semiconductor and semiconductor–dielectric interfaces, which may affect charge injection and charge trapping, introducing extrinsic behaviors to the measured transport characteristics. This chapter will review the methods of molecular monolayer assembly in FETs and the physics of FET behavior for various molecular monolayer chemistries and junction geometries that have been studied by the community for consideration in applications.

1 Background

The movement of charge in molecular compounds has traditionally been studied by the physical chemistry community through spectroscopic measurements in solution of charge transfer between donors and acceptors, separated by a relatively short σ or π bridge [2, 3]. Recent efforts to extend our understanding of charge transfer in molecular systems to the solid state have attracted attention for both fundamental and technological opportunities. While in this chapter we will focus on the FET, probing charge transport in the solid state necessitates the introduction of interfaces between the molecules and chemically and electronically different surfaces, such as metals and insulators. The solid state examines molecular systems between at least two electrodes under an applied bias, and in the FET, spaced by a dielectric layer from a third electrode. Interfacial charge transfer between a donor and an electrode has been investigated by probing molecular assemblies on metal surfaces through solution-phase electrochemical techniques [4–6] or photoemission spectroscopy [7, 8]. Research on charge transfer in solid-state molecular systems extends and marries the techniques, tooling, models, and theory of optical spectroscopy, electrochemistry, and charge transfer, common to the chemical community, with electrical measurements and charge transport, more common to mesoscopic physics. It is the interface spanning chemistry, physics, materials, and electrical disciplines that opens up the exploration of new chemical and physical phenomena unique to molecular systems. In this chapter, the aim is to bridge the languages used to describe the structures, the techniques used to probe charge transport, and the unique chemical and physical issues presented by charge transport in the molecular solid state.

2 Operation of the Molecular Monolayer Field-Effect Transistor

In the three-terminal junction of the FET (Fig. 1), the molecular monolayer forms the semiconducting channel extending between two electrodes, known as the source and the drain. A third electrode, the gate electrode, is separated from the charge-transporting monolayer channel by a dielectric layer, which may be fabricated on the underlying substrate, and/or may be a functional group synthesized in the design of the molecular species forming the monolayer. The influence of the gate on the channel can be understood as a capacitor, with the gate and the source-channel-drain representing parallel plates (Fig. 2). Applying a positive (negative) bias to the gate electrode may accumulate electrons (holes) in the semiconducting molecular channel. Increasing the magnitude of the applied bias increases the number of carriers accumulated in the channel at the interface with the gate dielectric layer. The number of accumulated carriers is proportional to the product of the

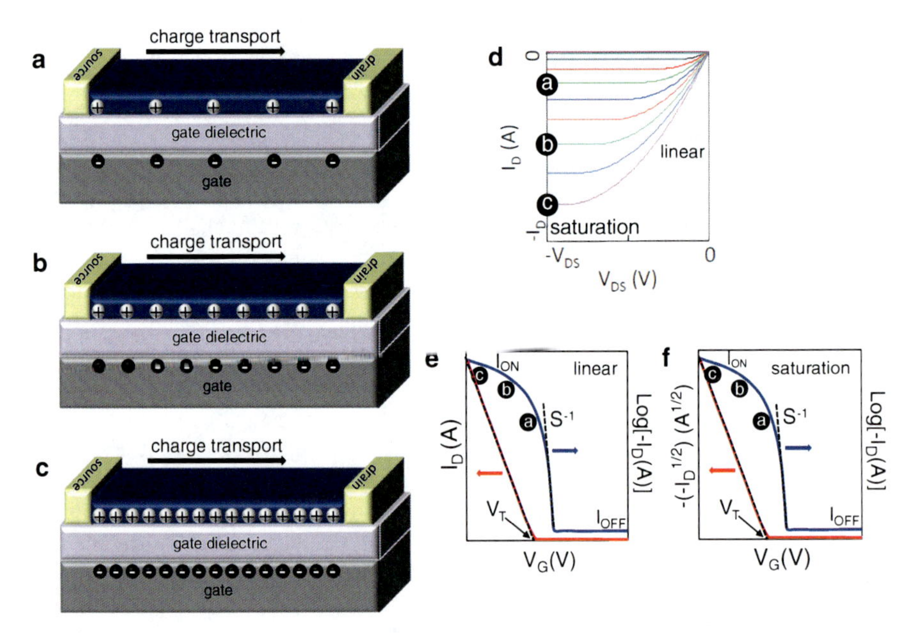

Fig. 2 Schematic of a p-type FET, where the semiconducting channel (*blue layer*) consists of a molecular monolayer. Application of an increasingly negative applied voltage to the gate [from (**a**) to (**c**)], with respect to the source, gives rise to an increase in the number of accumulated holes in the monolayer channel. (**d**) I_D–V_{DS} and (**e, f**) I_D–V_G curves are collected to characterize the FET behavior; mobility, threshold voltage (V_T), current modulation I_{ON}/I_{OFF}, and subthreshold slope (S^{-1}) in the linear and saturation regimes. Note: in the linear regime the current is linearly related to V_G and is lower, whereas in the saturation regime the current is related to the square of V_G and is comparably higher [as may be expected from (**d**)]. In the I_D–V_{DS} and I_D–V_G curves, the points (a–c) reflect the device schematics

capacitance of the dielectric layer and the applied gate bias, although not all induced carriers are free, as any traps in the material or at the device interfaces will be preferentially filled prior to creating mobile carriers. The threshold voltage describes the gate voltage required to fill the traps, and subsequently to induce free carriers in the channel. The threshold voltage may also be shifted by dopants in the semiconductor, even though most molecular semiconductors are unintentionally doped, and interface dipoles that develop upon formation of the metal–semiconductor interface. Applying a voltage between the source and drain electrodes results in currents that vary as the number of free carriers in the channel. The source and drain get their names, as commonly the source is grounded, and injects charge into the channel, which upon traversing the channel, exits at the drain. Figure 2a–c depicts the increase in channel hole density, as the applied gate voltage is made increasingly negative. By studying the thickness dependence of mobility in thin-film FETs, it has been shown in thin-film organic semiconductors that the charge-carrier concentration is modulated in only the first few nanometers of the semiconductor channel at the interface with the gate dielectric layer that gives rise to transistor action [9–15]. The "ultrathin" picture of the channel has spurred interest in studying charge transport and pursuing the physics and application of transistors having only a molecular monolayer as the semiconductor channel.

The gate-voltage-dependent FET behavior is characterized by two different measurements: (1) the current from source to drain (I_D) is measured as the voltage between drain and source (V_{DS}) is swept for various magnitudes of applied gate voltage (V_G) (Fig. 2d), yielding the "output" characteristics of the FET, and (2) I_D is measured as V_G is swept for constant V_{DS} (Fig. 2e, f), giving the "transfer" characteristics of the FET. It is in the I_D–V_G characteristics of the FET (plotted on a log scale) where its behavior as a switch "on" and "off" is seen. The measured device *I*–*V* characteristics are modeled by applying the same equations developed to describe the *I*–*V* behavior for the metal–oxide semiconductor field-effect transistor (MOSFET) and similarly for inorganic and organic thin film FETs. This analytical description of the FET assumes the field perpendicular to the channel produced by the gate is much greater than the field parallel to the channel from source-to-drain – this is known as the gradual channel approximation. For derivations of the analytical expressions, see references described in the context of MOSFETs [16, 17], inorganic [18, 19], and organic FETs [19–21]. The *I*–*V* characteristics of the FET at low V_{DS} is known as the linear regime, as I_D increases linearly with V_{DS}, as

$$I_{D,\text{linear}} = \frac{\varepsilon_i W}{t_i L} \mu_{\text{linear}} V_{DS} \left(V_G - V_T - \frac{V_{DS}}{2} \right), \qquad (1)$$

where ε_i and t_i are the dielectric constant and thickness of the insulating gate dielectric layer, W and L are the geometrical width and length of the FET, μ is the field-effect mobility of the semiconducting channel, and V_T is the device threshold voltage. In the linear regime, $V_{DS} << (V_G - V_T)$, and the last term may be neglected. At low voltages the accumulated carrier density in the channel is largely uniform from the source to the drain. Equation (1) is commonly used to extract the field-effect mobility (μ) for the semiconductor material in the device, and the device threshold voltage (V_T) from the slope and intercept, respectively, of the I_D–V_G characteristics (plotted on a linear scale) at low V_{DS} (approaching $V_{DS} \rightarrow 0$). At higher applied V_{DS}, referred to as the saturation regime, I_D saturates, as the accumulated carriers are depleted near the drain electrode, causing the channel to be "pinched-off." The FET behavior in the saturation regime is described by

$$I_{D,saturation} = \frac{\varepsilon_i W}{2t_i L} \mu_{sat} (V_G - V_T)^2 \tag{2}$$

and, upon plotting $I_D^{1/2}$ vs V_G, it is similarly used to extract μ and V_T for the accumulated carriers in the semiconducting channel. The field-effect mobility of molecular semiconducting thin film and monolayer channels is commonly reduced from that for an idealized layer, or single crystal, by disorder and defects that give rise to traps in the material and at the device interfaces. This is particularly important in studying charge transport in molecular monolayers, as disorder and defects in the single layer may dramatically affect the characteristics for different channel length FETs of the same molecular monolayer. As described above, V_T reflects carrier trapping, channel doping, and carrier injection at the metal–semiconductor interface. In the linear or saturation regimes of the I_D vs V_G characteristics, the current modulation between the "on" and "off" states (I_{ON}/I_{OFF}) and the subthreshold slope, that describes how quickly the FET turns on, are both important quantities to characterize for FET applications.

The gradual channel approximation (described above) may fail, as the channel length of the FET is shortened. The electrostatics of the FET limit $L > 1.5 \cdot t_i$ in a molecular FET, where the dielectric constant of the gate dielectric layer and semiconductor channel may be similar [22, 23]. This is particularly important in monolayer transistors, as many monolayer FETs studied have been limited to tens of nanometers channel length by the tens of nanometers size of ordered domains, and therefore require thin gate dielectric layers. Only recently (described below) have routes been shown to form more extended ordered molecular monolayers, allowing micron-scale FET channel lengths to be explored.

While thin-film organic semiconductors have been fabricated with p- or n-type or ambipolar (both electron and hole transport) behavior, molecular monolayer FETs demonstrated to-date (and described below) have only shown p-type characteristics. In thin-film organic semiconductors the electronic structure of the metal–semiconductor or semiconductor–dielectric interfaces are tailored by selecting the metal, the dielectric material, and the organic semiconductor, and profoundly affect the measured FET characteristics. Engineering the metal–semiconductor interface is critical to charge injection and the semiconductor–dielectric interface to preventing carrier trapping. For example, acene- and thiophene-based organic compounds typically have lower electron affinity and, in combination with high work-function metal electrodes, such as Au, form metal–semiconductor interfaces with lower barriers to hole injection, giving rise to FETs with p-type behavior [24–28]. In contrast, perylene and naphthalene diimides have higher electron affinity, and, fabricated with high workfunction metal electrodes, have metal–semiconductor interfaces with lower barriers to electron injection, and therefore form n-type FETs [29–31]. Recently it was reported that oxide dielectric materials, either from surface hydroxyl groups or water, act as electron traps [32]. Yet thermally oxidized Si wafers are a convenient platform upon which to fabricate molecular monolayer FET junctions, and SiO_2 and metal oxides provide good surfaces for chemisorption (see below). Polymer dielectric layers have been shown to eliminate the surface traps at oxide surfaces, allowing electron transport to be observed in thin-film materials previously thought to be "p-type." It has been shown that, by utilizing trap-free polymeric dielectric materials, and engineering the electronic structure and organization of molecules at the metal–semiconductor interface using self-assembled monolayers, p-type, n-type, and ambipolar organic thin FETs can be achieved [33–35]. While electron transport has not yet been observed in monolayer FETs, engineering the metal–monolayer and dielectric–monolayer interfaces, by either the choice of the metal and dielectric materials, or by synthesizing and assembling molecules to tailor the metal–molecule and dielectric–molecule interfaces, may provide a route forward to fabricate not only p-type, but n-type and ambipolar molecular monolayer FETs.

The monolayer channel offers unique advantages in FET performance: (1) the monolayer provides the channel for charge transport, achieving carrier mobilities similar to those in thin-film semiconductors and (2) carrier concentration is modulated only in about the first monolayer at the molecule–dielectric interface, allowing higher current modulation (I_{ON}/I_{OFF}) in monolayer FETs, which is limited by higher I_{OFF} in thin films by their greater thickness. Monolayer FETs also provide access to the channel modulated by the gate, important in probing the physics of charge transport in the molecular solid state, and to FET applications in electronics and chemical sensing.

3 Methods for Molecular Monolayer Organization on Device Surfaces

Three-methods have been pursued to organize monolayers from different organic compounds to form molecular monolayer FETs: (1) thermal evaporation of approximately monolayer thickness on the dielectric surface of FETs, (2) Langmuir–Blodgett assembly on the water surface and transfer to device surfaces, and (3) self-assembly of functionalized organic compounds on the surface of the gate, or gate dielectric layers of FETs.

3.1 *Thermal Evaporation*

Carefully controlled studies of the growth of physically evaporated organic semiconductors on surfaces have shown that near monolayers may be deposited.

The nature of the growth of the molecular layers, two-dimensional layer-by-layer, vs more three-dimensional island growth, depends on the chemistry of the underlying surface [36], the nature of the organic compound [37], and commonly changes from 2D to 3D growth with increasing film thickness [38]. The nature of growth and the development of charge transport, expressed by the increase and saturation in carrier mobility, have been studied on some of the highest carrier mobility, small-molecule thin-film semiconductors, such as pentacene, various derivatized oligothiophenes, and copper phthalocyanine. For example, pentacene and copper phthalocyanine FET studies show hole mobility saturates after six monolayers have been deposited [12, 39, 40]. Although in pentacene, only around two to three monolayers are believed complete, as island growth of pentacene is observed upon the first pentacene monolayer [41]. Asadi et al. studied the transport in high quality, near-single monolayer pentacene deposited by supersonic molecular beam deposition. Figure 3a shows an atomic force microscopy (AFM) image of the pentacene layer at the interface with one of the electrodes forming a bottom-contact, bottom-gate FET. The source and drain electrodes were modified with 1*H*,1*H*,2*H*,2*H*-perfluorooctanethiol (PFOT), which, as has been shown in thin films for a number of assembled thiolates, maintains the organization and orientation of large pentacene grains at and across the electrode interface, and improves device performance [42]. Figure 3b, c show well-behaved (b) output and (c) transfer characteristics for pentacene monolayer FETs. By modifying the FET source and drain electrodes, the FET hole mobility, calculated in the saturation regime using (2), was 0.015 cm^2/V s, a relatively high mobility for pentacene deposited on a bare SiO_2 surface, and the FET threshold voltage and hysteresis were also reduced. This work shows that surface modification, which has been used to improve electron and hole transport in thin-film organic semiconductors, provides a similar opportunity to improve the performance of monolayer FETs important in their application [33–35, 43–46].

In oligothiophene derivatives, hole mobilities are reported to saturate after only around one to two monolayers are deposited [10, 14]. Huang et al. studied vacuum-evaporated 5,5′-bis(4-hexylphenyl)-2,2′-bithiophene (6PTTP6) monolayers deposited by thermal evaporation (Fig. 4) [37]. Submonolayer (at 0.75 of a monolayer) platelets deposit, and some appear to coalesce, forming a near-complete monolayer prior to the next layer's growth (exemplified by the 1.27 monolayers and 1.31 monolayers). The hole mobility is seen to increase slightly as the first monolayer becomes nearly complete, and to increase by ~10^4, at 1.31 deposited monolayers (Fig. 4b). The hole mobility quickly saturates with further increase in 6PTTP6 deposition. Figure 4c depicts the expected physical underpinnings, as near-complete monolayers are deposited, but leave undesirable voids in the first layer (at 1.27 ML). Partial deposition of a second layer provides a pathway for charge transport, to bridge the voids in the underlying monolayer (at 1.31 ML), exemplifying the role of defects in limiting carrier mobility in monolayers. This near-monolayer 6PTTP6 semiconductor provides an attractive material system, as thick as the gate-modulated channel, that was further applied in chemical sensing applications (see below).

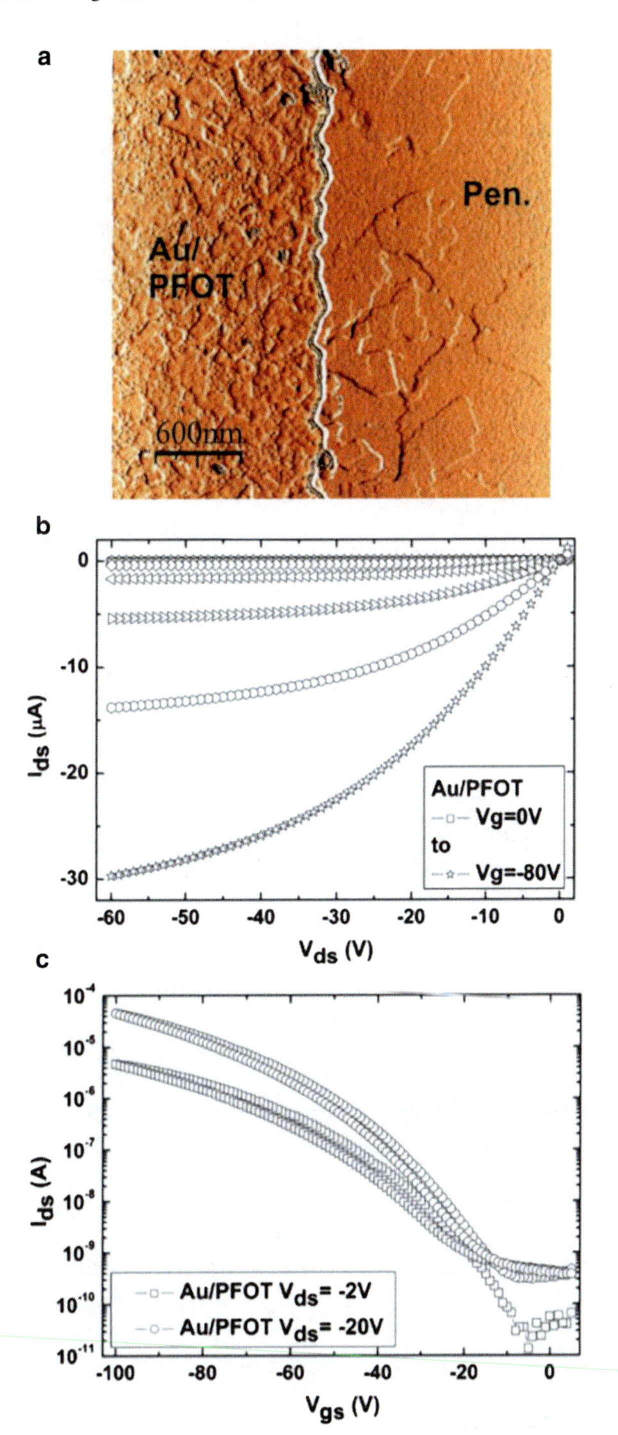

Fig. 3 Pentacene grown by supersonic molecular beam deposition to form near monolayer p-type FETs with thiolate monolayer modified Au source and drain contacts (**a**) visualized by atomic force microscopy and with well-behaved (**b**) I_D–V_{DS} and (**c**) I_D–V_G characteristics

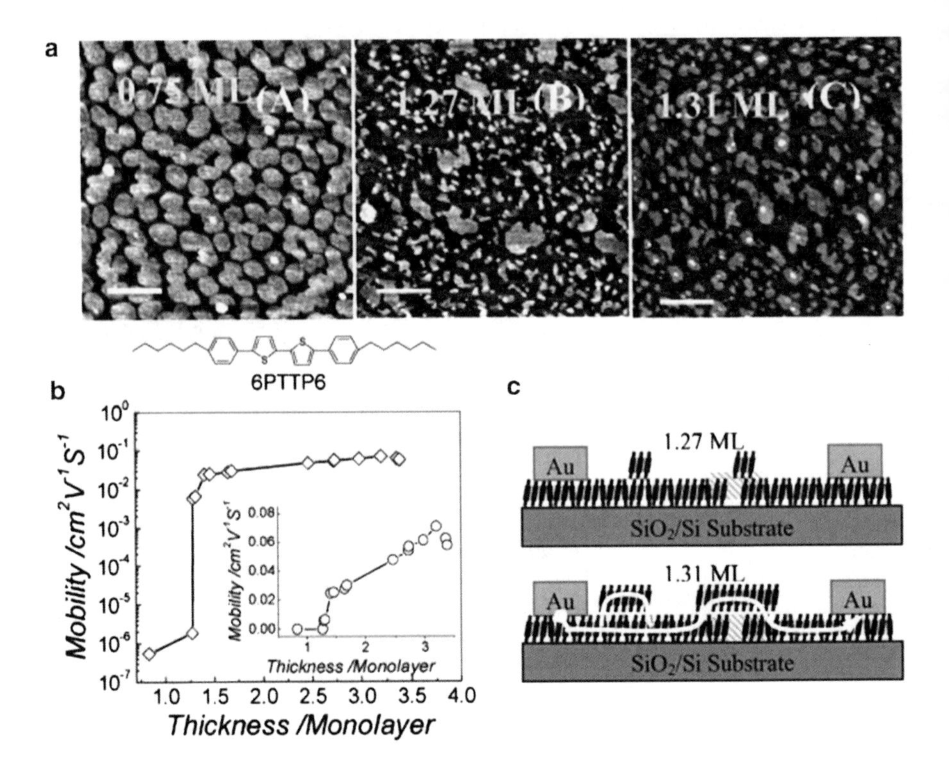

Fig. 4 Thermally evaporated (**a**) 0.75, 1.27, and 1.31 monolayers of 5,5′-bis(4-hexylphenyl)-2,2′-bithiophene [6PTTP6], which were incorporated in FETs and (**b**) show a dramatic increase in mobility at 1.31 monolayers. (**c**) The partial second layer is illustrated schematically to bridge voids in the underlying monolayer, providing pathways for increased charge transport

3.2 Langmuir–Blodgett Assembly

Langmuir–Blodgett (LB) assembly has been used to integrate monolayer (and multilayer) assemblies of a wide range of molecules, polymers, nanoparticles, lipids, and proteins on surfaces [47, 48]. Molecules, insoluble in the liquid subphase, are deposited and compressed into organized assemblies at the gas–liquid interface of a Langmuir–Blodgett trough. The organized assemblies are transferred to substrates by vertically extracting the substrate immersed in the liquid subphase, through the monolayer assembly, and into the gas phase using a robotic dipper. Typically, the liquid subphase is water, and the gas phase is air. Amphiphilic molecules are commonly used to orient molecules on the air–water interface and to transfer these assemblies to surfaces. For monolayer electronics, compounds with π-conjugated cores and aliphatic tail groups have been synthesized for LB assembly, although a number of unexpected compounds have also been found to assemble into LB films.

For example, Scott et al. derivatized pentacosa-10,12-diynoic acid by reaction of the terminal carboxylic acid with silanes and amines, and assembled these amphiphilic compounds at the air–water interface [49]. This compound is particularly interesting, as the middle of the compound contains diacetylene, which upon exposure to UV radiation (254 nm) was polymerized as a monolayer on the air–water interface, providing a strong, covalently-coupled conjugated pathway for charge transport. The polymerization can be followed as the absorption spectrum is red-shifted, converting the unpolymerized "blue phase" into the polymerized "red phase." The selection of the end group was shown to influence the assembly, polymerization, and film organization. The underivatized carboxylic acid showed hysteresis in its pressure–area isotherm (collected as the trough barrier compresses the monolayer and is taken back) and formed multilayers, instead of monolayers, upon transfer to substrates. The silane required a pH 4.5 water subphase to form silanol, but formed monolayers with poor crystallinity. The diethanolamine did not photopolymerize, as the bulkier side groups are expected to have limited the close approach of the compounds, preventing their photopolymerization. The ethanolamine derivative showed well-behaved reversible pressure–area isotherms, formed monolayers that were readily photopolymerized, and transferred as single monolayer polydiacetylenes to substrates. The ethanolamine was used to study charge transport by transferring the polydiacetylene monolayers to FET structures (Fig. 5). The measured I_D–V_{DS} characteristics show gate-modulated hole conductance with I_{ON}/I_{OFF} as high as 10^2, but the FET currents were <100 pA. The currents were constrained by (1) the device electrostatics, as the channel lengths studied (20–500 nm) were not substantially longer than the 100 nm SiO_2 gate dielectric layer, and (2) hydroxy-amide side groups, expected to give rise to high contact resistances at the metal–molecule interface, limiting carrier

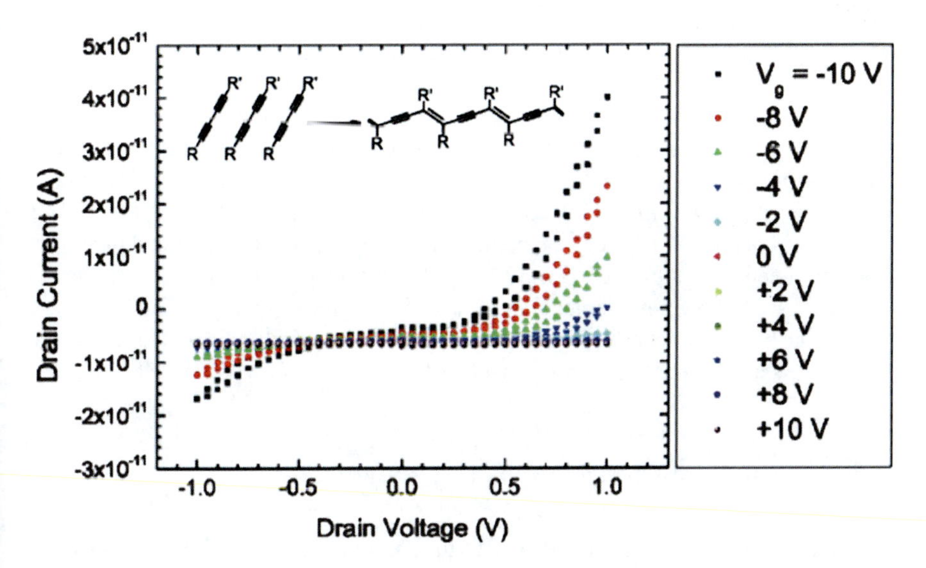

Fig. 5 I_D–V_{DS} characteristics for polydiacetylene monolayers, assembled and polymerized on the air–water interface (*inset*) and transferred to form bottom-contact, bottom-gate p-type FETs

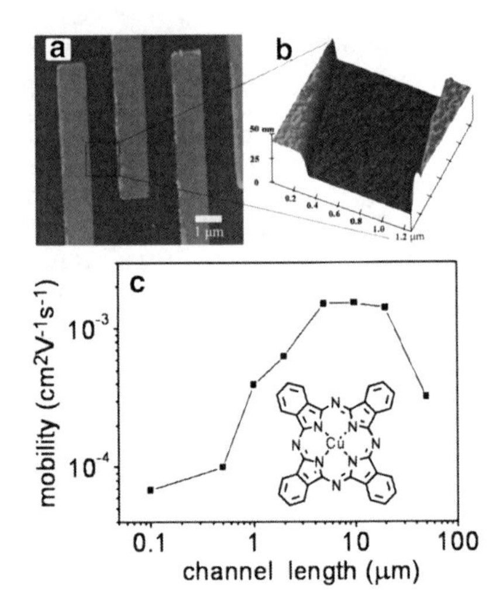

Fig. 6 (**a**) Scanning electron and (**b**) atomic force microscopy images of copper phthalocyanine Langmuir–Blodgett monolayer FETs. (**c**) Hole field-effect mobility as a function of the copper phthalocyanine channel length

injection. The polydiacetylene monolayers provide an intriguing opportunity to explore other derivatized diacetylenes to improve FET performance.

Copper phthalocyanine (CuPc) is an example of a molecular structure that is not a typical amphiphilic compound used in LB assembly, but has been reported by Wei et al. to form stable monolayers on the air–water interface of the LB trough [50]. The monolayers were transferred to the gate dielectric surface to form bottom-contact, bottom-gate FET structures. Figure 6 shows (a) scanning electron and (b) AFM images of FETs fabricated with monolayer CuPc semiconducting channels ranging from 100 nm to 50 μm in length. The CuPc monolayers form p-type FETs. The FET mobility, measured in the saturation regime, varied with channel length (Fig. 6c). The mobility first increased, as the channel length was enlarged from 100 nm to 5 μm, as high contact resistances, compared to the smaller channel resistances at shorter channel lengths, limited FET mobility, as has been observed in organic thin-film FETs with similar channel lengths [34, 51]; at the smaller channel lengths, the electrostatics (on a 100-nm SiO_2 gate dielectric layer) may restrict current modulation. The mobility in channels ranging from 5 to 20 μm were nearly constant at ~0.0015 cm^2/V s, but for the 50 μm channel-length devices the mobility again decreased, attributed to a large number of defects in the channel. The LB-transferred monolayer FETs were reported to have mobilities three times larger than those of 50 nm thick evaporated CuPc thin film FETs, attributed to achieving better orientation and crystallinity of the LB CuPc monolayers at the important semiconductor–dielectric interface of the FET.

3.3 *Chemically Directed Self-Assembly*

Organic compounds functionalized with groups that chemically bind to substrate surfaces will spontaneously self-assemble into close-packed, organized monolayers upon immersing the substrates into solutions of such compounds. Self-assembly on metal [52, 53] and oxide [53–55] surfaces has been utilized in molecular devices to control the surface wetting of deposited semiconductors [56, 57], to derivatize the metal and gate dielectric surfaces, to engineer the metal workfunction and the organization of organic semiconductors at metal–semiconductor interfaces [58–60], and to assemble ultrathin gate dielectrics [61–63]. Here, examples of self-assembled monolayers forming the active semiconductor channel of the FET, often referred to as the SAMFET (self-assembled monolayer FET), will be provided, as these self-assembled devices open an attractive route to produce low-cost, large-area electronics. To date, all of the SAMFETs have been fabricated with chemistries that bind to underlying SiO_2, metal oxide, and polymer-gate dielectric surfaces, but self-assembly of monolayers on metal [53] and doped-Si surfaces [64] provide an avenue for SAMFETs directly assembled on the FET gate.

The first demonstration of directly chemisorbing a semiconductor monolayer to the gate dielectric layer of an FET structure was demonstrated by Tulevski et al. [65]. They employed a catechol-derivatized tetracene which, using synchrotron X-ray reflectivity, was shown to form upright, close-packed assemblies on metal oxide surfaces such as Al_2O_3 (Fig. 7a), the catechol chelates Al_2O_3 forming an aluminum ester linkage. This chemistry was used to assemble tetracene-based monolayer semiconducting channels on 5-nm Al_2O_3 layers deposited on 5-nm thermally oxidized n+ Si wafers, forming the gate dielectric layer (Al_2O_3/SiO_2) and gate of the FETs. Au source and drain electrodes were defined by e-beam lithography, prior to assembling the monolayers, to form sub-100-nm monolayer channels (Fig. 7b, c).

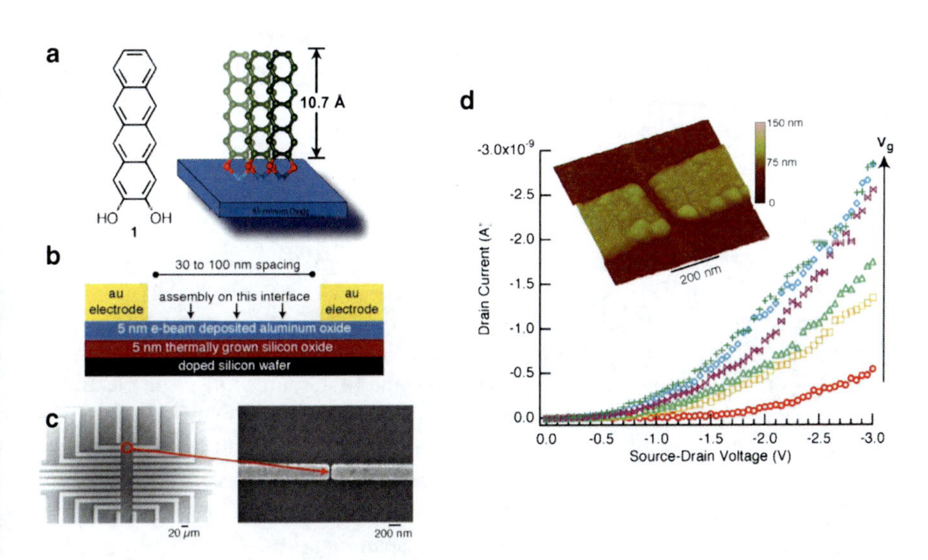

Fig. 7 (**a**) Catechol derivatized tetracenes self-assemble on metal oxide surfaces such as aluminum oxide. (**b**) Schematic and (**c**) scanning electron micrographs of FET structures fabricated with a 5-nm aluminum oxide layer on top of a 5-nm thermally oxidized Si wafer to allow self-assembly of the derivatized tetracene between sub-100 nm Au source and drain electrodes. (**d**) I_D–V_{DS} characteristics of the assembled tetracene monolayer FET for a 40 nm channel length showing hole modulation and (*inset*) an atomic force microscope image of the FET channel

Observation of gate-modulated hole currents (Fig. 7d) demonstrated the prospect of realizing SAMFETs, even though the FET characteristics were not well-behaved. These I–V characteristics were collected for device channel lengths $\leq$60 nm in >50% yield. The yield dropped sharply for longer channel lengths, as the monolayers were assembled on Al_2O_3 with ~40-nm grains, imaged by AFM, which is anticipated to give rise to disorder, limiting charge transport. The nonlinear behavior of the I–V characteristics may arise from high contact resistances, as the monolayer does not assemble on the Au surface, and device electrostatics, which are not as well-scaled for the shorter channel lengths.

More recently, Mottaghi et al. reported assembling monolayers of (1) quaterthiophene bonded to octanoic acid (4T) on ~100-nm Al_2O_3 gate-dielectric surfaces of FETs and (2) terthiophene added to preassembled long alkylsilanes (3T) on 10-nm SiO_2 gate dielectric surfaces of FETs [66]. The carboxylic acid group, provided by the octanoic acid of the 4T compound, is one of the first chemistries studied to assemble monolayers for molecular electronics [67]. Carboxylic acids, and other acids such as phosphonic and hydroxamic acids, assemble on more basic metal oxide surfaces, such as Al_2O_3 [52, 53]. The silane chemistry, used to functionalize the 3T compound and other examples described below, is widely used to derivatize hydroxylated surfaces, most commonly the SiO_2 surface, as surface water acts to catalyze the polymerization of polysiloxane, which connects to surface silanol groups [53]. The 3T and 4T FETs were fabricated with channel lengths of 50 nm to 1 μm, as longer channel lengths showed no current similarly expected to arise from disorder in the assembled monolayers. The device configuration selected for the 4T molecular monolayer FETs was bottom-contact, bottom-gate, whereas the 3T monolayer FETs were fabricated in top-contact, bottom-gate configuration. For the 4T devices, while the yield of devices was not high, out of 25 fabricated FETs, 4 showed nonlinear I_D–V_{DS} characteristics (akin to those in Fig. 7d), but remarkably 2 FETs showed well-behaved p-type FET characteristics (Fig. 8a). These FETs allowed hole mobilities to be extracted, with mobilities as high as 0.0035 cm^2/V s, only slightly lower than those of similar thin-film FETs. For 3T assemblies, 173 FETs were fabricated, of which 153 showed no current, 12 were short-circuited, but 7 samples (Fig. 8b) showed nonlinear but gate-modulated hole currents, estimated to have hole mobilities of 8×10^{-4} cm^3/V s. This work demonstrated the feasibility of attaining well-behaved molecular monolayer FETs for ordered assemblies that form good electrical contacts to the FET source and drain electrodes.

Instead of fabricating long-channel-length devices, that are often limited by disorder in self-assembled monolayers, Guo et al. formed ultrasmall ~2–6 nm source-drain channel lengths by oxidatively etching a gap in individual metallic single-walled carbon nanotubes (SWCNT) deposited across macroscopic metal electrodes (Fig. 9a) [68]. Ordered assemblies spanning the ultrasmall source-drain channels were achieved using columnar hexabenzocoronenes (HBCs), functionalized with acid chlorides, that chemisorbed on an underlying SiO_2 gate dielectric layer, grown on a highly doped Si wafer serving as the gate, to complete the FETs. An estimated column of 4–12 HBCs spanned the SWCNT junctions. Nanometer-scale monolayer and organic thin-film FETs typically show low current

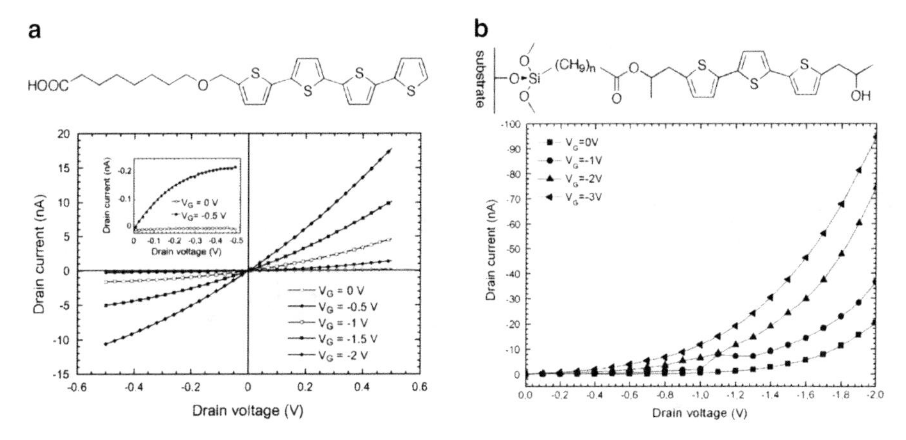

Fig. 8 (**a**) 8-Methylether-α,α′-quaterthiopheneoctanoic acid was assembled onto the ~100-nm Al_2O_3 gate dielectric surface of bottom-contact, bottom-gate FETs. I_D–V_{DS} characteristics of 2 of the 25 fabricated devices yielded well-behaved modulation of hole currents. (**b**) The vinyl group of assembled 10-undecenyl trichlorosilane monolayers assembled on 10-nm SiO_2 gate dielectric surfaces of FETs was oxidized allowing coupling of bis-(5,5″-(2-hydroxy 2-methylethyl))-2,2′:5′,2″-terthiophene by esterification. The FETs were fabricated in top-contact, bottom-gate configuration. Seven of 173 devices showed current modulation, seen in I_D–V_{DS} characteristics

modulation (I_{ON}/I_{OFF}) due to poor device electrostatics. Interestingly, using the SWCNT electrodes, the gate field is not as dramatically screened, allowing well-behaved monolayer HBC FETs to be fabricated. This is seen by the current saturation in the measured I_D–V_{DS} curves (Fig. 9b) and by I_{ON}/I_{OFF} ~10^5 collected in the I_D–V_G curve (Fig. 9c). Hole mobilities in excess of 1 cm^2/V s are calculated. Similarly short but wider channel Pt junctions assembled with the HBC do not show current modulation, as they effectively screen the gate field. In SWCNT FETs needle-like contacts were modeled to thin the Schottky barrier at the metal–semiconductor interface, giving rise to increased current levels at lower gate voltages [69]. The same enhanced field at the contacts may narrow the metal–molecule barrier in the HBC FETs, giving rise to the dramatically improved FET performance in SWCNT vs metal contacts, studied using Pt electrodes for the HBCs, and metals such as Au in the nanoscale monolayer FETs studied in the community. The SWCNT electrodes provide an exciting platform to study molecular monolayer FETs, and were further applied in a demonstration of chemical sensing (described below).

Recently, Smits et al. extended the length scale of monolayer ordering by synthesizing and assembling liquid-crystalline π-conjugated cores end-substituted with long alkyl chains bearing a terminal surface-anchoring group [70]. Quinquethiophene was selected as the π-conjugated core to maintain solubility, while having a large intermolecular π–π coupling. An undecane aliphatic chain was coupled to the quinquethiophene to promote liquid-crystalline ordering, and it was terminated with a silane to allow assembly on the SiO_2 gate dielectric surface to form SAMFETs. The quinquethiophene self-assembled monolayer was structurally characterized to form upright assemblies, ordered in-plane on the SiO_2 surface. The

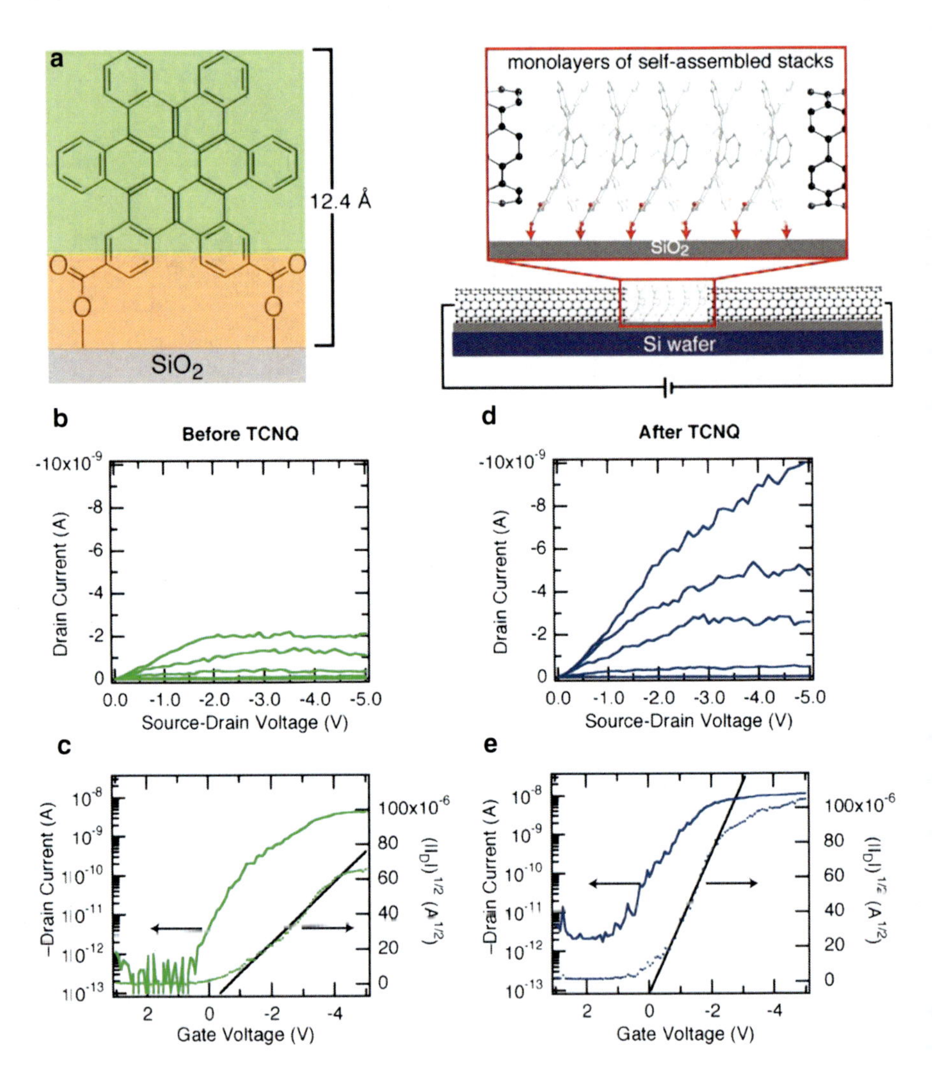

Fig. 9 (**a**) Acid chloride derivatized hexabenzocoronenes (HBC) assemble upright on the SiO_2 gate dielectric surface of heavily doped Si wafers (gate) and bridge ~2–6 nm gaps etched in metallic single-walled carbon nanotubes to form the source and drain of the FETs. (**b**) I_D–V_{DS} and (**c**) I_D–V_G characteristics of HBC monolayers as assembled and (**d**) I_D–V_{DS} and (**e**) I_D–V_G characteristics of the HBC FET upon exposure to TCNQ

longer-range ordering enabled by the liquid crystalline nature of the assemblies allowed micron-scale FETs to be fabricated with near unity yield [71]. Monolayer FETs characteristics are represented by the 40 μm channel-length FET (in a ring geometry) in Fig. 10. Linear and saturation regime I_D–V_G characteristics (Fig. 10a) were used to extract linear and saturation FET mobilities of 0.01–0.04 cm^2/V s and I_{ON}/I_{OFF} of 10^7 for the monolayer FETs. Both I_D–V_G and I_D–V_{DS} (Fig. 10b) characteristics were well-behaved, and showed nearly no hysteresis. The large current

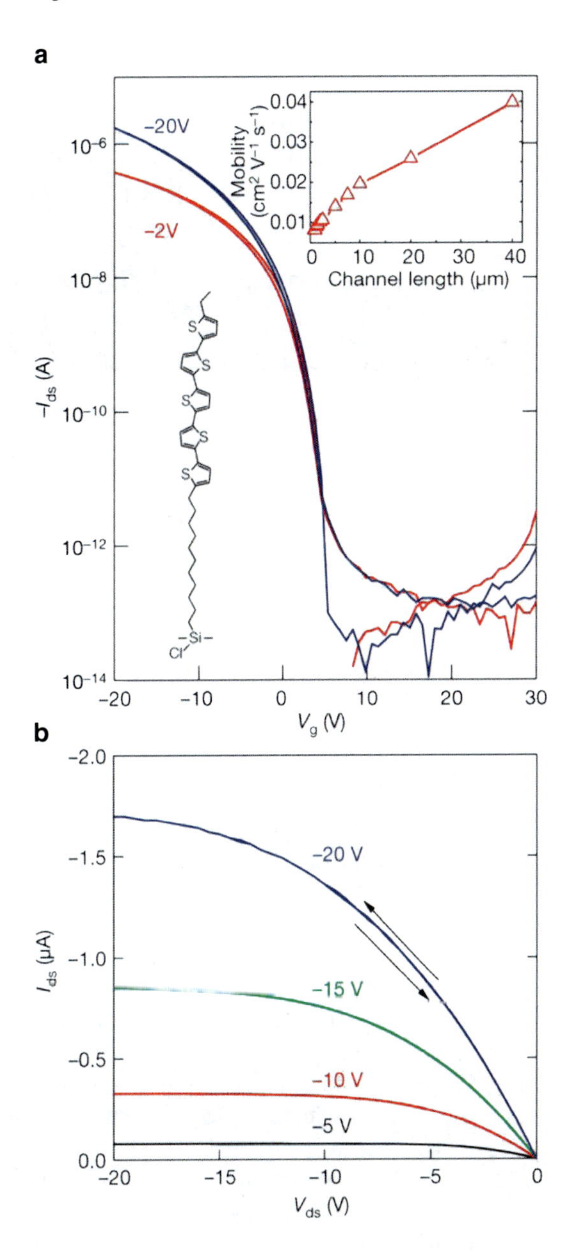

Fig. 10 (**a**) Transfer characteristics in the linear and saturation regimes for α-substituted quinquethiophene liquid crystalline monolayers assembled on the SiO_2 gate dielectric surface of a 40 μm channel length FET. (*Inset*) Field-effect mobility as a function of FET channel length. (**b**) Output characteristics of the FET. The transfer and output characteristics were scanned in both positive and negative directions in applied voltage

modulation I_{ON}/I_{OFF} exemplified the unique attributes achievable with molecular monolayer FETs.

The liquid crystalline, α-substituted quinquethiophene was used by the same group to translate this approach of self-assembly from the SiO_2 gate dielectric surface of FETs to assembly on organic dielectric surfaces, a first step toward fabricating flexible monolayer electronics [72]. Figure 11a provides a schematic of

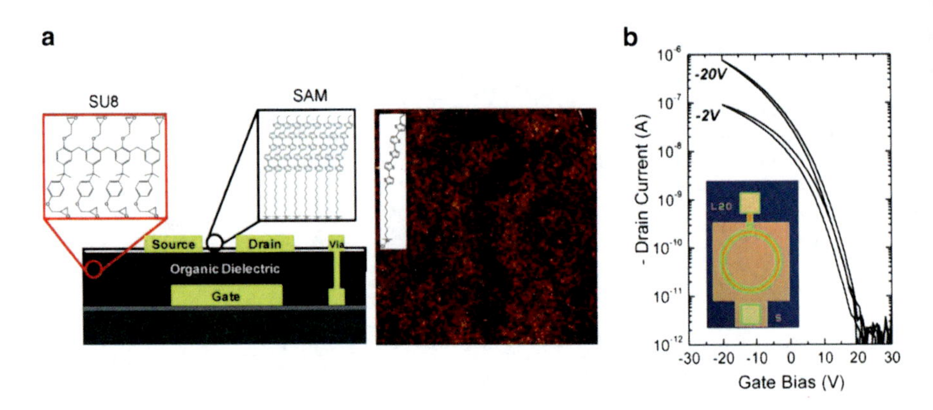

Fig. 11 (**a**) Schematic of an α-substituted quinquethiophene liquid crystalline monolayer assembled on an SU8 organic gate dielectric layer of an FET. (**b**) Transfer characteristics in the linear and saturation regimes for a 20 μm channel length monolayer FET fabricated (*inset*) in a ring geometry

the organic dielectric layer introduced into a bottom-contact, bottom-gate FET structure. The organic dielectric material selected was cross-linked SU8, a common negative photoresist. SU8 bears epoxides, that, upon exposure to oxygen plasma and hydrolysis, form hydroxyl groups that allow the self-assembly of the silane-functionalized alkylquinquethiophenes, with the same in-plane and upright structure as seen on the SiO_2 surface. The ordered packing of the monolayers is attributed to the liquid crystalline nature of the compound, provided by the long alkyl chains and π–π interactions between the thiophene cores. Figure 11b shows the I_D–V_G characteristics in the linear and saturation regimes for a 20-μm FET in a ring geometry. Field-effect mobilities of 0.02 cm^2/V s and I_{ON}/I_{OFF} ~10^6 were realized for the FETs having organic gate dielectric layers.

Fantastically, this same α-substituted quinquethiophene allowed the fabrication and characterization of a dual gate monolayer FET [73]. Using the bottom-gate monolayer FETs assembled on the SiO_2 gate dielectric surface, a top-gate dielectric and gate were fabricated by spin-coating an amorphous Teflon derivative (AF-1600, Sigma-Aldrich) and evaporating Au through a shadow mask, respectively (Fig. 12a). Dual-gate and wrap-around-gate FET architectures are being pursued in Si technology, as well as in the integration of one-dimensional nanoscale materials in FETs, to allow control of the carrier modulation in the entire semiconductor body. In molecular FETs gate modulation of the carrier concentration occurs in the first few nanometers of the gate dielectric interface (as described above). Unlike organic thin-film dual-gate FETs with semiconductor films >10 nm in thickness, which form two spatially separated channels, the dual-gate monolayer FET uniquely allows both the top and bottom FET gates to modulate overlapping semiconducting channels in the monolayer. The dual-gate FET structure is important in achieving higher currents and subthreshold slope (S^{-1} in Fig. 2e, f) and in tuning FET threshold voltage (V_T) to attain lower-noise circuitry. The top and bottom gate transfer characteristics (Fig. 12b) showed, by fixing the opposite gate

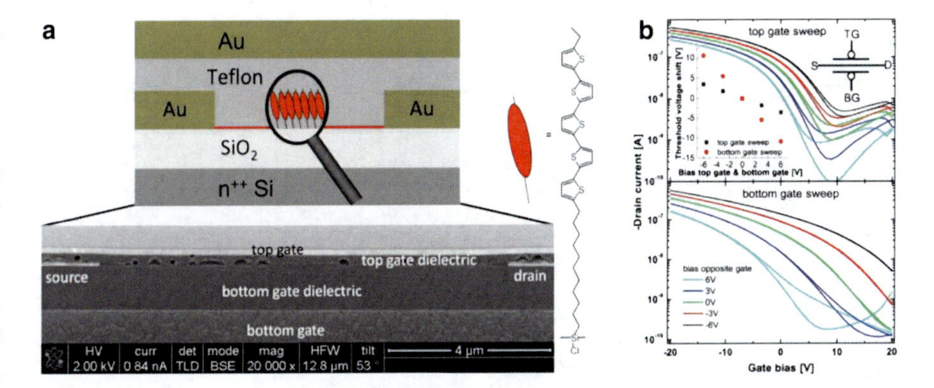

Fig. 12 (**a**) α-Substituted quinquethiophene liquid crystalline monolayer dual gate FET fabricated by assembly in a bottom gate FET configuration on an SiO_2 gate dielectric surface followed by spincoating a Teflon® top gate dielectric and evaporating a Au top gate. (**b**) Transfer characteristics of the top and bottom gate FETs and (*inset*) the threshold voltage at fixed gate voltages on the opposite gate

at voltages between −6 V and +6 V (Fig. 12b, inset), that V_T was linearly shifted. The V_T shift for the bottom gate is greater than that for the top gate, as the capacitance of the bottom SiO_2 gate dielectric layer is lower than that of the top Teflon gate dielectric layer.

4 Molecular Monolayer Circuitry

Integration of FETs into circuitry is one of the ultimate technology demonstrations, requiring many devices to be fabricated with near-uniform electrical characteristics. For the SAMFETs, the spontaneous nature of the assembly process renders it particularly attractive for low-cost, large-area electronics. The micron-scale channel α-substituted quinquethiophene SAMFETs were quickly implemented in different electronic circuits, with FETs having SiO_2 and SU8 gate dielectrics [70, 72].

While detailed circuit operation will not be described, these SAMFETs were demonstrated in three different circuit topologies. Figure 13a shows an optical micrograph of a unipolar p-type inverter and its transfer characteristics. The inverter is one of the simplest FET circuits, integrating two FETs, and is a building block of larger circuits. For the inverter, the dc transfer characteristics are described by the output voltage (V_{OUT}), measured as a function of input voltage (V_{IN}) for different supply voltages (V_{DD}). The action of the circuit is to invert the input signal, so a low V_{IN} gives rise to high V_{OUT} and a high V_{IN} gives rise to low V_{OUT}. The inverter can also act as a voltage amplifier. The gain of the monolayer inverters is shown inset for different supply voltages. Combining multiple inverters with a buffer was used to form (Fig. 13b) seven-stage ring oscillators that on the SU8 gate dielectric oscillate at 2 kHz. A 4-bit code generator, building on the oscillator, was demonstrated, requiring

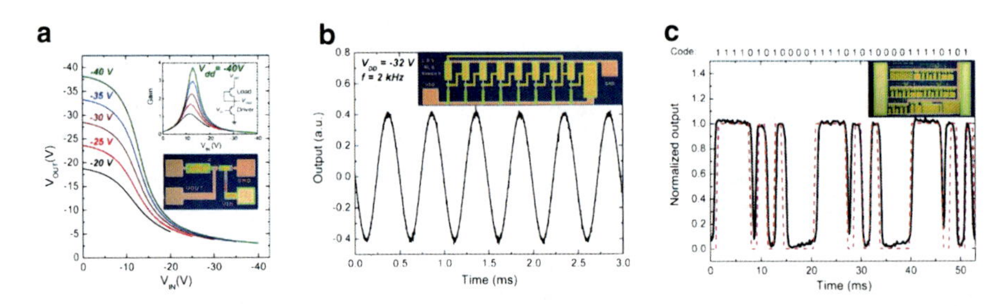

Fig. 13 Monolayer FETs assembled on SU8 organic gate dielectric layers configured to form different circuits. (**a**) dc transfer characteristics of an inverter at different supply voltages (V_{dd}) (**b**) output characteristics of a 7-stage ring oscillator, and (**c**) output of a 4-bit code generator (*continuous line*) and the preprogrammed code (*dotted line*)

over 100 SAMFETs with uniform enough electrical characteristics (Fig. 13c). These monolayer circuits have performances similar to those of organic thin-film circuits, and demonstrate the uniformity in performance that may be achieved for hundreds of FETs, necessary to realize monolayer FET integrated electronics.

5 Molecular Monolayer Chemical Sensors

The access of analytes to the gate-modulated semiconductor channel that is achievable in molecular monolayer FETs makes this device architecture particularly attractive for applications in chemical sensing. Both vacuum-evaporated near-monolayer and self-assembled-monolayer molecular FETs have been applied in chemical sensing. The near-1.3 monolayer thermally evaporated FETs of 6PTTP6, described above, were used with an added 4-nm blend layer of 6PTTP6 and HO6OPT (5,50-bis (4-(6-hydroxyhexyloxy)phenyl)-2,20-bithiophene) [37]. The HO6OPT introduces surface hydroxyl groups that act as receptors of dimethyl methylphosphonate (DMMP), a nerve agent. The FET mobility reversibly decreased by 70% after 40 s exposure to 5 ppm DMMP, and was recovered by flowing N_2 gas over the device (Fig. 14).

SAMFETs have also been used in chemical sensing. The α-substituted quinquethiophene SAMFETs were covered with a ~10-nm pinhole-riddled iron^{III} tetraphenylporphyrin chloride layer, that acts as a receptor to nitric oxide (NO), an important biomarker [74]. The threshold voltage, measured by the FET transfer characteristics with the porphyrin receptor shifts upon increased exposure to NO. Annealing the monolayer FET in vacuum restores the initial FET behavior. Also, in the single monolayer HBC assembled FETs between metallic SWCNT source and drain electrodes increased current levels were measured in I_D–V_{DS} and I_D–V_G characteristics (Fig. 9) upon exposure to solutions of the electron acceptor TCNQ [68]. While the mechanism of response is not known, TCNQ has an affinity for coronene, and likely gives rise to charge transfer between electron-deficient TCNQ

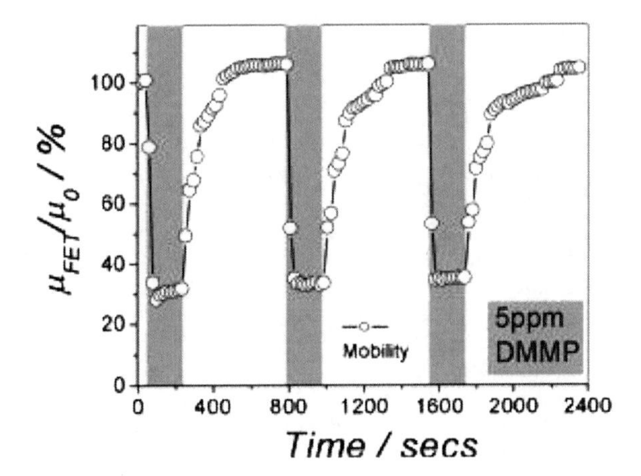

Fig. 14 Normalized field-effect mobility recorded as a function of time upon (*gray*) exposure to 5 ppm DMMP and (*white*) flowing fresh N_2 gas

and electron-rich HBC. The exposure to TCNQ is observed as a positive shift in V_T of the FETs. Application of molecular monolayer FETs in chemical sensing has only recently been reported. The mechanisms of interaction between the monolayer and the analyte may vary with the system under study, and give different output signatures, such as changes in mobility or threshold voltage. Monolayer FETs are expected to provide the ultimate sensing performance, with faster response times and higher sensitivity than for thin-film semiconductor FETs, as already reducing the thickness of organic thin films has allowed the demonstration of higher-performance chemical sensors [75].

6 Conclusion

Molecular monolayer FETs have rapidly been advanced from first demonstrations to integration in electronic circuitry and chemical sensors. The order and organization in molecular monolayers on the length scale of the device is critical to realizing high-performance monolayer electronics, and may be lengthened through molecular design. The flexibility afforded by synthetic chemistry promises a wider range of molecular materials. Functionalization is important both to our understanding of charge transport in the molecular solid state and to electronic and sensing applications. Understanding how to manipulate charge injection and charge transport, and preventing undesirable charge trapping, may lead not only to the p-type molecular monolayer FETs demonstrated to date, but to p-type, n-type, and ambipolar FETs desirable for molecular-monolayer-like CMOS (complementary metal–oxide semiconductor FET) electronics. The first steps have been taken to transfer molecular assembly from inorganic to organic surfaces, providing a pathway for flexible, molecular, monolayer electronics.

References

1. Aviram A, Ratner MA (1974) Molecular rectifiers. Chem Phys Lett 29:277–283
2. Davis WB, Ratner MA, Wasielewski MR (2001) Conformational gating of long distance electron transfer through wire-like bridges in donor-bridge-acceptor molecules. J Am Chem Soc 123:7877–7886
3. Lakowicz JR (2006) Principles of fluorescence spectroscopy, vol 1, 3rd edn. Springer, New York
4. Chidsey CED (1991) Free energy and temperature dependence of electron transfer at the metal-electrolyte interface. Science 251:919–922
5. Sikes HD, Smalley JF, Dudek SP, Cook AR, Newton MD, Chidsey CED, Feldberg SW (2001) Rapid electron tunneling through oligophenylenevinylene bridges. Science 291:1519–1523
6. Creager S, Yu CJ, Bandad C, O'Connor S, MacLean T, Lam E, Chong Y, Olsen GT, Luo J, Gozin M, Kayyem JF (1999) Electron transfer at electrodes through conjugated 'molecular wire' bridges. J Am Chem Soc 121:1059–1064
7. Zhu X-Y (2004) Charge transport in metal-molecule interfaces: a spectroscopic view. J Phys Chem B 108:8778–8793
8. Cahen D, Kahn A, Umbach E (2005) Engergetics of molecular interfaces. Mater Today 8:32–41
9. Loiacono MJ, Granstrom EL, Frisbie CD (1998) Investigation of charge transport in thin, doped sexithiophene crystals by conducting probe atomic force microscopy. J Phys Chem B 102:1679–1688
10. Dinelli F, Murgia M, Levy P, Cavallini M, Biscarini F, de Leeuw DM (2004) Spatially correlated charge transport in organic thin film transistors. Phys Rev Lett 92:116802–116803
11. Horowitz G, Haijlaoui R, Delannoy P (1995) Temperature dependence of the field-effect mobility of sexithiophene, determination of the density of traps. J Phys III 5:355–371
12. Ruiz R, Papadimitratos A, Mayer AC, Malliaras GG (2005) Thickness dependence of mobility in pentacene thin film transistors. Adv Mater 17:1795–1798
13. Park B-N, Seo S, Evans PG (2007) Channel formation in single-monolayer pentacene thin film transistors. J Phys D 40:3506–3511
14. Dodabalapur A, Torsi L, Katz HE (1995) Organic transistors – 2-dimensional transport and improved electrical characteristics. Science 268:270–271
15. Muck T, Wagner V, Bass U, Leufgen M, Geurts J, Molenkamp LW (2004) In situ electrical characterization of DH4T field-effect transistors. Synth Met 146:317–320
16. Sze SM, Ng KK (2007) Physics of semiconductor devices, 3rd edn. Wiley, Hoboken, NJ
17. Pierret RF (1990) Field effect devices, vol 4, 2nd edn. Prentice Hall, Reading, MA
18. Kanicki J (1992) Amorphous and microcrystalline semiconductor devices, vol 2: materials and device physics. Artech House, Boston, MA
19. Kagan CR, Andry P (2003) Thin-film transistors. Marcel Dekker, New York, NY
20. Zaumseil J, Sirringhaus H (2007) Electron and ambipolar transport in organic field-effect transistors. Chem Rev 107:1296–1323
21. Klauk H (2010) Organic thin-film transistors. Chem Soc Rev 39:2643–2666
22. Kagan CR, Afzali A, Martel R, Gignac LM, Solomon PM, Schrott AG, Ek B (2002) Evaluations and considerations for self-assembled monolayer field-effect transistors. Nano Lett 3:119–124
23. Frank DJ, Dennard RH, Nowak E, Solomon PM, Taur Y, Wong H-SP (2001) Device scaling limits of Si MOSFETs and their application dependencies. Proc IEEE 90:259–288
24. Asadi A, Svensson C, Willander M, Inganas O (1988) Field-effect mobility of poly(3-hexylthiophene). Appl Phys Lett 53:195–197
25. Garnier F, Yassar A, Hajlaoui R, Horowitz G, Deloffre F, Servet B, Ries S, Alnot P (1993) Molecular engineering of organic semiconductors: design of self-assembly properties in conjugated thiophene oligomers. J Am Chem Soc 115:8716–8721

26. Lin YY, Gundlach DJ, Nelson SF, Jackson TN (1997) Stacked pentacene layer organic thin-film transistors. IEEE Electron Dev Lett 18:606–608
27. Payne MM, Parkin SR, Anthony JE, Kuo CC, Jackson TN (2005) Organic field-effect transistors from solution-deposited functionalized acenes with mobilities as high as 1 cm2/V-s. J Am Chem Soc 127:4986–4987
28. Kagan CR, Afzali A, Graham TO (2005) Operational and environmental stability of pentacene thin film transistors. Appl Phys Lett 86:193505
29. Ling MM, Erk P, Gomez M, Koenemann M, Locklin J, Bao ZN (2007) Air-stable n-channel organic semiconductors based on perylene diimide derivatives without strong electron withdrawing groups. Adv Mater 19:1123
30. Jones BA, Facchetti A, Wasielewski MR, Marks TJ (2008) Effects of arylene diimide thin film growth on n-channel OFET performance. Adv Funct Mater 18:1329–1339
31. Katz HE, Lovinger AJ, Johnson J, Kloc C, Siegrist T, Li W, Lin Y-Y, Dodabalapur A (2000) A soluble and air-stable organic semiconductor with high electron mobility. Nature 404:478–481
32. Chua L-L, Zaumseil J, Chang J-F, Ou EC-W, Ho PK-H, Sirringhaus H, Friend RH (2005) General observation of n-type field-effect behaviour in organic semiconductors. Nature 434:194–1999
33. Saudari SR, Frail PR, Kagan CR (2009) Ambipolar transport in solution-deposited pentacene transistors enhanced by molecular engineering of device contacts. Appl Phys Lett 95:023301
34. Saudari SR, Lin Y-J, Lai Y, Kagan CR (2010) Device configurations for ambipolar transport in flexible pentacene transistors. Adv Mater 44:5063–5068
35. Cheng XY, Noh YY, Wang JP, Tello Frisch MJ, Blum RP, Vollmer A, Rabe JP, Koch N, Sirringhaus H (2009) Controlling electron and hole charge injection in ambipolar organic field-effect transistors by self-assembled monolayers. Adv Funct Mater 19:2407–2415
36. Meyer zu Heringdorf F-J, Reuter MC, Tromp RM (2001) Growth dynamics of pentacene thin films. Nature 412:517–520
37. Huang J, Sun J, Katz HE (2008) Monolayer-dimensional 5,5'-bis(4-hexylphenyl)-2,2'-bithiophene transistors and chemically responsive heterostructures. Adv Mater 20:2567–2572
38. Venables JA, Spiller GDT, Hanbucken M (1984) Nucleation and growth of thin-films. Rep Prog Phys 47:399–459
39. Gao J, Xu JB, Zhu M, Ke N, Ma D (2007) Thickness dependence of mobility in CuPc thin film on amorphous SiO_2 substrate. J Phys D 40:5666–5669
40. Kiguchi M, Nakayama M, Fujiwara K, Ueno K, Shimada T, Saiki K (2003) Accumulation and depletion layer thickness in organic field effect transistors. Jpn J Appl Phys 42:L1408–L1410
41. Asadi K, Wu Y, Gholamrezaie F, Rudolf P, Blom PWM (2009) Single-layer pentacene field-effect transistors using electrodes modified with self-assembled monolayers. Adv Mater 21:4109–4114
42. Kymissis I, Dimitrakopoulos CD, Purushshothaman S (2001) High-performance bottom electrode organic thin-film transistors. IEEE Trans Electron Dev 48:1060–1064
43. Gundlach DJ, Jia LL, Jackson TN (2001) Pentacene TFT with improved linear region characteristics using chemically modified source and drain electrodes. IEEE Electron Dev 22:571–573
44. Tulevski GS, Miao Q, Afzali A, Graham TO, Kagan CR, Nuckolls C (2006) Chemical complementarity in the contacts for nanoscale organic field-effect transistors. J Am Chem Soc 128:1788–1789
45. Bock C, Pham DV, Kunze U, Kafer D, Witte G, Woll C (2006) Improved morphology and charge carrier injection in pentacene field-effect transistors with thiol-treated electrodes. J Appl Phys 100:114517
46. Hamadani BH, Corley DA, Ciszek JW, Tour JM, Natelson D (2006) Controlling charge injection in organic field-effect transistors using self-assembled monolayers. Nano Lett 6:1303–1306
47. Tredgold RH (1987) The physics of Langmuir-Blodgett films. Rep Prog Phys 50:109–128
48. Peterson IR (1990) Langmuir Blodgett films. J Phys D 23:379–395

49. Scott JC, Samuel JDJ, Hou JH, Rettner CT, Miller RD (2006) Monolayer transistor using a highly ordered conjugated polymer as the channel. Nano Lett 6:2916–2919
50. Wei Z, Cao Y, Ma W, Wang C, Xu W, Guo X, Hu W, Zhu D (2009) Langmuir-Blodgett monolayer transistors of copper phthalocyanine. Appl Phys Lett 95:033304
51. Klauk H, Schmid G, Radlik W, Weber W, Zhou LS, Sheraw CD, Nichols JA, Jackson TN (2003) Contact resistance in organic thin film transistors. Solid State Electron 47:297–301
52. Nuzzo RG, Allara DL (1983) Adsorption of bifunctional organic disulfides on gold surfaces. J Am Chem Soc 105:4481–4483
53. Ulman A (1996) Formation and structure of self-assembled monolayers. Chem Rev 96:1533–1554
54. Sagiv J (1980) Organized monolayers by adsorption. 1. Formation and structure of oleophobic mixed monolayers on solid surfaces. J Am Chem Soc 102:92–98
55. Allara DL, Nuzzo RG (1985) Spontaneously organized molecular assemblies. 1. Formation, dynamics, and physical-properties of normal-alkanoic acids adsorbed from solution on an oxidized aluminum surface. Langmuir 1:45–52
56. Chen C-Y, Wu K-Y, Chao Y-C, Zan H-W, Meng H-F, Tao Y-T (2011) Concomitant tuning of metal work function and wetting property with mixed self-assembled monolayers. Org Electron 12:148–153
57. Kagan CR, Breen T-L, Kosbar LL (2001) Patterning organic-inorganic thin-film transistors using microcontact printed templates. Appl Phys Lett 79:3536–3538
58. de Boer B, Hadipour A, Mandoc MM, van Woudenbergh T, Blom PWM (2005) Tuning of metal work functions with self-assembled monolayers. Adv Mater 17:621–625
59. Mathijssen SGJ, van Hal PA, van den Biggelaar TJM, Smits ECP, de Boer B, Kemerink M, Janssen RAJ, de Leeuw DM (2008) Manipulating the local light emission in organic light-emitting diodes by using patterned self-assembled monolayers. Adv Mater 20:2703–2706
60. Miozzo L, Yassar A, Horowitz G (2010) Surface engineering for high performance organic electronic devices: the chemical approach. J Mater Chem 20:2513–2538
61. Collet J, Tharaud O, Chapoton A, Vuillaume D (2000) Low-voltage, 30 nm channel length, organic transistors with a self-assembled monolayer as gate insulating films. Appl Phys Lett 76:1941–1943
62. Klauk H, Zschieschang U, Pflaum J, Halik M (2007) Ultralow-power organic complementary circuits. Nature 445:745–748
63. DiBenedetto SA, Facchetti A, Ratner MA, Marks TJ (2009) Molecular self-assembled monolayers and multilayers for organic and unconventional inorganic thin film transistor applications. Adv Mater 21:1407–1433
64. Cohen YS, Vilan A, Ron I, Cahen D (2009) Hydrolysis improves packing density of bromine-terminated alkyl-chain, silicon-carbon monolayers linked to silicon. J Phys Chem C 113:6174–6181
65. Tulevski GS, Miao Q, Fukuto M, Abram R, Ocko B, Pindak R, Steigerwald ML, Kagan CR, Nuckolls C (2004) Attaching organic semiconductors to gate oxides: in situ assembly of monolayer field effects transistors. J Am Chem Soc 126:15048–15050
66. Mottaghi M, Lang P, Rodriguez F, Rumyantseva A, Yassar A, Horowitz G, Lenfant S, Tondelier D, Vuillaume D (2007) Low-operating-voltage organic transistors made of bifunctional self-assembled monolayers. Adv Funct Mater 17:597–604
67. Mann B, Kuhn H (1971) Tunneling through fatty acid salt monolayers. J Appl Phys 42:4398–4405
68. Guo X, Myers M, Xiao S, Lefenfeld M, Steiner R, Tulevski GS, Tang J, Baumert J, Leibfarth F, Yardley JT, Steigerwald ML, Kim P, Nuckolls C (2006) Chemoresponsive monolayer transistors. PNAS 103:11452–11456
69. Heinze S, Tersoff J, Martel R, Derycke V, Appenzeller J, Avouris PH (2002) Carbon nanotubes as Schottky barrier transistors. Phys Rev Lett 106801:1–4
70. Smits ECP, Mathijssen SGJ, van Hal PA, Setayesh S, Geuns TCT, Mutsaers KAHA, Cantotore E, Wondergem HJ, Werzer O, Resel R, Kemerink M, Kirchmeyer S, Muzaarov AM,

Ponomarenko SA, deBoer B, PBlom PWM, deLeeuw DM (2008) Bottom-up organic integrated circuits. Nature 455:956–959

71. Mathijssen SGJ, Smits ECP, van Hal PA, Wondergem HJ, Ponomarenko SA, Moser A, Resel R, Bobbert PA, Kemerink M, Janssen RAJ, de Leeuw DM (2009) Monolayer coverage and channel length set the mobility in self-assembled monolayer field-effect transistors. Nat Nanotechnol 4:674–679
72. Gholamrezaie F, Mathijssen SGJ, Smits ECP, Geuns TCT, van Hall PA, Ponomarenko SA, Flesch H-G, Resel R, Cantatore E, Blom PWN, de Leeuw DM (2010) Ordered semiconducting self-assembled monolayers on polymeric surfaces utilized in organic integrated circuits. Nano Lett 10:1998–2002
73. Spojkman M, Mathijssen SGJ, Smits ECP, Kemerink M, Blom PWM, de Leeuw DM (2010) Monolayer dual gate transistors with a single charge transport layer. Appl Phys Lett 96:143304
74. Andringa A-M, Spijkman M-J, Smits ECP, Mathijssen SGJ, van Hal PA, Setayesh S, Willard MP, Borshchev OV, Ponomarenko SA, Blom PWM, de Leeuw DM (2010) Gas sensing with self-assembled monolayer field-effect transistors. Org Electron 11:895–898
75. Yang RD, Gredig T, Colesniuc CN, Park J, Schuller IK, Trogler WC, Kummel AC (2007) Ultrathin organic transistors for chemical sensing. Appl Phys Lett 90:263506

Top Curr Chem (2012) 312: 239–274
DOI: 10.1007/128_2011_177

Published online: 28 July 2011

Issues and Challenges in Vapor-Deposited Top Metal Contacts for Molecule-Based Electronic Devices

Masato M. Maitani and David L. Allara

Abstract Metal vapor deposition to form ohmic contacts is commonly used in the fabrication of organic electronic devices because of significant manufacturability advantages. In the case of single molecular layer devices, however, the extremely small thickness, typically ~1–2 nm, presents serious challenges in achieving good contacts and device integrity. This review focuses on recent scientific aspects of metal vapor deposition on monolayer thickness molecular films, particularly self-assembled monolayers, ranging across mechanisms of metal nucleation, metal-molecular group interactions and chemical reactions, diffusion of metal atoms within and through organic films, and the correlations of these and other factors with device function. Results for both non-reactive and reactive metal deposition are reviewed. Finally, novel strategies are considered which show promise for providing highly reliable and durable metal/organic top contacts for use in metal–molecule–metal junctions for device applications.

Keywords Metal vapor deposition · Metal-organic interface · Molecular devices · Molecular electronics · Self-Assembled Monolayers

Contents

M.M. Maitani
Department of Applied Chemistry, Tokyo Institute of Technology, Tokyo 152-8552, Japan

D.L. Allara (✉)
Department of Chemistry, Pennsylvania State University, University Park, PA 16802, USA

Department of Materials Science and Engineering, Pennsylvania State University, University Park, PA 16802, USA
e-mail: dla3@psu.edu

1 Introduction and Scope of the Review

Metal vapor deposition to form ohmic contacts is commonly used in fabrication of electronic devices with organic films serving as the active functional component and is one of the means of preparing research devices based on single molecular monolayers. In terms of the manufacturability of a molecule-based device with top metal contacts there are significant advantages to using vapor deposition, including a clean environment, simple integration into a standard fab process, the potential ease of making highly uniform contacts over large area device wafers, the flexibility of choosing from a large variety of possible metals, and the added advantage that for certain types of devices the use of electropositive metals can result in charge injection to form a favorable interface dipole to assist charge transport across the junction [1–10].

In the case of single molecular layer devices, however, the extremely small thickness, at the 1–2 nm scale dictated by the end-to-end dimensions of a molecule, presents serious challenges in achieving good contacts and device integrity. In contrast to inorganic thin films, which have strong chemical bonding throughout the atomic lattice and can form highly uniform, dense coatings via deposition methods such as atomic layer deposition (ALD), soft matter molecular films are bound together by weak intermolecular interactions, mostly van der Waals in origin, which typically allow defects, such as pinholes caused by missing molecules and misoriented molecules, to arise after deposition. The soft character of the films frequently leads to damage by thermal exposure, for example via impact of depositing metal atoms with high kinetic energies when generated from molten sources at high temperatures (>1,000 K typically). The penetration of only a few atoms stacked axially in a defect hole of ~1–2 nm length can lead to a local short in the junction. Further, the rich π-electron character of typical molecules of interest for device activity can pose issues with respect to adverse chemical interactions for many metals, ranging from simple donor–acceptor complexation to vigorous degradation of the molecular backbone for those metals which can form strong C–metal or O–metal bonds, corresponding to inorganic carbide and oxide products. With these and related factors in mind, it is clear that there is a significant challenge in meeting the key objective of creating an organic/metal contact for a given organic film and metal atom combination which produces the desired device behavior with

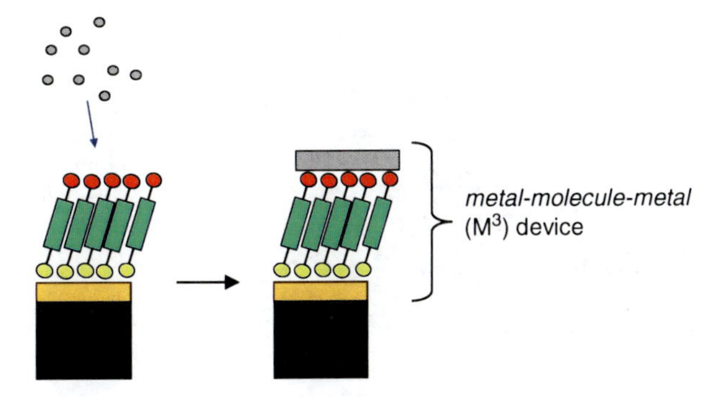

Fig. 1 Schematic of an M^3 device fabricated with a vapor-deposited top metal contact

good junction properties that last through millions of device cycles. To meet this challenge a range of scientific and engineering knowledge is needed, including mechanisms of metal nucleation on specific organic surfaces, metal-molecular group interactions and reactions, diffusion of metal atoms within and through organic films, and the correlations of these and many other factors with device function.

The focus in this chapter is on the scientific issues and challenges in the application of vapor deposited top metal contacts to devices fabricated from self-assembled monolayers (SAMs) and related films such as Langmuir Blodgett (LB) monolayers. Specifically the chapter will review the types of possible interactions of nascent metal atoms with SAMs, with a view to the effects on top contact behavior, and then review examples from the literature on specific metals and SAMs. A schematic of a metal/molecule/metal or M^3 device is shown in Fig. 1 for reference.

2 General Structural Features of Organized Molecular Monolayers

In order to understand the effects of metal vapor deposition on supported molecular monolayers it is necessary to define and understand the basic elements of their structure. For reference, the general features are shown in Fig. 2. This diagram also applies to LB films and we will use the term SAM to cover both cases unless otherwise specified. SAMs can be divided into two limiting classes according to the *headgroup* attachment at the *substrate surface* to form the *headgroup–substrate interface*: (1) headgroups chemically attached to the substrate by donor-acceptor bonding, e.g., thiol groups (–SH) chemisorbed to Au as thiolate (–S) species, covalent bonding, e.g., C–Si bonds in the case of alkylsilanes grafted onto hydride Si surfaces, and ionic types of bonding, e.g., carboxylic acids ($-CO_2H$) chemisorbed as carboxylate ions ($-CO_2^-$) on oxide-covered Ag and (2)

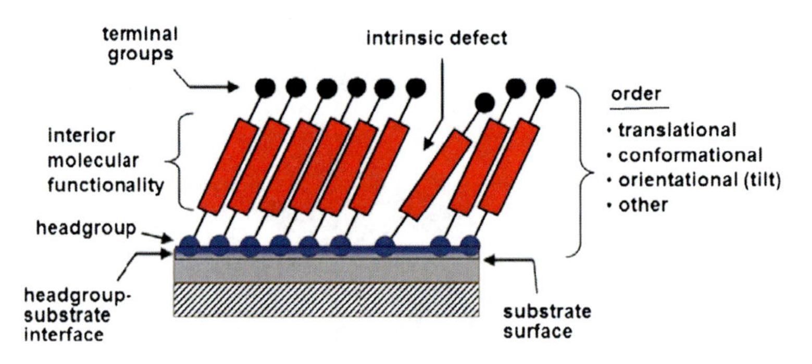

Fig. 2 Schematic of a SAM showing the constituent features of interest for applications

headgroups solvated by a thin layer of water (~1–3 monolayers) adsorbed at the substrate surface, e.g., typical LB films drawn from an air–water interface onto hydrophilic substrates and octadecyl trisiloxane (ODS) films self-assembled on highly hydrated silicon oxide surfaces. The first class consists of chemically attached films on a substrate template and the second is essentially a surfactant monolayer on a supported ultrathin water film. In general, organized molecular monolayer structures will fall somewhere between these two limits depending on the specific molecules, substrates and preparation conditions. In the case of a crystalline substrate with strong bonding sites the translational (the two-dimensional arrangement) organization of a SAM is driven by a matching of the molecular sizes and headgroup positions to an ordered arrangement of substrate surface atoms. In the case of weaker bonding substrate sites and those with no preferential sites (e.g., a water layer), the intrinsic intermolecular forces tend to drive the molecules into an ordered packing (essentially a surfactant layer self-assembling on a liquid surface). The orientational arrangements of the molecules (tilting and twisting) and the conformational sequences are driven by secondary interactions of the interior molecular structure or "backbone" which guide the molecules into tighter contact, thus lowering the potential energy of the system. In the case of M^3 devices, the interior molecular groups are functional in terms of moderating charge transport through the molecule, thus they are termed *interior molecular functionality* and their exact orientational and conformational ordering can be important in the functioning of the device. Finally, the *terminal groups* of the SAM, exposed to the ambient interface (or the top electrode in the fabricated device) will be organized generally according to the headgroup-substrate and interior molecular ordering.

The static defects in a SAM are of considerable importance in making high quality vapor deposited metal top contact layers since metal atoms are ~3 Å in diameter and can easily penetrate through molecular scale defects. In general, because of the statistical difficulty of annealing planar, two-dimensional films, the translational ordering of a molecular monolayer is typically not long range with translational correlation lengths of ordered domains for well-organized SAMs of the order of ~10–20 nm on single crystal substrates, which gives rise to defects in

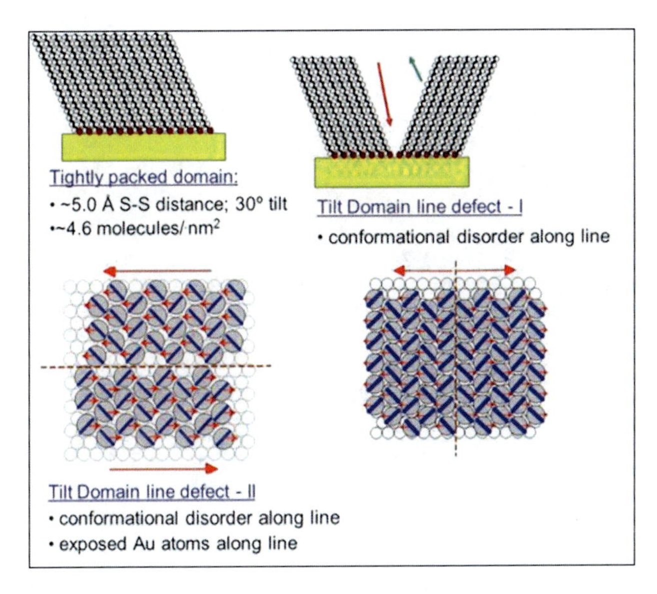

Fig. 3 Schematic of a SAM showing different types of molecular defects

the monolayer between the domains. Further, if the substrates are polycrystalline defects, they are readily introduced at the grain boundaries on the underlying substrate surface. The main molecular defects consist of missing molecules, which constitutes the smallest "pinhole," depressions in the SAM due to missing atom defects in the substrate in an otherwise defect free substrate surface terrace (also termed "pit defects"), and various defects at the edges of molecular domains, as illustrated in Fig. 3 for the example of an alkanethiolate on Au(111) SAM. Overall, there are a number of different types of intrinsic static defects that can play a role in producing defects in vapor deposited top metal contacts in terms of defeating the production of highly uniform, sharp interface contacts with no leakage paths between the electrodes.

Dynamic defects can also arise in certain types of SAM and are particularly important in organothiolate SAMs on Au surfaces [11–13]. These types of defects arise from the intrinsic local thermal motion of the headgroup laterally around the pinning site on the substrate. Coupling with transverse motions of the molecules in a local region can result in transient formation of a molecular diameter scale pinhole in the film, even for highly organized domains on a crystalline pinning lattice. These transient holes in turn can provide channels for metal atom transport through the film to the substrate. Transient defects can be expected for any SAM in which the displacement potential for a headgroup motion around the pinning site is of the order of the thermal energy (~kT) in the system and are particularly notable for organothiolates on Au(111). An illustration of the transient pinhole for an organothiolate adsorbate on Au(111) is given in Fig. 4 [11–13]. Extensive consideration of such effects has been made in a study of O atom penetration into various Langmuir-Blodgett films by Naaman et al. [14].

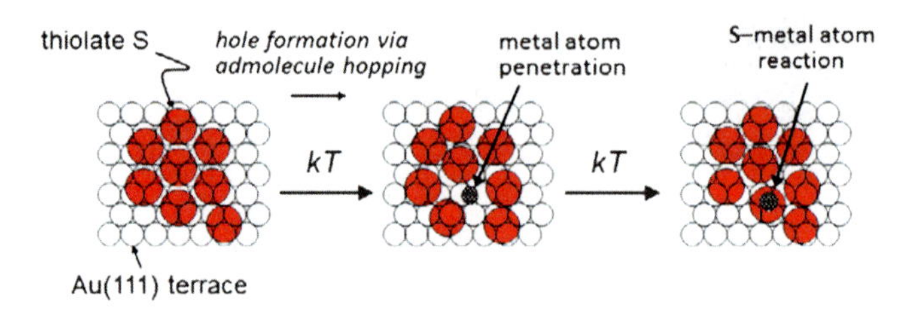

Fig. 4 Schematic illustrating the dynamic mechanism of metal atom penetration into a SAM by means of transient opening of holes by lateral motion of the adsorbate molecules

Finally, another class of important defects involves impurity materials included in the junction during processing, typically water, which is often hard to remove if a component of the junction has some hydrophilic character and the processing includes steps with air exposure or even cooling of samples under vacuum, e.g., during metal vapor deposition, where the vacuum chamber background pressure allows residual water vapor to condense on the sample prior to metal flux exposure. In the case of LB films drawn from the air/water interface with salts in the aqueous subphase, water layers and salts remain in the film/substrate interface and survive during vacuum processing. Thus in order to fabricate molecular junctions for device applications, contamination is an important issue [15–20].

3 General Aspects of Metal Vapor Deposition

Vacuum deposition of metal films is typically done by exposure of a sample to metal atom fluxes typically produced by three major methods: (1) evaporation from a heated boat or filament containing the pure metal (thermal evaporation), (2) evaporation from a grounded, conductive boat in which a tightly focused electron beam on an isolated region of a metal charge contained in the crucible is heated, causing metal vaporization (electron-beam or e-beam evaporation), and (3) ejection of metal atoms from a metal target into the vapor phase by means of energetic inert gas ions striking the target (sputtering). In the first two cases the energy for vaporization of the metal comes from the thermal energy delivered to the metal source. Given the high vaporization energies of most metals, this process places the metal source at temperatures well above 1,000 K and can result in thermal damage to an exposed SAM by radiation heating and/or by conduction heat transfer from the impinging thermally excited metal atoms. In the case of sputtering the presence of the energetic ions and neutral species used to bombard the source target and the excess energy of the metal vapor itself can also lead to sample damage. The main advantage of sputtering is the extremely fast rates of deposition, useful for example in producing micrometer-thick films. For any method, damage to the sample during deposition can be reduced by cooling the sample, for example, to liquid nitrogen temperatures, and for evaporation processing minimizing sample exposure to

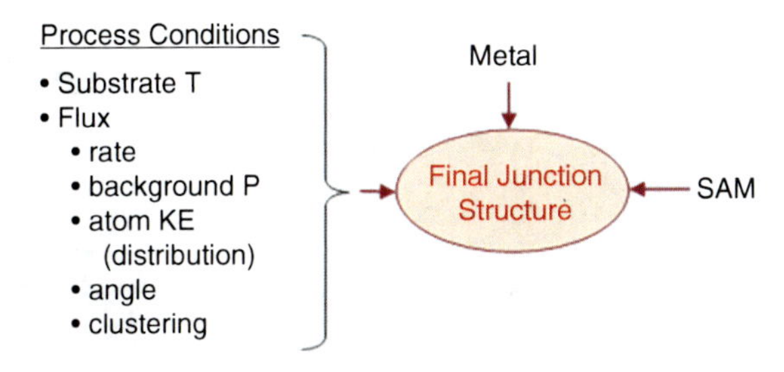

Fig. 5 Diagram showing the major aspects of metal atom vapor deposition on a SAM

radiation from the metal source. In all three methods the nascent atoms are delivered ballistically (line of sight) onto the surface. Since this mode of deposition cannot coat conformally surfaces with roughness features larger than the metal film thickness or with interior cavities, the final films may be highly non-uniform with incomplete coverage across the surface.

A general summary of the variables involved in producing good junctions are depicted in the schematic in Fig. 5. The main challenge is to choose the best metal and process conditions for a given SAM with the potential function of interest for device operation. The metal can vary widely depending upon the factors such as reactivity with the SAM and some desired molecule/metal interface electrical property. Once the metal has been chosen, the variables that can be controlled most readily are the rate of deposition, background pressure, deposition angle, and substrate temperature. More difficult to control directly are the average atom kinetic energy (KE), which is typically set by the source temperature of the vaporizing metal, and the degree of clustering of gaseous metal atoms, which can be a complex parameter of several of the other variables. There will be some discussion of these factors in later sections but in general these sorts of details are beyond the scope of this chapter and can be found elsewhere in many standard references and texts.

4 Issues in Vapor Deposited Top Metal Electrodes for M^3 Devices

A perfect top contact would have a highly uniform layer of metal with complete, conformal contact at the SAM surface to form a sharp interface in which there is good overlap of the electronic states of the metal surface and the molecule terminal groups. This ideal situation is not to be expected in real device processing as a number of physical and chemical defects can easily arise, as summarized in the schematic in Fig. 6.

With respect to the sharpness of the top metal/molecule interface, thermodynamically, given the vastly different lattice energies of the soft-condensed matter SAM and the hard metal layer, one would expect the interface to be near the limit of no interfacial mixing (essentially a large chi mixing parameter, $\chi \gg 1$) and the interface to approach an infinitely sharp condition. An example of the complexity that can be introduced in this simple picture is illustrated by considering the case of

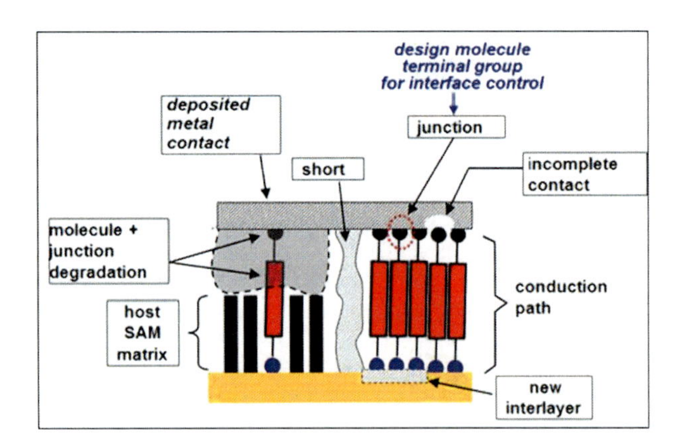

Fig. 6 Schematic showing the main features and technical issues of a SAM undergoing metal vapor deposition

chemical bond formation between the terminal groups and the condensed metal atoms, favorable for good electronic overlap in the junction, In this case an interfacial layer of metal–molecule reaction product is formed so the interface has new electronic character introduced, where, in the limit of highly exothermic reactions, such as can occur with vigorously reactive metals such as titanium, the reaction products can reach the limit of inorganic carbides, oxides, or nitrides, for example, and result in new inorganic layers introduced in the interfacial region, with likely adverse effects on charge transport. The obvious trade-off is that increased chemical interaction favors uniform metal nucleation, which results in uniform lateral growth and high contact integrity, whereas extensive chemical interaction can continue into the interior of the molecule and destroy the intended device functionality for the given molecule.

The presence of lateral density defects in the SAM, e.g., static pinholes resulting from missing molecules or dynamic formation of transient diffusion channels (see Sect. 5), can result in metal atom penetration into the SAM interior. On a thermodynamic basis, if there is no chemical reaction with the interior groups of the SAM, the metal atoms will cluster if space is available. Further, those atoms which can reach the underlying bottom metal/headgroup interface region may have a highly favorable interaction with the underlying metal phase, ranging from insertion into the headgroup/metal bond, to formation of an interface metal alloy and/or even to penetrating further into the bottom metal phase to form a bulk phase alloy. The simplest issue is the formation of metal filaments which leads to local shorts in the contact, essentially a set of very low resistance, parallel paths across the junction. In the case of formation of interlayer phases at the bottom contact, adverse changes in junction impedance may result.

In addition to these static problems with M^3 junctions, once a device is operating, dynamic perturbations may arise during a given area of device operation, as illustrated in Fig. 7. The major cause of perturbations would be the onset of extremely high electric fields across the junction, e.g., ~5×10^8 V m^{-1} for an

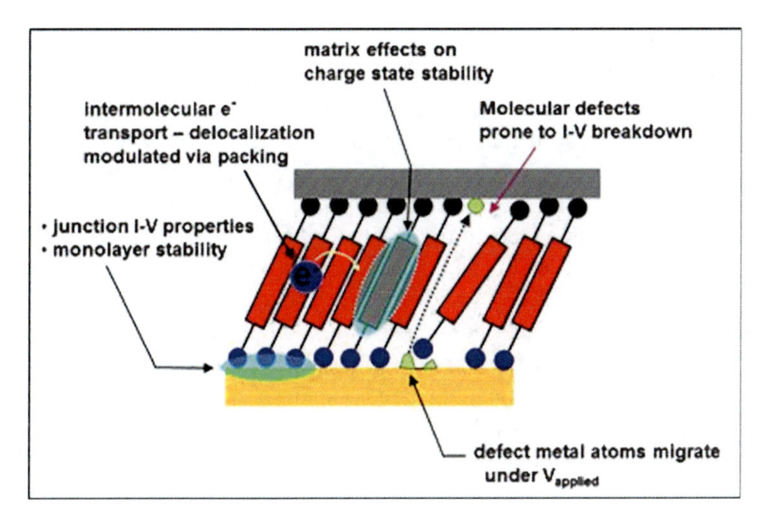

Fig. 7 Schematic of an M^{3}device with illustrations of features critical to the device operation

~2 nm long molecule in a junction at a 1 V bias. At these high fields there is sufficient energy to cause migration of metal atoms, either those intrinsic to the electrode or impurities, across the junction, in some cases even causing filament formation with resultant shorting. In addition, the fields can exert large electrostatic forces capable of twisting molecules bearing dipoles into new configurations with permanent changes in charge transport behavior. In some cases, if controlled, the latter effects can be the basis of switching or memristor effects but when these effects are spurious they lead to device failure over cycling. Other effects can arise, such as electrochemistry if the junction contains traces of water and salts and thermally driven effects arising from junction heating during operation.

Overall, from this discussion, amplified with the pictorial illustrations in Figs. 6 and 7, it is clear that a number of adverse perturbations can arise during top metal processing and operation of an M^3 device. The examples given in the following sections will demonstrate that these effects indeed arise and that considerable design and processing strategies are needed to overcome these problems along with a deep base of fundamental knowledge of the physical and chemical structures and operational mechanisms underlying the behavior of the M^3 devices.

5 Specific Fundamental Studies for Different Metals and Device Application

5.1 Overview and General Considerations

Since SAMs for device application typically involve organothiolate molecules assembled on an Au substrate, this review will focus on these types of SAMs for the most part. Extensions to other SAMs can be made by comparing the attachment chemistries and substrate characteristics and making projections based on the differences. The most critical issue for any SAM is metal atom leakage to the substrate with formation of filaments which can cause shorts. This problem typically arises from the presence of intrinsic static defects caused by the imperfection of self-assembly and atomic level morphologies of Au substrates, step edges and kinks, and from transient, dynamic defects caused by high frequency phonon type lateral motions of the headgroup moiety around the pinning site on the substrate [11, 12, 14]. Limited filament formation will interfere with the desired level of device operation, and continued formation eventually leads to complete device failure. In the case of penetration of nascent metal atoms through dynamic, transient defects, insertion into S–Au bonds can occur, and if the new S–metal–Au bonding is stronger than S–Au bonds this can result in diminishing of the lateral fluctuations, thereby stopping formation of the temporal vacancy channels, thus limiting the extent of penetration. On the other hand, this freezing process of the lateral positional fluctuation may also lead to new static defects because of the continuing disordering of the SAM molecule positions driven by the incommensurate non-uniform insertion of deposited metal into the SAM-substrate interface [20, 21].

In order to avoid introducing chemical reaction products into the junction via reactions of the deposited metal with the SAM and to avoid oxidation or corrosion of the top contact, inert or low reactive metal species, such as Au, Ag, Cu, and Pd, have been used [22–28]. Nonreactive metal atoms, however, due to their inertness with the SAM molecules, can easily penetrate wherever defects arise. In the case of dynamic defects which arise uniformly across the surface, the penetration can even continuously produce an under layer at the SAM/substrate interface. For example, in the case of inert aliphatic thiolate SAM molecules on Au, deposited Au atoms subsequently diffuse around the ambient surface and penetrate through both static and dynamic channels of SAMs, resulting in a "floating" SAM structure [11–13, 29–31]. To avoid the penetration through SAMs, reactive functional groups, such as –COOH, $-COOCH_3$, –OH, $-COCH_3$, –CN, –SH, etc., have been introduced to the SAM backbone, generally as terminal groups, as "traps" for the deposited atoms at the ambient surface. A variety of results with such reactive terminal groups will be discussed.

On the other hand, highly reactive metals can result in only top deposition on the SAM without penetration, though this takes place under conditions where reaction occurs, often resulting in serious degradation and corrosion of the SAM backbone. For example, Ti metal deposition will seriously degrade simple aliphatic types of thiolate SAMs to produce oxide and carbide species [32–35]. Another issue of reactive metal is reaction with adventitious contamination incorporated in the film during formation. In cases with water trapped, such as in LB films, deposition of reaction metal atoms such as Ti can result in formation of metal oxide species [17, 36]. The metal oxide produced during top metal deposition severely affects the electronic characteristics of the M^3 junction, although the coincident metal oxide species can actually be utilized as the materials basis for a memorister device [37, 38]. Even background oxygen gas in vacuum chambers can cause slow oxidation. For example, exposure of deposited Al to background pressures of ~10^{-9} Torr for extended periods can result in aluminum oxide formation at the exposed Al metal surfaces [20].

In the following literature review, advantages and disadvantages of both low reactive (inert) and highly reactive metal species are summarized with a few categories of metal species. Since the combination of metal and functional groups of SAMs is a key factor to determine the final metal contact structure, it is important to understand the fundamental mechanistic pathways of impinging metal atom adsorption on organic film surfaces. Therefore some fundamental studies of metal–SAM reactions are summarized briefly along with associated surface analysis methods used to characterize the film structures, including IRS, XPS, ToF-SIMS, SPM, spectroscopic ellipsometry, and related techniques. Readers, should refer to more comprehensive reviews for detailed discussions in this area [22–28]. The main focus of this review is on recent experimental studies designed to elucidate the fundamental issues in constructing M^3 molecular junctions metal vapor top contacts. Finally, a few promising alternate types of metal deposition processes are summarized which look to be useful.

5.2 *Low and Non-Reactive Metals*

5.2.1 Au

Room Temperature Deposition

Gold is by far the most common metal used in top metal contacts for devices based on SAMs. It is preferred in terms of its high electrical conductivity and near complete inertness towards typical environmental species such as oxygen and water vapor, as well as even towards more reactive gases such as acid vapors. Although well-made alkanethiolate SAMs on Au{111} are closely packed with excellent barrier characteristics, e.g., as observed in electrochemical measurements [39], vapor deposited Au atoms can easily penetrate through the SAMs [29–31, 40, 41], generally accepted as due to a dynamic defect mechanism [11–13]. Since Au is a highly inert metal, very few terminal groups are effective as trap groups to prevent metal penetration. Gold does interact with highly polarizable molecules and atoms, e.g., I, S, Se, and pi-electron rich aromatic molecules, to form moderately strong donor–acceptor bonding (e.g., ~30–60 kJ/mol). Based on this type of interaction, SAMs with thiol terminal groups have been used to trap depositing gold atoms [29, 40–43], typically for M^3 junctions with rigid dithiol molecules such as oligophenyldithiol [44]. In contrast, though, filament growth with Au deposition on an alkanedithiol SAM has been observed by conducting probe

AFM (cp-AFM).[1] This specific measurement detects highly local and isolated penetration spots caused by either static pinholes or local blocking of a thiol terminal group at the SAM/vacuum interface due to some local disorder in the SAM.

To avoid penetration and filament formation via static and randomly scattered pinholes, one approach is to diminish the area of the junction until statistically the presence of a pinhole defect is near vanishing, as might be calculated by a Poisson distribution. An example of this strategy is the use of a nanopore junction of ~50 nm diameter, though in this case the device fabrication yields were still reported to be quite low, down to a few percent [16].

Another strategy is the use of terminal groups which are cohesively bound together to prevent the surface from following the dynamic diffusion channel formation from the substrate interface, thus closing down penetration through the surface layer. One example is the introduction of strong ionic bonding between carboxylic acid terminal groups by titration with potassium vapor to form $K^+(-CO_2)^-$ species. While this strategy does prevent the standard result of a floating SAM, cp-AFM images reveal short-circuit filament growth, presumably due to static defects of SAM, especially at step edges of the Au substrate [45, 46].

Cryogenic Deposition

One potential approach to avoiding rapid diffusion of Au atoms across the SAM/vacuum surface and into isolated penetration channels is to cool the SAM to cryogenic temperatures during deposition. In addition, the low temperatures should avoid any side reactions with metals that are more reactive than Au [22]. It is reported, however, that Au cryogenic deposition does not achieve reliable reproducibility [16, 47], presumably because of the presence of intrinsic static defects which provide filament growth channels, either during deposition or upon warmup. The use of a nanopore structure may decrease the probability of static defect formation, as described above, but is reported to provide only low yields as well as varying I-V (current-voltage) characteristics due to filaments. It also has been confirmed that cryogenic deposition of Au at ~10 K rarely prevents short-circuit filament growth through alkanethiolate and even thiol-terminated SAMs in large area (~a few micrometers) contacts [44, 48]. This could be because of a low reactivity of the metal with the terminal functional groups of the SAMs (or more properly, kinetic barriers to the trapping interaction), as Tarlov reported [22]. Under this circumstance the metal atoms could diffuse into the intrinsic static defects before being trapped by thiol terminal groups at cryogenic conditions. Another possibility is that metal clusters and atoms could be loosely trapped in SAM domains and subsequently become mobile and penetrate through the SAM upon sample warmup [22].

---

[1] Our preliminary cp-AFM results of Au thermal deposition on octanedithiol SAM under UHV condition at room temperature and 10 K revealed short-circuit filament generation.

5.2.2 Cu and Ag

Deposition on Non-Reactive SAMs

The general outcome of using Cu and Ag metals in top contact deposition on non-reactive SAMs, e.g., *n*-alkanethiolate SAMs/Au, gives very similar results to Au, including dominant formation of floating SAMs at room temperature and formation of metal filaments at cryogenic sample temperatures [22–27, 32, 33, 49]. Since Ag atoms are more mobile than Au atoms on an inert surface, penetration is expected to be quite rapid into pinholes. Arenethiolate/Au types of SAMs appear to lead to top layer deposition, possibly due to Ag atom/phenyl group interactions via the aromatic π-electrons [33], but further analysis, for example by cp-AFM, is needed to conclude that there is no penetration. In the case of Cu, an interaction between a phenyl ring and Cu atoms has been concluded based on the appearance of $Cu–C_6H_5^+$ species in time-of-flight secondary ion mass spectrometry (TOF-SIMS) spectra [33].

Deposition on Reactive SAMs

There have been a number of studies with Cu and Ag deposition on SAMs with varied terminal groups including –COOH, $–COOCH_3$, $–OCH_3$, –OH, –CN, and –SH [23–27, 50–52]. While interactions with Ag appear minimal, Cu can interact with O-containing functionality, e.g., –COOH interactions can lead to formation of inorganic phases such as CuO, Cu_2O, and $Cu(OH)_2$ [23–27, 52, 53] and interaction between Cu and $COOCH_3$ can result in formation of carboxylate [23–27, 52]. In spite of the interaction in the latter case, some fraction of Cu atoms still penetrate through the SAM to produce an under layer with the overall result of disordering of the SAM. In the case of SAMs with –CN terminal groups, only weak interactions occur, which leads to significant penetration with formation of an under layer at the SAM-substrate interface. Similarly, for Ag and Cu on $-OCH_3$ terminated SAMs, some complexation of the metal atoms with the OCH_3 groups occurs, as evidenced by TOF-SIMS, and considerable penetration through the SAMs occurs [23, 32, 33, 49, 52], though Cu appears to interact more strongly [52]. Overall, it is clear that Ag and Cu are not promising candidates for preparing top metal electrodes for M^3 junctions for most organiothiolate/Au SAMs under typical deposition conditions unless highly reactive terminal groups are present, such as demonstrated for a phenylthiolate type SAM with a thiol terminal group which appears to exhibit a sufficiently strong interaction with Cu atoms to prevent penetration [51].

5.2.3 Pd

It remains a challenge to stop completely the penetration of Pd into SAMs by vapor deposition. One might expect problems given the high vaporization temperature of the metal which leads to high kinetic energies of the nascent atoms and opportunities for thermal damage and activated penetration rates upon adsorption. In one study of constructing a nanopore junction device, Pd was deposited on 1,4-phenylene diisocyanide SAM on a Pd substrate with the idea of creating a strong trapping interaction with the –N≡C terminal group to give a good junction yield. The results showed significant hopping dominated charge transport characteristics which indicate a nonuniform top metal/SAM interface and unidentified defect components [44].

Although direct vapor deposition of Pd on SAMs with various types of trapping functional groups and/or under cryogenic condition could potentially reduce the extent of short-circuits through static pinhole defects and the level of thermally induced damage, the indirect deposition method proposed by Cahen and co-workers appears to be an attractive solution as shown by the good device yields (5–50% in yield) [54]. Further, less penetration was observed with indirect vapor deposition of Pd on inert alkanethiolate SAMs under cryogenic condition (150–200 K) as compared with Au, which is not surprising given the stronger interaction of Pd towards many common molecular groups compared to Au. However, it has also been proposed that the difference of device yields of Pd compared to Au arises because of the difference in growth mechanisms of Au and Pd; Pd quickly coalesces into a continuous film before penetrating into SAMs, while the Au atoms are less bound to the SAM, making them more mobile for diffusion into the SAM before coalescence [47, 54]. A systematic study of Pd indirectly deposited on SAMs on GaAs with various terminal functional groups (–$N(CH_3)_2$, –OCH_3, –CH_3, –H, –F, –CF_3, –Br, –CN, –NO_2, –CN) revealed low damage depositions, whereas direct deposition induced degradation of the SAMs [55]. Overall, the indirect deposition appears to be a promising process to construct top Pd contacts on SAMs.

5.2.4 Pb

While Pb is typically not a useful choice for device top contacts because of its low melting point and soft mechanical characteristics, it is of considerable interest for its superconducting characteristics. There appear to be, however, no reports on Pb deposition on SAMs. In our own preliminary experiments of Pb deposition on C16 alkanethiolate SAM on Au/mica from UHV AFM imaging, we observed complete penetration with no top surface cluster formation and continuing penetration into the underlying Au lattice.[2]

[2]Our preliminary AFM observation of Pb thermal deposition on hexadecanedithiol SAM on Au/mica under UHV condition at room temperature revealed continuous penetration of Pb as unpublished results.

5.3 *Reactive Metals*

5.3.1 Overview

Reactive metals are of interest for two primary reasons: (1) reaction with the uppermost part of the SAM which can drive uniform nucleation with no penetration and (2) for electropositive metals, injection of electrons into the SAM to create a favorable dipole at the metal/SAM interface for device operation. With respect to the first, as opposed to the results with non-reactive metal deposition, some reports of reactive metal deposition appear to show prevention of metal penetration with the avoidance of short-circuits across the M^3 junction. In general, serious concerns remain that some of metal atoms react destructively with the SAM backbone to produce inorganic species, e.g., carbides and oxides in the case of aggressive metals such as titanium.

5.3.2 Ti

Reactivity with SAMs

Titanium is a highly reactive metal with most organics, especially in the form of nascent atoms, due to the very favorable thermochemistry to form inorganic oxides and carbides, and even hydrides, at ambient temperatures. It has been reported for normally inert *n*-alkanethiolate and aromatic ring thiolates/Au SAMs that deposited Ti atoms do form top metal overlayers but the initial atoms depositing react vigorously to form multiple layers of inorganic carbides which continue from the top surface down into the SAM [23, 33–35, 40, 41, 56]. Studies made for the series of $-CH_3$, $-OH$, $-COOH$, $-COOCH_3$, and $-CN$ reveal extensive reactions and the penetration depth of Ti into the SAM molecular backbone appears to correlate somewhat with the oxygen component of the terminal group following the series $-CH_3 < -OH < -COOCH_3 < -COOH$ [23, 35]. In the case of the $-OCH_3$ group, it has been reported that Ti deposition results in parallel reactions with both the $-OCH_3$ terminus and $-CH_2-$ backbone groups to degrade the SAM extensively [33, 56]. Extensive degradation to form carbides also has been observed for aromatic SAMs [33, 40, 41].

Based on cp-AFM evidence for the simple case of an *n*-alkanethiolate/Au SAM, the M^3 structures show no evidence for penetration of metal to form conducting filaments that can cause shorts. The resultant junctions, however, do show extensive formation of reaction product layers with complex chemical compositions which may lead to unfavorable characteristics for molecular device operation. Indeed, in recent reports the use of Ti deposition on LB films, which contain water and inorganic salts at the bottom Pt electrode/LB film interface, leads to formation of inorganic titanium oxide type species in the junction but these complex inorganic layers have also been reported to impart fortuitously quite useful device characteristics, somewhat independent of the specific molecule in the junction [17]. Following this strategy, reliable memristor devices have been by produced by deliberate fabrication of ~1–2 nm metal–TiO_x–metal junctions [36–38].

5.3.3 Mg, Ca, Na, K

Deposition of the highly electropositive alkaline earth metals, e.g., Ca and Mg, and alkali metals, e.g., Na and K, on organic films have strong tendencies to inject electrons into the organic moiety. For the alkaline earth metals this effect can be useful for generating low work function cathode metal electrodes in organic light emission diodes (OLEDs) and organic electro luminescence display (OELDs). Alkaline earth metals, however, are not only active in injecting electrons into conjugated molecules but also are chemically reactive and can form reaction products with many organic groups [56, 57]. In general, the alkali metals are not used for commercial applications but are useful to study for their simple behavior as one electron injection atoms.

Reactivity with Inert SAMs

For inert SAMs such as *n*-alkanethiolates/Au, alkaline earth and alkali metal deposition on inert SAMs tends to exhibit low sticking coefficients of the nascent metal atoms due to quite weak interactions with the $-CH_3$ terminus; sometimes $<10^{-6}$ of the impinging metal atoms stick to the surface while the rest scatter off the surface [23, 58]. Bammel and co-workers observed quite slow penetration of Na through this inert SAM [59]. In the case of Mg and Ca depositions on *n*-alkanethiolate SAMs it was observed that while Mg does not react it does undergo continuous penetration thorough the SAM. In contrast, Ca does react to some extent resulting in calcium carbide species formation [56, 57]. In the case of K on an *n*-alkanethiolate SAM the results are more complicated. For example, at ~10 K atoms per SAM molecule, it has been reported that half of the deposited metal penetrates to the SAM/Au interface while the remainder is claimed to remain embedded within the SAM matrix [60], though such space is not available theoretically in a dense SAM.

Reactivity with Reactive SAMs

Both alkali metal and alkaline earth atoms are reactive with oxygen-bearing terminal groups but the latter appear to be far more destructive. K deposition on $-COOCH_3$ and –COOH terminated SAMs results in a 1:1 reaction to form $-COO^-$ species, which in the ester case requires cleavage of the $-CH_3$ group [45, 46, 58, 59, 61]. Walker and co-workers have reported that Ca deposition on $-OCH_3$

functionalized alkanethiolate SAMs results in selective reaction with the $-OCH_3$ terminus in the initial stage of deposition with continued deposition leading to the onset of reaction with the $-(CH_2)-$ backbone and degradation of the SAM, similar to the case of Ti deposition [56]. The interactions of Mg and Ca atoms on $-COOH$, $-OH$, $-OCH_3$, and $-CO_2CH_3$ functional groups have been studied. Both Mg and Ca react with these groups with insertion of metal into C–O bonds with cleavage [57]. Continuing deposition of Mg forms a top metal layer [62, 63], although further analysis of filament generation still has to be performed in order to show reliability of this deposition for M^3 junction fabrication. Ahn and co-workers have reported the complete penetration of K on thiophene terminated alkane thiolate/Au SAMs. The thiophene rings remained intact even after deposition of large amounts of K [64].

5.3.4 Al

Aluminum, a highly electropositive metal similar to the alkaline earth metals, can be useful in deposition of low work function cathode metal electrodes in OLEDs and OELDs, and is preferred for many electronics applications because of its low cost and high conductivity, although it is somewhat prone to electro migration [8, 65]. Since Al is an important metal for device applications its deposition on SAMs has been extensively studied [20, 21, 23, 32, 33, 41, 50, 65–69]. The critical problem for Al, however, is its facility in penetration along with filament growth.

Penetration and Top Deposition Transition on Inert SAM

Al deposition on relatively inert *n*-alkanethiolate/Au SAMs tends to proceed with partitioning between some complexation with surface terminal groups and penetration to the underlying Au–S interface. Though Al atoms are highly reactive with a number of functional groups and are thermochemically driven to form oxides, carbides, and hydrides, kinetic control typically tends to drive the Al atoms to undergo penetration with stabilization at the underlying S/Au interface. As the reactivity of the terminal group increases, however, the partitioning to form surface overlayers of Al metal increases, often with unexpected metal morphologies such as porous films [20, 21, 41, 49, 68, 69].

Two specific illustrative cases of the extreme limits of behavior are given by Al deposition on $-CH_3$ and $-COOCH_3$ terminated hexadecanethiolate/Au SAMs [20, 21]. The $-CH_3$ terminated SAM case shows a spectrum of deposition modes, including penetration to the Au-SAM interface and ambient surface overlayer formation. The penetration was explained in terms of Al atoms diffusing into dynamically formed temporal vacancies in the SAM (see Sect. 2) caused by fluctuations of Au–thiolate moieties around their equilibrium positions on the Au substrate [11–13]. Once the Al atoms arrive at the substrate, energetically favorable insertion into S–Au bonds can occur. This in turn can result in strongly decreased positional fluctuations of the adsorbate, thereby shutting off the temporal vacancy channels. In the case of aromatic thiolate/Au SAMs, the penetration of the Al is somewhat retarded and localized to a depth of ~1 nm due to some complexation with the phenyl rings, resulting in a more selective top layer formation [41].

Reactivity with Functional Groups

It has been observed that Al penetration can be avoided by introducing reactive functional groups such as $-CO_2H$, $-OCH_3$, $-CH_2OH$, $-CO_2CH_3$, and –SH, thereby resulting in a more precise Al/SAM/Au layered structure [20, 21, 23, 32, 33, 41, 49, 65–69]. Note, however, that the trapping of the Al atoms at the ambient interface, for example by –COOH or $-COOCH_3$, arises because of chemical reactions which produce various organometallic and oxide species at the top surface, as concluded from infrared reflection spectroscopy, X-ray photoelectron spectroscopy, ToF-SIMS, and spectroscopic ellipsometry [20, 21, 32, 33, 49, 67–69]. More information on minor penetration channels can be obtained from application of AFM and cp-AFM to image both the morphology and the electrical character of the deposited metal surfaces as a function of deposition coverage and such studies in the case of the $-COOCH_3$ terminated SAM reveal scattered, isolated metal filaments, likely formed by static pinholes [20]. Thus, even if the terminal group reactions are successful in shutting down the dynamic penetration channels, it appears that static pinhole channels may still be available. Thus formation of ideal M^3 junctions may require more than the presence of surface trap groups. Another example of Al deposition on reactive functional groups is given by the report of deposition on a thiophene-terminated alkanethiolate/Au SAM. Spectroscopic analyses reveals formation of a top overlayer, presumably created by a strong Al-thiophene interaction which leads to top Al deposition with an insulating interface [65, 66].

5.3.5 Cr, Ni

Deposition of Cr can produce degradation with formation of inorganic species for terminal functional groups. For example, Cr on $-COOCH_3$ terminal groups shows evidence for formation of inorganic carbide and oxide species [24, 28]. Surprisingly, the deposition on the –COOH group does not show this feature. As the coverage of Cr increases, the interaction between carboxyl group and Cr is reported to vanish, although a chromium oxide species was detected, probably due to the background residual oxygen. No evidence for penetration was observed in these experiments, which suggests that Cr on a –COOH terminated SAM may be useful for fabrication of M^3 junctions [24, 28].

On the other hand, Ni shows intermediate or low reactivity with –COOH, $-COOCH_3$, –SH, and –OH groups, but penetration has not been observed

[24, 28, 70, 71]. Ni and Cr deposition on –CN terminated alkanethiolate/Au SAMs both under room temperature and modest cryogenic conditions (173 K) are reported to result in penetration, although at cryogenic temperatures the deposition is slightly diminished [23]. Further, Cr typically reveals less penetration than Ni, probably due to the difference in reactivity with –CN. Nitride and carbide species were observed for Cr at high coverage, similar to typical observations for Cr deposition on polymer surfaces, while no evidence of carbide production was observed for Ni. Aromatic thiolate/Au SAMs appear to give selective top deposition of Cr [33, 72]. In this case, degradation side reactions were not observed. Ni deposition, however, showed evidence of penetration though aromatic SAMs whereas cross-linked aromatic SAMs, induced by electron beam irradiation, give selective top metal deposition [73]. Ni deposition on cross-linked aromatic SAMs was confirmed to result in a completely electrically isolated top Ni layer by electrochemical analyses [74].

5.3.6 Fe

Although Fe deposition has been less studied than other metals, some reports are available. Deposition on fluorinated organic films, such as partially fluorinated decanethiolate/Au SAMs, shows extensive defluorination of the molecular group to form Fe(II) fluorides, with no evidence for reactions with the C atoms of the chains [75]. Further, for an aromatic thiolate/Au SAM, no degrading side reactions were observed [33].

5.4 Strategies for Producing High Quality Top Contacts with No Shorts from Deposition of Low Reactivity Metals

As reviewed above, direct vapor deposition of metal onto SAMs has still not been developed to a point where the process could be used to fabricate large quantities of M^3 junctions with sufficiently high yields for practical device applications, though, clearly, small quantities of junctions can be produced for scientific research into the study of molecular conduction and switching, for example. In order to utilize vapor phase deposition, challenges remain. Achieving high yields, reaching well beyond 99%, of non-shorted junctions operating within extremely small tolerances in their electronic characteristics for millions of junctions, requires key factors, such as illustrated earlier in Figs. 6 and 7, to be precisely controlled. From this standpoint, there are several novel techniques proposed recently which could serve as alternate or additional methods to improve fabrication yields of M^3 devices instead of using standard metal vapor top deposition. In this section, some of these new strategies are reviewed.

5.4.1 Terminal Group Chemical Trap Reactions

Donor–Acceptor Interactions with Selected SAMs Terminal Groups

As described in the earlier sections, introducing reactive terminal groups has been utilized in attempts to stop the penetration of metal into SAMs with varying success for different metal functional group combinations. One of the notable issues was the presence of static pinholes which often continue to allow some penetration at these local defect sites even though the dominant mechanism for metal atom deposition was via trapping at the terminal groups. Since it is virtually impossible to self-assemble perfect SAMs without any static defect, the trapping strategy is a challenge for generating the highest quality M^3 devices. An excellent demonstration is given by cp-AFM observations with Al on $-COOCH_3$ [20] and Au on –SH terminated alkanethiolate SAMs for which short-circuit filament features were typically observed at the step edges of the Au substrate where the SAM typically has a significant number of static defects. For some metals even the simple phenyl ring substituents in aromatic thiolate types of SAMs can show selective top deposition, likely via some metal atom–phenyl ring interactions which may be sufficient to reduce pinhole diffusion [33, 65, 66, 72, 73]. Therefore aromatic types of SAMs with appropriate molecular structures bearing reactive terminal functional groups could be promising candidates for preparing M^3 molecular junction devices, though cp-AFM analysis still has to be employed with these systems to confirm no filament growth in SAMs [20, 46, 48]. This strategy is good since the typical device type molecules have aromatic rings with rich π-electron character.

5.4.2 Chemical Vapor Deposition

Chemical vapor deposition (CVD) on SAMs could be a very promising method to deposit metal contacts on SAMs for M^3 molecular junctions. The limitation of CVD processes is the limited variation and availability of the organometallic precursors which have to be volatile, reactive with respect to decomposition at the SAM surface, and applicable at relatively low surface temperature, well below the stability threshold of typical SAMs, for example, at ~70 °C [76]. Since CVD processes are often based on the decomposition of precursor by specific reaction with functional groups on the SAM, the decomposed precursor organic products must be volatile, leaving only the metal film on the surface. In spite of these limitation, various metal species, e.g., Au, Pd, Al, Cu, and V, have been successfully deposited on SAMs. There is a potential clear advantage for CVD in the cases of refractory metals with high vaporization temperatures (well above 1,000 K) where the metal atoms arrive at the SAM surface at quite high kinetic energies, a condition that can cause considerable damage to the SAM, whereas for CVD the precursor vapor generally arrives at ambient temperatures which in principle should reduce the rate of diffusion across the surface and favor surface metal cluster formation, thus reducing penetration. Reports are available which show the possibility of using CVD to deposit metal films on SAMs [77–79]. Wöll and Fishcher and co-workers have reviewed wide variations of metal deposition on SAMs, including CVD, which can provide reproducible metal-organic interfaces, as characterized by spectroscopic studies [80]. It should pointed out, however, that degradation of a SAM also can occur due to the intrinsic reactivity of metal atoms that form, regardless of CVD or vapor deposition. For example, Wöll and co-workers have observed for CVD on a SAM that the catalytic reactivity of Pd results in the degradation of the SAMs upon exposure to air. The top metal layer can form without penetration of metal and damage by the use of H_2 exposure to reduce Pd on the surface [78]. Finally, since CVD processes require very specific surface chemical reactions, the metal deposition is very surface sensitive and the deposition site can be controlled by the use of surface functional groups, which can also lead to a patterning strategy, in principle.

Au

For Au CVD an organogold complex, (trimethylphosphine)methylgold(I) ((Me3P) $AuCH_3$), has been used with the result of selective deposition on $-SH$ and $-CH_3$ terminated SAMs maintained at 70 °C [81–83]. Further, CVD at 70 °C on SAMs with $-SH$, $-CH_3$, and $-OH$ terminal groups was studied by spectroscopic and microscopic analyses with the result that no penetration was observed for the–SH surface but was observed for the $-CH_3$ and –OH surfaces. Standard thermal deposition of Au on the –SH surface showed, in contrast to CVD, that the penetration cannot be completely prevented [83].

Cu

It has been reported that Cu deposition can be carried out via organometallic precursors at elevated temperatures onto –SH and $-SO_3H$ terminated SAMs, typically on non-metallic substrates. In the case of silane SAMs on SiO_2, top deposition of metallic Cu has been reported for substrate temperatures >150 °C, as confirmed by XPS, X-ray diffraction, and SEM [84, 85]. The surface functionality apparently changes the activation energy of nucleation; $-SO_3H$ appears to show the lowest activation energy, likely due to protonation of ligands leading to more rapid nucleation of Cu [86].

Pd

Cyclopentadienyl-allyl-palladium (Cp-(allyl)Pd) precursors have been typically used for Pd deposition on SAMs at room temperature. Successful CVD has been reported for –SH terminated SAMs [87], though upon exposure to air degradation results. This can be prevented apparently by H_2 exposure [78]. Since Cp-(allyl)Pd on both –SH and non-functionalized aromatic SAM is not selective, –OH terminal groups were introduced to achieve selectivity. Hydroxyl group-terminated aromatic SAMs functionalized with trimethylamine alane as a seed layer activates the reaction with Pd precursor, resulting in Pd deposition with surface selectivity relative to an inert alkanethiolate SAM.

Al

Trimethylamine alane has been used for Al deposition on various SAMs functionalized with –COOH, –OH, and –CH_3 terminal groups. No reaction was observed with –CH_3 but reactions were observed with –OH and –COOH. In the latter cases room temperature deposition created Al oxide and Al layers under conditions of N_2 purging and UHV, respectively [79, 88, 89].

5.4.3 Atomic Layer Deposition

ALD via deposition of organometallic complexes has been used to prepare the top contact between metal and SAMs [90–93]. The concept of ALD is quite similar to CVD in terms of using organometallic precursors rather than a metal atom vapor. Specifically, the ALD process can be distinguished from CVD by the use of a self-limiting deposition process in ALD, in contrast to a continual process in CVD. In ALD the chemical reaction between precursor and substrate surface can be limited to exactly one monolayer, regardless of the presence of a large excess of precursor vapor, whereas CVD continues to grow films and the achievement of a single uniform monolayer is extremely difficult. On the other hand, the disadvantage of ALD is the slow rate of growth. In particular, each layer to be formed requires: (1) exposure to the precursor gas to form the layer, (2) complete evacuation of the precursor gas, and (3) a decomposition process, such as heating, hydrogen, or other chemical regent exposure for reduction to generate the metal species from precursors. The ALD process for SAMs has been applied to deposition of metals such as Ti, Ru, Hf, and Pt. The complex, tetrakis(dimethylamido)titanium has been used to deposit Ti on SAMs with –OH, –NH_2, and –CH_3 terminal groups [94–96]. A two-step-ALD process of Cu deposition on –CH_3 and –COOH SAMs on silicon substrates has been reported in which the second step was followed by H_2 exposure. The final samples, characterized by XPS, IRS, AFM, and Hg drop current-voltage measurements, showed no damage of the SAM in general, but did show reaction of the –COOH group with Cu [97]. SAMs with terminal functional groups of –NH_2, –OH, and –COOH appear to be free of Cu metal penetration, although the –CH_3 terminated SAM sometimes allows the penetration during successive reaction processes. It should be emphasized that ALD provides self-limiting deposition by ligand exchange of the organometallic complex. Therefore some systems can have different stoichiometries, for example, 1:1 or 1:2 reactions between metal precursors depending on the terminal functional groups. Finally, the major challenges in ALD for standard large area M^3 top metal contacts with thicknesses in the range of ~200 nm are the slow deposition rate with time consuming processing and how to develop the more difficult growth of metals rather than oxides for reactive metals such as Ti and Al.

5.4.4 Cluster Deposition

Cluster Deposition Mechanisms

Cluster deposition is a method in which clusters of metal atoms of selected sizes and distributions are deposited on a surface. This process prevents individual atoms from impinging on the surface and thus provides a great reduction in the diffusion of metals across the surface and into pores because of the large size of the diffusing species, though if the cluster size is too large the metal layer can become grainy with poor conduction and mechanical properties. Some information is available in several review papers [55, 76, 98, 99]. Cluster deposition can been performed by different methods involving formation of the clusters directly before they impinge on the surface or after surface adsorption. In the former case, several methods have been used such as extremely high deposition fluxes of metal vapors which allow extensive metal atom-metal atom collisions, plasma sputtering or laser ablation of clusters from a metal target, and subjecting a metal vapor flux from a thermal source (e.g., e-beam hearth) to scattering from an introduced inert gas or even an inert surface such as a Teflon coated plate, which cools the vapor and reduces the mean free path, thereby encouraging metal atom coalescence from increased metal–metal atom collisions. There are a few studies with quite small metal clusters, down to just a few atoms, which provide information on the intrinsic properties of isolated metal cluster and their size effects [100, 101], but here we focus primarily on relatively large cluster deposition processes which have been applied to fabricate M^3 molecular device junctions. The second general method involves impinging the metal atom vapor directly onto a cryogenically cooled inert buffer layer on the SAM surface. In this process the KE of the metal atoms is instantly quenched and coalescence of the atoms occurs, creating clusters directly on the surface.

Cluster Deposition via Inert Gas Scattering

Metzger and co-workers and Cahen and co-workers have both proposed formation of clusters via a decrease of the KE of atoms in a metal vapor by introducing Ar gas into the vacuum deposition chamber to lower the base pressure during deposition from typical values of ~10^{-7} Torr or lower to several orders of magnitude higher pressures. This condition in principle should produce not only clusters but cooler temperatures of the depositing metal. In addition, an advantage can be realized by facing the SAM surface away from the metal thermal source to reduce radiation heating of the surface from the source. Further cooling can be achieved by sample cooling down to 77 K using liquid nitrogen. In the case of the Cahen group, the sample temperature was maintained at 150 ~ 200 K and another cold finger (at 77 K) was placed next to the sample to collect all contaminants or residual gas to avoid contamination of the cooled sample surface by condensation of background gas. This process was reported to improve significantly the device yield for Au, typically a highly penetrating metal, presumably due to minimizing the surface diffusion length of metal atoms on SAM surface [47, 54, 55, 76, 98, 99, 102–104].

Cluster Deposition via High Deposition Rates

Thermal deposition of Cu on alkanethiolates/Au SAMs at high deposition rates (~10 nm/s) gives top deposition of the metal, in contrast to the typical floating SAMs with extensive pinholes, but does result in irreversible disordering of the SAM. A series of analyses after Cu deposition involving electrochemistry, XPS, and IRS confirmed the disordering and showed that it is similar to simple heating of the SAM to elevated temperatures but the electrochemical analyses revealed no pinholes [50, 51]. Gold cluster deposition on SAMs of benzenedimethanethiol on Au has been carried out with low KE clusters produced by laser vaporization to give precisely controlled sizes, ~250 atoms (2.6 nm in diameter). The depositions resulted in top metal layers with no floating SAM, pinholes, or migration of the SAMs onto the deposited clusters [43]. On the other hand, a recent scanning tunneling microscope (STM) investigation reports the opposite results. In the case of nanometer scale Au clusters ($d = 1.5 \sim 3$ nm) deposited onto a dodecanethiolate/Au SAM, the clusters were observed to diffuse to the SAM/substrate interface resulting in buried Au clusters below the SAM [105]. Consequently, it is important to confirm independently that there is no filament growth short-circuiting the M^3 junctions by scanning probe microscopic analyses for local information. Although coulomb blockade characteristics are sometimes considered to be an indicator of isolated metal on top of SAMs [101], Sheng and co-workers have suggested that a coulomb blockade feature can be observed with a small fraction of Au clusters which remains on top of inert alkane thiolate SAM after the majority of deposited Au penetrates to the SAM/substrate interface [31]. Deng and co-workers also confirmed that the device yield was significantly improved by applying cluster deposition to prepare the top contact of a nanopore molecular junction device, with a maximum yield ~40% or possibly even better [106]. The large variation of chip-to-chip yield, however, indicates that achieving high yields with the cluster deposition process is still a large challenge.

5.4.5 Low Temperature Depositions with Soft-Landing

Since high kinetic energies of the impinging metal atoms or clusters can have deleterious effects, including damage to the SAMs, accelerating undesired chemical reactions, diffusion across the SAM surface to find penetration channels for either static or dynamic penetration, the main strategy of soft-landing process is to freeze impinging metal atoms completely in an inert buffer layer. The thicker the layer, the more one can expect the metal film to be formed without contact with the SAM until the buffer layer is removed by evaporation on warm up. This buffer layer assisted growth (BLAG) strategy was proposed originally by Weaver and co-workers with rare gas buffer layers such as Xe or Kr on the target surface held at cryogenic temperatures by liquid helium below the condensation point of the gas [107–109]. Following the soft-landing process step the sample is heated sufficiently to allow the buffer layer to sublime into the vacuum, leaving a top layer of immobile metal nanoclusters to make contact with the SAM. In our own work we have observed a high yield of large area (many micrometers) M^3 devices which show good tunneling characteristics through the SAM. The less than expected conductance by about one order of magnitude, however, suggests that the contact of the nanocluster assembly with the SAM surface is only at specific points likely arising from the spatial distribution of clusters directly at the SAM surface. This situation presumably could be improved by controlling the surface morphology of buffer layers through variations in the process parameters [48].

5.5 *Various Alternate Methods to Prevent Shorts in M^3 Devices*

5.5.1 Plugging of Defect Pinholes with Polymers

Since it is extremely difficult to establish the ideal abrupt organic/metal interface, de Boer and co-workers have proposed the use of conductive organic diffusion barrier layers inserted between the SAM and top metal electrode, thus preventing metal atom diffusion to the bottom electrode [110, 111]. They demonstrated organic barrier layers via spin coating of a conductive polymer, poly-ethylenedioxythiophene derivatives (PEDOT). The Au top electrode was prepared by vapor deposition of Au on top of the PEDOT layer. Since PEDOT can act as a conducting material, direct tunneling characteristics through the molecular junction were observed, similar to those in other M^3 architectures [112–115]. This process looks promising to prevent penetration of vaporized metal penetration and provide high yields of molecular devices for large scale device manufacturing, although some inconsistencies in electronic characteristics are pointed out that arise from a comparison between polymer/SAM and Au/SAM interfaces [116, 117]. Further studies of the conductive polymer/SAM interface is desired in order to understand better the dependencies of the interface characteristics, morphology, and conductance on the SAM functional groups [118].

5.5.2 Oxidation of Conducting Metal Filaments to form Dielectric Material

As has been noted for Al deposition on inert alkanethiolate SAMs, the metal atoms easily penetrate through both static and dynamic defects of the SAMs. Al insertion in S–Au bonds, however, closes the dynamic penetration channel due to formation of more stable S–Al bonds preventing sulfur positional fluctuation, though static defects may actually increase via the freezing in of some molecular displacements at the Au–Al–S interface, as is indicated by the presence of shorts at higher Al coverages as observed by cp-AFM while under vacuum. After continued exposure to the vacuum condition for a few hours, however, the shorts vanish [20]. This is undoubtedly due to the oxidation of the Al metal in the filaments by residual oxygen species in the UHV chamber background gas to form insulating aluminum oxide filaments. This prompts a strategy for closing both dynamic and static penetration channels for metals with strong tendencies to form strong metal–S bonds at the Au–SAM interface of an organothiolate SAM and to undergo oxidation of any filaments to form insulating filaments. Although this strategy has not yet been extended to chip level device arrays, this process could be one to achieve high device yield of M^3 molecular junction for large scale device applications.

5.5.3 Metal Deposition on SAM/Non-Metal Substrate Structures

Since typical SAMs are usually prepared on metal substrates, mostly Au using divalent organosulfur chemistry, most studies of metal deposition on SAMs have been performed on metal substrates. Nevertheless extensive work is also available for SAMs on non-metal substrates. Though typically this is of little interest for molecular electronic devices, there are a few relevant examples and this system as well can teach us about the fundamentals of top metal layer formation. Since SAMs can have both intrinsic static defects due to imperfections in self-assembly and dynamic defects for thermal fluctuation of molecular positions at the pinning sites, non-metal substrates, which often provide strong bonding with SAM molecules, may actually increase static defects since annealing of positions during assembly may be difficult while decreasing dynamic defects due to the inability of thermally driven positional fluctuations. SAMs of interest include molecules with siloxane bonding on oxide substrates, such as silicon oxide, tin oxide, and indium tin oxide. Issues for these systems are: (1) difficulty of SAM characterization, (2) difficulty of introducing terminal groups with varied chemistry, and (3) difficulty in detecting filament formation or metal penetration by electrical measurements since the substrate is usually a wide bandgap insulator. In one study, Al deposition on various types of aromatic silane derivatives on silicon oxide covered silicon substrates revealed Fermi level pinning at the metal/SAM interface, which provides a rectification behavior [119]. In another study, vapor deposition of Au and Ag on functionalized alkyl and alkylsiloxane SAMs on Si–H and Si–O substrates, respectively, revealed different characteristics. Although Au shows a clear penetration feature to the SAM/substrate interface, Ag remains on top of SAMs. This was understood by the difference of Ag and Au in reactivity with Si substrate [120]. On the other hand, another study reported that Ag deposition on fluorinated alkyl silane SAMs on an SiO_2 substrate revealed destructive reaction between Ag and fluorinated alkyl moiety [121]. Vapor deposition of Al on alkanethiolate SAMs on a conductive indium tin oxide substrate revealed resistive tunneling type I-V characteristics rather than a short-circuited junction, although it is not clear whether this feature arises from the molecular junction or a resistive Al oxide layer [122]. Overall such studies show that it is important to extend the area of fundamental study of SAMs to different substrates, such as oxides and semiconductors, as well as various transition metals to explore conditions for preparing uniformly assembled steady SAMs without static defects and the reduction or elimination of transient dynamic defects [123, 124].

6 Future Developments and Needs

The area of molecular electronics as a competitive technology for standard semiconductor devices has experienced drastic fluctuations between boom and disappointment in this past decade. Overwhelmingly the disappointments can be attributed to a lack of rigor in device fabrication, the neglect of critical process issues, and the difficulty of independent characterization of the devices other than the standard I-V testing. For M^3 devices the challenge has been to avoid shorts and erratic behavior of the devices, a standard result once the top contact has been deposited. Recently extensive efforts have revealed undesirable features arising, such as heterogeneity of the top metal/SAM and SAM/substrate interfaces for deposition of a large variety of metal species. Based on this fundamental knowledge, recent research into new types of vapor phase deposition processes and some alternate processes have provided promising results with acceptable yields and device reproducibility. There still remains long way to go, however, to produce reliable molecular electronic devices for mass production. Looking ahead, we summarize the future needs to achieve the ultimate goal of the production of high quality M^3 molecular electronic devices and the related research needs:

1. To avoid misleading results, the electronic characteristics of a device always have to be analyzed in a proper statistical manner to characterize the results of the process in terms of device structure details, yield, electrical performance, and durability.
2. Since the deviation of device characteristics may come from the local heterogeneity of metal/molecule interfaces, the local characterization of the molecular structure and interfaces is necessary to learn how to alter processes and materials in order to achieve high yield, stable process condition, and low deviations of device characteristics. A few researchers have proposed the use of the scanning probe microscopic technique [conducting probe AFM, surface potential

(Kelvin-probe) AFM, and STM] to analyze the local phenomena and structure after top contact construction. Development of new techniques such as surface plasmon resonance Raman, for example, could be applied to advantage for junctions fabricated with plasmonically active metals such as silver [125] or gold [126].
3. Since the static intrinsic defects and dynamic defects are critical problem for SAMs in M^3 junctions, improvement of SAM assembly is still desired to achieve more closely packed stable SAMs. Engineering of interactions, such as headgroup–substrate, headgroup–headgroup, and intermolecular interaction with the backbone structure and the terminal groups, remains a frontier area since most SAMs studied are still limited to a few types of organothiolate molecules on noble metal substrates. A few studies have extended SAMs to non-metal substrates, e.g., SiO_2, GaAs, and ITO. Therefore continuation of the development of new types of SAMs on different substrates needs to be studied with an emphasis on characterization of the fundamental aspects of orientation, bonding characteristics, packing features, intermolecular interactions, thermal stability, and molecular mobility, which ultimately leads to methods for minimizing the intrinsic static and dynamic defects.
4. While recent developments have opened a door to new types of top contacts, for example, using conductive polymer/SAM layers, as promising strategies, this approach is largely an unknown area and questions of how the conductive polymer/SAM interface contributes to the electronic characteristics of M^3 junction as a faction of chemical functional groups needs study. In particular, the influence of the interface has to be studied from various aspects such as surface dipole, bonding, hydrophilic–hydrophobic characteristics, morphologies, overall electronic characteristics, and yield and operating stability of the device.
5. Lastly, and ultimately the most important, is evaluation of the applicability of the metal vapor top contact process for insertion into fabrication lines for mass production. At this point, significant improvements are required, well beyond the present state-of-the art, in order to achieve large cost efficient, high yield, high throughput fabrication. Hopefully, in the next decade, many of these issues will be resolved by the integrated research results from workers worldwide.

References

1. Scott JC (2003) Metal-organic interface and charge injection in organic electronic devices. J Vac Sci Technol A 21(3):521–531
2. Dimitrakopoulos CD, Malenfant PRL (2002) Organic thin film transistors for large area electronics. Adv Mater 14(2):99–117
3. Faupel F, Willecke R, Thran A (1998) Diffusion of metals in polymers. Mater Sci Eng R 22 (1):1–55
4. Horowitz G (1998) Organic field-effect transistors. Adv Mater 10(5):365–377
5. Mantooth BA, Weiss PS (2003) Fabrication, assembly, and characterization of molecular electronic components. Proc IEEE 9(11):1785–1802

6. Mitschke U, Bauerle P (2000) The electroluminescence of organic materials. J Mater Chem 10(7):1471–1507
7. Pesavento PV, Chesterfield RJ, Newman CR, Frisbie D (2004) Gated four-probe measurements on pentacene thin-film transistors: contact resistance as a function of gate voltage and temperature. J Appl Phys 96(12):7312–7324
8. Shen C, Kahn A, Schwartz J (2001) Chemical and electrical properties of interfaces between magnesium and aluminum and tris-(8-hydroxy quinoline) aluminum. J Appl Phys 89 (1):449–459
9. Watkins NJ, Yan L, Gao Y (2002) Electronic structure symmetry of interfaces between pentacene and metals. Appl Phys Lett 80(23):4384–4386
10. Durr AC, Schreiber F, Kelsch M, Carstanjen HD, Dosch H, Seeck OH (2003) Morphology and interdiffusion behavior of evaporated metal films on crystalline diindenoperylene thin films. J Appl Phys 93(9):5201–5209
11. Bhatia R, Garrison BJ (1997) Phase transitions in a methyl-terminated monolayer self-assembled on Au{111}. Langmuir 13(4):765–769
12. Bhatia R, Garrison BJ (1997) Structure of c(4x2) superlattice in alkanethiolate self-assembled monolayers. Langmuir 13(15):4038–4043
13. Alkis S, Cao C, Cheng HP, Krause JL (2009) Molecular dynamics simulations of Au penetration through alkanethiol monolayers on the Au(111) surface. J Phys Chem C 113:6360–6366
14. Paz Y, Trakhtenberg S, Naaman RJ (1994) Reaction between O(^{3}P) and organized organic thin films. Phys Chem 98(51):13517–13523
15. Long DP, Lazorcik JL, Mantooth BA, Moore MH, Ratner MA, Troisi A, Yao Y, Ciszek JW, Tour JM, Shashidhar R (2006) Effects of hydration on molecular junction transport. Nat Mater 5(11):901–908
16. Wang W, Lee T, Reed MA (2005) Electronic transport in molecular self-assembled monolayer devices. Proc IEEE 93(10):1815–1824
17. Stewart DR, Ohlberg DAA, Beck PA, Chen Y, Williams RS, Jeppesen JO, Nielsen KA, Stoddart JF (2004) Molecule-independent electrical switching in Pt/organic monolayer/Ti devices. Nano Lett 4:133–136
18. Lai YS, Tu CH, Kwong DL, Chen JS (2005) Bistable resistance switching of poly(N-vinylcarbazole) films for nonvolatile memory applications. Appl Phys Lett 87 (12):122101–122103
19. Tu CH, Lai YS, Kwong DL (2006) Electrical switching and transport in the Si/organic monolayer/Au and Si/organic bilayer/Al devices. Appl Phys Lett 89(6):062105–062113
20. Maitani MM, Daniel TA, Cabarcos OM, Allara DL (2009) Nascent metal atom condensation in self-assembled monolayer matrices: coverage-driven morphology transitions from buried adlayers to electrically active metal atom nanofilaments to overlayer clusters during aluminum atom deposition on alkanethiolate/gold monolayers. J Am Chem Soc 131 (23):8016–8029
21. Hooper A, Fisher GL, Konstadinidis K, Jung D, Nguyen H, Opila R, Collins RW, Winograd N, Allara DL (1999) Chemical effects of methyl and methyl ester groups on the nucleation and growth of vapor-deposited aluminum films. J Am Chem Soc 121:8052–8064
22. Tarlov MJ (1992) Silver metalization of octadecanethiol monolayers self-assembled on gold. Langmuir 8:80–89
23. Jung DR, Czanderna AW (1994) Chemical and physical interactions at metal/self-assembled organic monolayer interfaces. Crit Rev Solid State Mater Sci 19:1–54
24. Jung DR, Czanderna AW, Herdt GC (1996) Interactions and penetration at metal/self-assembled organic monolayer interfaces. J Vac Sci Technol A 14:1779–1787
25. Herdt GC, King DE, Czanderna AW (1997) Penetration of deposited Au, Cu, and Ag overlayers through alkanethiol self-assembled monolayers on gold or silver. Z Phys Chem 202:163–196

26. Herdt GC, Jung DR, Czanderna AW (1995) Weak interactions between deposited metal overlayers and organic functional groups of self-assembled monolayers. Prog Surf Sci 50:103–129
27. Herdt GC, Czanderna AW (1997) Metal overlayers on organic functional groups of self-organized molecular assemblies. 7. Ion scattering spectroscopy and X-ray photoelectron spectroscopy of Cu/CH_3 and Cu/$COOCH_3$. J Vac Sci Technol A 15:513–519
28. Herdt GC, Czanderna AW (1999) Metal overlayers on organic functional groups of self-assembled monolayers: VIII. X-Ray photoelectron spectroscopy of the Ni/COOH interface. J Vac Sci Technol A 17:3415–3418
29. Ohgi T, Sheng H-Y, Nejoh H (1998) Au particle deposition onto self-assembled monolayers of thiol and dithiol molecules. Appl Surf Sci 130–132:919–924
30. Ohgi T, Sheng H-Y, Dong Z-C, Nejoh H (1999) Observation of Au deposited self-assembled monolayers of octanethiol by scanning tunneling microscopy. Surf Sci 442:277–282
31. Wang B, Xiao X, Sheng P (2000) Growth and characterization of Au clusters on alkanethiol self-assembled monolayers. J Vac Sci Technol B 18:2351–2358
32. Walker AV, Tighe TB, Reinard MD, Haynie BC, Allara DL, Winograd N (2003) Solvation of zero-valent metals in organic thin films. Chem Phys Lett 369:615–620
33. Haynie BC, Walker AV, Tighe TB, Allara DL, Winograd N (2003) Adventures in molecular electronics: how to attach wires to molecules. Appl Surf Sci 203–204:433–436
34. Tighe TB, Daniel TA, Zhu Z, Uppili S, Winograd N, Allara DL (2005) Evolution of the interface and metal film morphology in the vapor deposition of Ti on hexadecanethiolate hydrocarbon monolayers on Au. J Phys Chem B 109:21006–21014
35. Konstadinidis K, Zhang P, Opila RL, Allara DL (1995) An in-situ X-ray photoelectron study of the interaction between vapor-deposited Ti atoms and functional groups at the surfaces of self-assembled monolayers. Surf Sci 338:300–312
36. Blackstock JJ, Stickle WF, Donley CL, Stewart DR, Williams RS (2007) Internal structure of a molecular junction device: chemical reduction of PtO_2 by Ti evaporation onto an interceding organic monolayer. J Phys Chem C 111:16–20
37. Borghetti J, Snider GS, Kuekes PJ, Yang JJ, Stewart DR, Williams RS (2010) 'Memristive' switches enable 'stateful' logic operations via material implication. Nature 464:873–876
38. Yang JJ, Pickett MD, Li X, Ohlberg DAA, Stewart DR, Williams RS (2008) Memristive switching mechanism for metal/oxide/metal nanodevices. Nat Nanotechnol 3:429–433
39. Porter MD, Bright TB, Allara DL, Chidsey CED (1987) Spontaneously organized molecular assemblies. 4. Structural characterization of n alkyl thiol monolayers on gold by optical ellipsometry, infrared spectroscopy, and electrochemistry. J Am Chem Soc 109:3559–3568
40. Walker AV, Tighe TB, Stapleton J, Haynie BC, Uppili S, Allara DL, Winograd N (2004) Interaction of vapor-deposited Ti and Au with molecular wires. Appl Phys Lett 84:4008–4010
41. deBoer B, Frank MM, Chabal YJ, Jiang W, Garfunkel E, Bao Z (2004) Metallic contact formation for molecular electronics: interactions between vapor-deposited metals and self-assembled monolayers of conjugated mono- and dithiols. Langmuir 20:1539–1542
42. Ohgi T, Sheng HY, Dong ZC, Nejoh H, Fujita D (2001) Charging effects in gold nanoclusters grown on octanedithiol layers. Appl Phys Lett 79:2453–2455
43. Vandamme N, Snauwaert J, Janssens E, Vandeweert E, Lievens P, Van Haesendonck C (2004) Visualization of gold clusters deposited on a dithiol self-assembled monolayer by tapping mode atomic force microscopy. Surf Sci 558:57–64
44. Chen J, Wang W, Klemic J, Reed MA, Axelrod BW, Kaschak DM, Rawlett AM, Price DW, Dirk SM, Tour JM, Grubisha DS, Bennett DW (2002) Molecular wires, switches, and memories. Molecular electronics II, Ann NY Acad Sci 960:69–99
45. Zhu Z, Allara DL, Winograd N (2006) Chemistry of metal atoms reacting with alkanethiol self-assembled monolayers. Appl Surf Sci 252(19):6686–6688

46. Zhu Z, Daniel TA, Maitani M, Cabarcos OM, Allara DL, Winograd N (2006) Controlling gold atom penetration through alkanethiolate self-assembled monolayers on Au{111} by adjusting terminal group intermolecular interactions. J Am Chem Soc 128:13710–13719
47. Haick H, Ghabboun J, Cahen D (2005) Pd versus Au as evaporated metal contacts to molecules. Appl Phys Lett 86:042113
48. Maitani MM, Allara DL, Ohlberg DAA, Li Z, Williams RS, Stewart DR (2010) High integrity metal/organic device interfaces via low temperature buffer layer assisted metal atom nucleation. Appl Phys Lett 96(17):173109–173113
49. Walker AV, Tighe TB, Cabarcos OM, Reinard MD, Haynie BC, Uppili S, Winograd N, Allara DL (2004) The dynamics of noble metal atom penetration through methoxy-terminated alkanethiolate monolayers. J Am Chem Soc 126:3954–3963
50. Colavita PE, Doescher MS, Molliet A, Evans U, Reddic J, Zhou J, Chen D, Miney PG, Myrick ML (2002) Effects of metal coating on self-assembled monolayers on gold. 1. Copper on dodecanethiol and octadecanethiol. Langmuir 18(22):8503–8509
51. Colavita PE, Miney PG, Taylor L, Priore R, Pearson DL, Ratliff J, Ma S, Ozturk O, Chen DA, Myrick ML (2005) Effects of metal coating on self-assembled monolayers on gold. 2. Copper on an oligo(phenylene-ethynylene) monolayer. Langmuir 21(26):12268–12277
52. Nagy G, Walker AV (2006) Dynamics of the interaction of vapor-deposited copper with alkanethiolate monolayers: bond insertion, complexation, and penetration pathways. J Phys Chem B 110:12543–12554
53. Dake LS, King DE, Czanderna AW (2000) Ion scattering and X-ray photoelectron spectroscopy of copper overlayers vacuum deposited onto mercaptohexadecanoic acid self-assembled monolayers. Solid State Sci 2(8):781–789
54. Haick H, Ambrico M, Ghabboun J, Ligonzo T, Cahen D (2004) Contacting organic molecules by metal evaporation. Phys Chem Chem Phys 6(19):4538–4541
55. Haick H, Niitsoo O, Ghabboun J, Cahen D (2007) Electrical contacts to organic molecular films by metal evaporation: effect of contacting details. J Phys Chem C 111:2318–2329
56. Walker AV, Tighe TB, Haynie BC, Uppili S, Winograd N, Allara DL (2005) Chemical pathways in the interactions of reactive metal atoms with organic surfaces: vapor deposition of Ca and Ti on a methoxy-terminated alkanethiolate monolayer on Au. J Phys Chem 109:11263–11272
57. Nagy G, Walker AV (2007) Dynamics of reactive metal adsorption on organic thin films. J Phys Chem C 111:8543–8556
58. Balzer F, Bammel K, Rubahn HG (1993) Laser investigation Na atoms deposited via inert spacer layers close metal surfaces. J Chem Phys 98:7625–7635
59. Bammel K, Ellis J, Rubahn HG (1993) Two-photon laser observation of diffusion of Na atoms through self-assembled monolayers on a Au surface. Chem Phys Lett 201 (1–4):101–107
60. Ge Y, Weidner T, Ahn H, Whitten JE, Zharnikov M (2009) Energy level pinning in self-assembled alkanethiol monolayers. J Phys Chem C 113(11):4575–83
61. Zhu Z, Haynie BC, Winograd N (2004) Static SIMS study of the behavior of K atoms on -CH_3, -CO_2H and -CO_2CH_3 terminated self-assembled monolayers. Appl Surf Sci 231–232:318–322
62. Walker AV, Tighe TB, Cabarcos OM, Haynie BC, Allara DL, Winograd N (2007) Dynamics of interaction of magnesium atoms on methoxy-terminated self-assembled monolayers: an example of a reactive metal with a low sticking probability. J Phys Chem C 111:765–772
63. Zhou C, Nagy G, Walker AV (2005) Toward molecular electronic circuitry: selective deposition of metals on patterned self-assembled monolayer surfaces. J Am Chem Soc 127:12160–12161
64. Ahn H, Whitten JE (2009) Potassium deposition on a thiophene-terminated alkanethiol monolayer. J Electron Spectros Relat Phenomena 172(1–3):107–113

65. Dannetun P, Boman M, Stafstrom S, Salaneck WR, Lazzaroni R, Fredriksson C, Bredas JL, Zamboni R, Taliani C (1993) The chemical and electronic structure of the interface between aluminum and polythiophene semiconductors. J Chem Phys 99(1):664–672
66. Ahn H, Whitten JE (2003) Vapor-deposition of aluminum on thiophene-terminated self-assembled monolayers on gold. J Phys Chem B 107(27):6565–6572
67. Fisher GL, Hooper A, Opila RL, Jung DR, Allara DL, Winograd N (1999) The interaction between vapor-deposited Al atoms and methylester-terminated self-assembled monolayers studied by time-of-flight secondary ion mass spectrometry, X-ray photoelectron spectroscopy and infrared reflectance spectroscopy. J Electron Spectrosc Relat Phenom 98–99:139–148
68. Fisher GL, Hooper AE, Opila RL, Allara DL, Winograd N (2000) The interaction of vapor-deposited Al Atoms with CO_2H groups at the surface of a self-assembled alkanethiolate monolayer on gold. J Phys Chem B 104(14):3267–3273
69. Fisher GL, Walker AV, Hooper AE, Tighe TB, Bahnck KB, Skriba HT, Reinard MD, Haynie BC, Opila RL, Winograd N, Allara DL (2002) Bond insertion, complexation, and penetration pathways of vapor-deposited aluminum atoms with HO- and CH_3O-terminated organic monolayers. J Am Chem Soc 124(19):5528–5541
70. Tai Y, Shaporenko A, Eck W, Grunze M, Zharnikov M (2004) Abrupt change in the structure of self-assembled monolayers upon metal evaporation. Appl Phys Lett 85(25):6257–6259
71. Tai Y, Shaporenko A, Noda H, Grunze M, Zharnikov M (2005) Fabrication of stable metal films on the surface of self-assembled monolayers. Adv Mater 17(14):1745–1749
72. Wacker D, Weiss K, Kazmaier U, Woll C (1997) Realization of a phenyl-terminated organic surface and its interaction with chromium atoms. Langmuir 13(25):6689–6696
73. Tai Y, Shaporenko A, Grunze M, Zharnikov M (2005) Effect of irradiation dose in making an insulator from a self-assembled monolayer. J Phys Chem B 109(41):19411–19415
74. Noda H, Tai Y, Shaporenko A, Grunze M, Zharnikov M (2005) Electrochemical characterizations of nickel deposition on aromatic dithiol monolayers on gold electrodes. J Phys Chem B 109(47):22371–22376
75. Carlo SR, Wagner AJ, Fairbrother DH (2000) Iron metalization of fluorinated organic films: a combined X-ray photoelectron spectroscopy and atomic force microscopy study. J Phys Chem B 104(28):6633–6641
76. Bittner AM (2006) Clusters on soft matter surfaces. Surf Sci Rep 61(9):383–428
77. Hermes S, Zacher D, Baunemann A, Woll C, Fischer RA (2007) Selective growth and MOCVD loading of small single crystals of MOF 5 at alumina and silica surfaces modified with organic self-assembled monolayers. Chem Mater 19:2168–2173
78. Rajalingam K, Strunskus T, Terfort A, Fischer RA, Woll C (2008) Metallization of a thiol-terminated organic surface using chemical vapor deposition. Langmuir 24(15):7986–7994
79. Weiß J, Himmel H-J, Fischer RA, Wöll C (1998) Self-terminated CVD-functionalization of organic self-assembled monolayers (SAMs) with trimethylamine alane (TMAA). Chem Vap Deposition 4(1):17–21
80. Zacher D, Shekhah O, Woll C, Fischer RA (2009) Thin films of metal-organic frameworks. Chem Soc Rev 38(5):1418–1429
81. Kashammer J, Wohlfart P, Weiß J, Winter C, Fischer R, Mittler-Neher S (1998) Selective gold deposition via CVD onto self-assembled organic monolayers. Opt Mater 9(1–4):406–410
82. Wohlfart P, Weiß J, Kashammer J, Winter C, Scheumann V, Fischer RA, Mittler-Neher S (1999) Selective ultrathin gold deposition by organometallic chemical vapor deposition onto organic self-assembled monolayers (SAMs). Thin Solid Films 340(1–2):274–279
83. Winter C, Weckenmann U, Fischer RA, Kashammer J, Scheumann V, Mittler S (2000) Selective nucleation and area-selective OMCVD of gold on patterned self-assembled organic monolayers studied by AFM and XPS: a comparison of OMCVD and PVD. Chem Vap Deposition 6(4):199–205

84. Semaltianos NG, Pastol J-L, Doppelt P (2004) Copper nucleation by chemical vapour deposition on organosilane treated SiO_2 surfaces. Surf Sci 562(1–3):157–169
85. Semaltianos NG, Pastol J-L, Doppelt P (2004) Copper chemical vapour deposition on organosilane-treated SiO_2 surfaces. Appl Surf Sci 222(1–4):102–109
86. Liu X, Wang Q, Wu S, Liu Z (2006) Enhanced CVD of copper films on self-assembled monolayers as ultrathin diffusion barriers. J Electrochem Soc 153(3):C142–C145
87. Rajalingam K, Bashir A, Badin M, Schroeder F, Hardman N, Strunskus T, Fischer RA, Woll C (2007) Chemistry in confined geometries: reactions at an organic surface. ChemPhysChem 8:657–660
88. Lu P, Demirkan K, Opila RL, Walker AV (2008) Room-temperature chemical vapor deposition of aluminum and aluminum oxides on alkanethiolate self-assembled monolayers. J Phys Chem C 112(6):2091–2098
89. Wohlfart P, Weiß J, Käshammer J, Kreiter K, Winter C, Fischer RA, Mittler-Neher S (1999) MOCVD of aluminum oxide/hydroxide onto organic self-assembled monolayers. Chem Vap Deposition 5(4):165–170
90. Chen R, Bent SF (2006) Chemistry for positive pattern transfer using area-selective atomic layer deposition. Adv Mater 18:1086
91. Chen R, Kim H, McIntyre PC, Bent SF (2005) Investigation of self-assembled monolayer resists for hafnium dioxide atomic layer deposition. Chem Mater 17:536
92. Park MH, Jang YJ, Sung-Suh HM, Sung MM (2004) Selective atomic layer deposition of titanium oxide on patterned self-assembled monolayers formed by microcontact printing. Langmuir 20:2257–2260
93. Park KJ, Doub JM, Gougousi T, Parsons GN (2005) Microcontact patterning of ruthenium gate electrodes by selective area atomic layer deposition. Appl Phys Lett 86:051903
94. Killampalli A, Aravind S, Ma PF, Engstrom JR (2005) The reaction of tetrakis (dimethylamido)titanium with self-assembled alkyltrichlorosilane monolayers possessing -OH, -NH2, and -CH3 terminal groups. J Am Chem Soc 127(17):6300–6310
95. Dube A, Chadeayne AR, Sharma M, Wolczanski PT, Engstrom JR (2005) Covalent attachment of a transition metal coordination complex to functionalized oligo(phenylene-ethynylene) self-assembled monolayers. J Am Chem Soc 127(41):14299–14309
96. Dube A, Sharma M, Ma PF, Ercius PA, Muller DA, Engstrom JR (2007) Effects of interfacial organic layers on nucleation, growth, and morphological evolution in atomic layer thin film deposition. J Phys Chem C 111:11045–11058
97. Seitz O, Dai M, Aguirre-Tostado FS, Wallace RM, Chabal YJ (2009) Copper-metal deposition on self assembled monolayer for making top contacts in molecular electronic devices. J Am Chem Soc 131(50):18159–18167
98. Haick H, Cahen D (2008) Contacting organic molecules by soft methods: towards molecule-based electronic devices. Acc Chem Res 41(3):359–366
99. Metzger RM, Xu T, Peterson IR (2001) Electrical rectification by a monolayer of hexadecyl-quinolinium tricyanoquinodimethanide measured between macroscopic gold electrodes. J Phys Chem B 105:7280–7290
100. Binns C (2001) Nanoclusters deposited on surfaces. Surf Sci Rep 44(1–2):1–49
101. Andres R, Bein T, Dorogi M, Feng S, Henderson J, Kubiak C, Mahoney W, Osifchin R, Reifenberger R (1996) "Coulomb staircase" at room temperature in a self-assembled molecular nanostructure. Science 272:1323–1325
102. Okazaki N, Sambles JR, Jory MJ, Ashwell GJ (2002) Molecular rectification at 8 K in an Au/$C_{16}H_{33}$Q-3CNQ LB film/Au structure. Appl Phys Lett 81(12):2300–2302
103. Xu T, Peterson IR, Lakshmikantham MV, Metzger RM (2001) Rectification by a monolayer of hexadecylquinolinium tricyanoquinodimethanide between gold electrodes. Angew Chem Int Ed 40(9):1749–1752
104. Haick H, Ambrico M, Ligonzo T, Cahen D (2004) Discontinuous molecular films can control metal/semiconductor junction. Adv Mater 16:2145

105. Lando A, Lauwaet K, Lievens P (2009) Controlled nanostructuring of a gold film covered with alkanethiol SAM by low energy cluster implantation. Phys Chem Chem Phys 11 (10):1521–1525
106. Deng J, Hofbauer W, Chandrasekhar N, Shea SJ (2007) Metallization for crossbar molecular devices. Nanotechnology 18(15):155202
107. Waddill GD, Vitomirov IM, Aldao CM, Weaver JH (1989) Cluster deposition on GaAs (110): formation of abrupt, defect-free interfaces. Phys Rev Lett 62(13):1568
108. Weaver JH, Waddill GD (1991) Cluster assembly of interfaces: nanoscale engineering. Science 251(5000):1444–1451
109. Bruch LW, Diehl RD, Venables JA (2007) Progress in the measurement and modeling of physisorbed layers. Rev Mod Phys 79(4):1381–1454
110. Akkerman HB, Blom PWM, de Leeuw DM, de Boer B (2006) Towards molecular electronics with large-area molecular junctions. Nature 441(7089):69–72
111. van Hal PA, Smits ECP, Geuns TCT, Akkerman HB, De Brito BC, Perissinotto S, Lanzani G, Kronemeijer AJ, Geskin V, Cornil J, Blom PWM, de Boer B, de Leeuw DM (2008) Upscaling, integration and electrical characterization of molecular junctions. Nat Nanotechnol 3(12):749–754
112. Salomon A, Cahen D, Lindsay S, Tomfohr J, Engelkes VB, Frisbie CD (2003) Comparison of electronic transport measurements on organic molecules. Adv Mater 15(22):1881–1890
113. Engelkes VB, Beebe JM, Frisbie CD (2004) Length-dependent transport in molecular junctions based on SAMs of alkanethiols and alkanedithiols: effect of metal work function and applied bias on tunneling efficiency and contact resistance. J Am Chem Soc 126 (43):14287–14296
114. Wang W, Lee T, Reed MA (2003) Mechanism of electron conduction in self-assembled alkanethiol monolayer devices. Phys Rev B 68(3):035416
115. Akkerman HB, de Boer B (2008) Electrical conduction through single molecules and self-assembled monolayers. J Phys Condensed Matter 20(1):013001
116. Wang G, Yoo H, Na SI, Kim TW, Cho B, Kim DY, Lee T (2009) Electrical conduction through self-assembled monolayers in molecular junctions: Au/molecules/Au versus Au/molecule/PEDOT:PSS/Au. Thin Solid Films 518(2):824–828
117. Wang G, Kim Y, Choe M, Kim T-W, Lee T (2011) A new approach for molecular electronic junctions with a multilayer graphene electrode. Adv Mater 23(6):755–60. doi:10.1002/adma.201003178
118. Akkerman HB (2008) Large area molecular junctions. Ph.D. Thesis, University of Groningen, The Netherlands, ISBN 978-90-367-3441-7
119. Lenfant S, Guerin D, TranVan F, Chevrot C, Palacin S, Bourgoin JP, Bouloussa O, Rondelez F, Vuillaume D (2006) Electron transport through rectifying self-assembled monolayer diodes on silicon: Fermi-level pinning at the molecule-metal interface. J Phys Chem B 110 (28):13947–13958
120. Hacker CA, Richter CA, Gergel-Hackett N, Richter LJ (2007) Origin of differing reactivities of aliphatic chains on H-Si(111) and oxide surfaces with metal. J Phys Chem C 111 (26):9384–9392
121. Zuo J, Keil P, Valtiner M, Thissen P, Grundmeier G (2008) Deposition of Ag nanoparticles on fluoroalkylsilane self-assembled monolayers with varying chain length. Surf Sci 602 (24):3750–3759
122. Kolipaka S, Aithal RK, Kuila D (2006) Fabrication and characterization of an indium tin oxide-octadecanethiol-aluminum junction for molecular electronics. Appl Phys Lett 88 (23):233104–233113
123. Ulman A (1996) Formation and structure of self-assembled monolayers. Chem Rev 96 (4):1533–1554
124. McGuiness CL, Blasini D, Masejewski JP, Uppili S, Cabarcos OM, Smilgies D, Allara DL (2007) Molecular self-assembly at bare semiconductor surfaces: characterization of a homologous series of n-alkanethiolate monolayers on GaAs(001). ACS Nano 1(1):30–49

125. Ioffe Z, Shamai T, Ophir A, Noy G, Yutsis I, Kfir K, Cheshnovsky O, Selzer Y (2008) Detection of heating in current-carrying molecular junctions by Raman scattering. Nat Nanotechnol 3(12):727–732
126. Yoon HP, Maitani MM, Cabarcos OM, Cai L, Mayer TS, Allara DL (2010) Crossed-nanowire molecular junctions: a new multi-spectroscopy platform for conduction-structure correlations. Nano Lett 10:2897–2902

Top Curr Chem (2012) 312: 275–302
DOI: 10.1007/128_2011_223

Published online: 2 August 2011

Spin Polarized Electron Tunneling and Magnetoresistance in Molecular Junctions

Greg Szulczewski

Abstract This chapter reviews tunneling of spin-polarized electrons through molecules positioned between ferromagnetic electrodes, which gives rise to tunneling magnetoresistance. Such measurements yield important insight into the factors governing spin-polarized electron injection into organic semiconductors, thereby offering the possibility to manipulate the quantum-mechanical spin degrees of freedom for charge carriers in optical/electrical devices. In the first section of the chapter a brief description of the Jullière model of spin-dependent electron tunneling is reviewed. Next, a brief description of device fabrication and characterization is presented. The bulk of the review highlights experimental studies on spin-polarized electron tunneling and magnetoresistance in molecular junctions. In addition, some experiments describing spin-polarized scanning tunneling microscopy/spectroscopy on single molecules are mentioned. Finally, some general conclusions and prospectus on the impact of spin-polarized tunneling in molecular junctions are offered.

Keywords Magnetoresistance, Moleculer junctions · Tunneling

Contents

G. Szulczewski (✉)
Department of Chemistry, University of Alabama, Tuscaloosa, AL 34587, USA
e-mail: gjs@ua.edu

1 Introduction

Although tunneling is a well-established theoretical concept in quantum mechanics, experimental studies of tunneling in metal/insulator/metal structures were largely absent until the invention of the transistor. In the 1950s and 1960s experimental studies on tunneling had a renaissance. Driven largely to verify experimentally the Bardeen–Copper–Schreefer (BCS) theory of superconductivity, tunneling through M–I–M structures, where M is a metal and I is an electrical insulator, was a very active research area at Bell Labs and General Electric. An excellent book that describes many of the seminal activities during these decades was edited by Burstein and Lundqvist [1]. The importance of this era was recognized when the 1973 Nobel Prize in Physics was awarded to Leo Esaki and Ivar Giaever for "their experimental discoveries regarding tunneling phenomena in semiconductors and superconductors" and to Brian Josephson "for his theoretical predictions of the properties of a supercurrent through a tunnel barrier, in particular those phenomena which are generally known as the Josephson effects." The importance of tunneling in real-world devices cannot be understated. In 2007, Intel announced that it would replace the SiO_2 gate dielectric with HfO_2 in order to decrease the leakage current due to tunneling in field-effect transistors. To paraphrase Gordon Moore, one of Intel's founders, the transition away from SiO_2 marked the biggest change in transistor technology since the late 1960s.

The development of advanced electronic devices is confronted by imposing technical limits and costs set by lithographic tools, so scientists and engineers have been exploring new paradigms to design future computing and memory technologies. One example that has reached commercialization is solid-state memory. Today there are inexpensive portable "flash" drives that contain as much memory as hard-disk drives did only a few years ago. Currently there is great interest in creating a "universal" memory from solid-state structures with no moving parts and low power consumption. One of the most appealing strategies is to combine transistor logic with magnetic tunnel junctions. The theoretical prediction [2, 3] and experimental verification [4] of spin-torque-induced switching allows the magnetization direction of a thin magnetic layer in a nanojunction to be controlled with an electrical current. In this way an electrically addressable magnetic memory element can be placed at every transistor, to create magnetic random-access memory.

Advances in the synthesis and processing of organic molecules and polymers have led to the development of organic semiconductor-based devices that exhibit a multitude of functionalities, e.g., photovoltaics, memory/logic, and displays;

some commercial organic semiconductor devices have already appeared on the market. The primary motivation to use organic materials is the prospect of low-cost and low-temperature processing methods, such as roll-to-roll and ink-jet printing, that enable devices to be fabricated on flexible substrate. The market for such products is expected to reach US $50 billion by 2020 [5]. To expand the versatility of molecular materials; there has been an effort to manipulate the spin degrees of freedom of electrical carriers (mainly electrons) in organic semiconductors. It is envisioned that, in addition to the charge-transport property, spin-dependent properties, such as the recombination of electron-hole pairs to yield light, can be controlled to yield a new class of multifunctional materials. To this end, there are a few basic issues that must be addressed. First, undoped organic semiconductors have no free carriers, so electrons (or holes) must be supplied by a source, usually a metal. To observe spin-dependent effects, it is natural to use magnetic metals or alloys, which possess an imbalance of spin-polarized electrons. Second, a device that can discriminate the current carried by "up" and "down" spin electrons must be available. Fortunately the fate of spin-polarized electron transport in ferromagnetic metals has been well studied over the past three decades and the Nobel Prize in Physics in 2007 was award to Albert Fert and Peter Grünberg "for the discovery of giant magnetoresistance." Magnetoresistance was originally observed in multilayer structures with alternating magnetic and nonmagnetic layers [6]. In a trilayer structure, often called a spin-valve, two magnetic electrodes are separated by a very thin nonmagnetic layer, which can be conducting or insulating.

This chapter focuses on spin-polarized tunneling through molecules, which results in tunneling magnetoresistance (TMR). This is in distinct contrast to another form of "organic magnetoresistance," or OMAR, observed in organic semiconductor devices without magnetic electrodes [7–10]. There are several theories to explain OMAR, and the interested reader is referred to the literature for the details [11, 12]. This chapter also excludes the magnetoresistance observed in carbon nanotubes [13], graphene [14], and organic superconductors [15]. Furthermore, there is no discussion of the theoretical approaches developed to model the organic magnetoresistance phenomena. Section 2 presents a brief tutorial on the concepts of spin-polarization, superconductors, and band-structure of ferromagnetic metals. In Sect. 3 basic experimental methods are discussed. Section 4 summarizes the results of several experiments. Section 5 gives a brief synopsis of the field. In general, this chapter is written from a chronological perspective, covering the years 2004 through 2010.

2 Basic Concepts in Tunneling with Ferromagnetic Metals and Superconductors

2.1 Spin Polarization of Ferromagnetic Metals

Majority- and minority-spin electrons are often referred to as "up" or "down" electrons, respectively. Consider a simplified drawing of the bulk electronic band

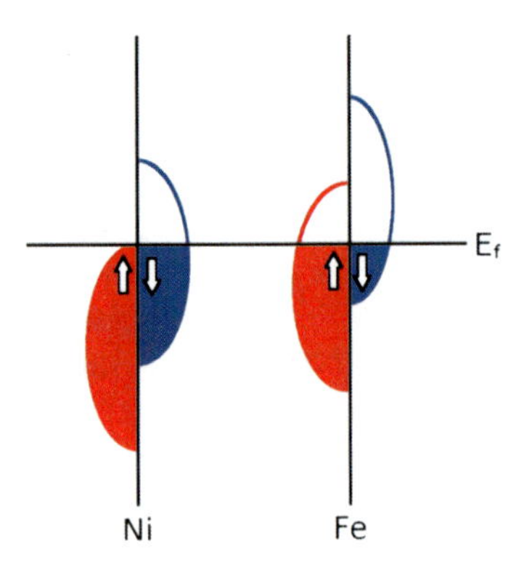

Fig. 1 Schematic diagram depicting the bulk spin-dependent electronic structure of Fe and Ni

structure for Fe and Ni shown in Fig. 1. It is well known that the *d*-bands split due to the exchange interaction [16]. For Ni there is a larger density of states for the minority electrons at the Fermi level, while Fe has a larger density of states for the majority electrons. Not shown in Fig. 1 is the *sp*-hybridized band that crosses the Fermi energy: this has important consequences in the transport studies described below. The band structure depicted in Fig. 1 suggests that the majority and minority bands for Fe are reversed with respect to Ni. However, tunneling experiments that utilize a superconducting electrode as a spin-detector reveal a difference picture. Tedrow and Meservey pioneered measurement of the spin-polarization of a ferromagnetic thin film by tunneling spectroscopy. The details of the technique will not be described here, but the interested reader can consult their excellent review article [17]. In brief, a thin (~4 nm) Al electrode is deposited onto an insulating substrate, followed by brief oxidation, to produce a very thin Al_2O_3 layer. Next, a ferromagnetic metal or alloy can be deposited on top to form a tunnel junction. When the junction is cooled below ~2 K, the Al becomes superconducting. As a result, magnetic fields up to about 4 T can be applied that split the degenerate density of states by the Zeeman energy. In this configuration, the superconducting Al electrode functions as a spin-detector. When a small bias (~1 mV) is applied to the junction, no current flows, due to the energy gap in the superconductor. Once the gap energy is exceeded by the bias, two peaks are observed in the differential conductance spectrum, $G(V) = dI/dV$. These peaks correspond to the current from the "up" and "down" spin electrons in the ferromagnetic metal. To a first approximation, the difference in the amplitude of these peaks can be used to estimate the spin-polarization of the ferromagnetic layers. Using a more rigorous theoretical description of the superconductor, it is possible to determine both the sign and magnitude of the spin polarization [18]. The results of many experiments reveal that the sign of the polarization is positive for Fe, Ni, and Co, which does not seem to support the simple band structure shown in Fig. 1. The origin of the positive spin-polarization in Fe, Ni, and Co results from the fact that the tunneling probability is higher for *s* electrons than for *d* electrons because they have a higher velocity at the Fermi energy.

2.2 Tunneling Magnetoresistance and the Jullière Model

In 1975 Jullière measured the TMR ratio for a Co/Ge/Fe trilayer at 4.2 K, and proposed a simple way to estimate the magnitude of the effect [19]. The model is essentially a combination of Mott's two-current model in ferromagnetic metals and the Meservey–Tedrow estimate of the effective density of states measured by spin-polarized tunneling. In the Jullière model, the up and down spin electrons are thought to pass the barrier in parallel channels, without spin scattering. A schematic drawing is shown in Fig. 2. Under these assumptions, the tunneling conductance between the ferromagnetic electrodes can be written as the sum of the conductance for the two channels. Thus, for parallel electrode alignment of the electrode magnetization, the up-spin electrons on the left side ($\uparrow_l$) are still up-spin on the right ($\uparrow_r$) after tunneling through the barrier. Conversely, the down-spin electrons on the left ($\downarrow_l$) remain down-spin on the right ($\downarrow_r$). In contrast, for antiparallel alignments of the electrode magnetization, electrons that are locally up-spin find themselves in a region of opposite magnetization, which means that they are locally down-spin. Consequently, down-spins on the left become up-spins on the right. In this model it is straightforward to derive the formula for TMR. In the parallel and antiparallel configuration of the electrode magnetization the equations for DC conductance (G_p and G_{ap}, respectively) are

$$G_p = G(\uparrow_l, \uparrow_r) + G(\downarrow_l, \downarrow_r) \tag{1}$$

and

$$G_{ap} = G(\uparrow_l, \downarrow_r) + G(\uparrow_r, \downarrow_l). \tag{2}$$

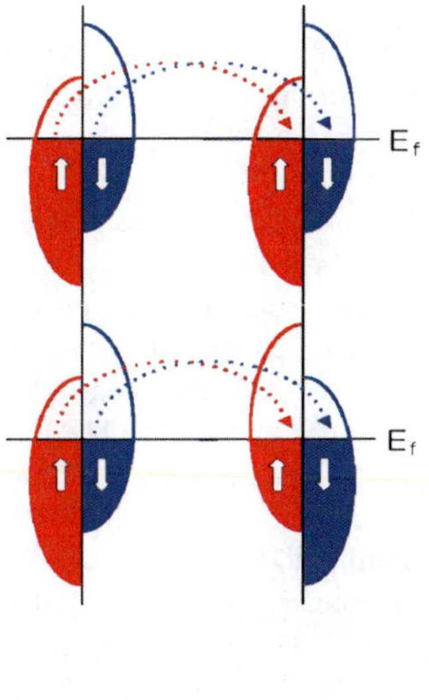

Fig. 2 Schematic diagram of spin-conserved tunneling via the Jullière model. The *top* and *bottom panels* indicate parallel and antiparallel alignment of the magnetization of the top and bottom electrodes, respectively

Jullière assumed that the conductance is proportional to the density of states of the left and right electrodes. The TMR is usually defined as the ratio of the change in conductance to the minimum conductance by the equation

$$\mathrm{TMR} = (G_{\mathrm{p}} - G_{\mathrm{ap}})/G_{\mathrm{ap}}. \tag{3}$$

Furthermore, if the polarizations of the left and right electrodes are defined as

$$P_{\mathrm{L}} = [\mathrm{N}(\uparrow_{\mathrm{l}}) - \mathrm{N}(\downarrow_{\mathrm{l}})]/[\mathrm{N}(\uparrow_{\mathrm{l}}) + \mathrm{N}(\downarrow_{\mathrm{l}})] \tag{4}$$

and

$$P_{\mathrm{R}} = [\mathrm{N}(\uparrow_{\mathrm{r}}) - \mathrm{N}(\downarrow_{\mathrm{r}})]/[\mathrm{N}(\uparrow_{\mathrm{r}}) + \mathrm{N}(\downarrow_{\mathrm{r}})] \tag{5}$$

respectively, where N is proportional to the number of up and down spin electrons, then the equation for TMR becomes

$$\mathrm{TMR} = 2P_{\mathrm{L}}P_{\mathrm{R}}/(1 - P_{\mathrm{L}}P_{\mathrm{R}}). \tag{6}$$

Equation (6) shows that TMR can be expressed in terms of the spin polarization of the left and right electrodes. Although this formula is frequently used to analyze data from TMR experiments, it must be applied with great care. First, the polarization should not be interpreted as the spin-polarization of the density of states at the Fermi energy. The reason for this is that the spin polarization of the tunneling current can only be measured when electrons tunnel between a ferromagnetic electrode and a superconducting electrode. Second, even though Ni (and Co) has more minority electrons than majority electrons at the Fermi energy, it is the majority electrons that tunnel with a higher probability. Third, the sign of the TMR has been observed to reverse when the barrier layer is replaced, even though the ferromagnetic electrodes remain the same. As a result, the Jullière formula is most useful when comparing TMR for junctions with different electrodes, while keeping the barrier material constant.

In contrast to the discussion above with amorphous barriers, it is possible to use first-principles electron-structure calculations to describe TMR with crystalline tunnel barriers. In the Jullière model the TMR is dependent only on the polarization of the electrodes, and not on the properties of the barrier. In contrast, theoretical work by Butler and coworkers showed that the transmission probability for the tunneling electrons depends on the symmetry of the barrier, which has a dramatic influence on the calculated TMR values [20]. In the case of Fe(100)/MgO(100)/Fe (100) the majority of electrons in the Fe are spin-up. They are derived from a band of delta-symmetry. In 2004 these theoretical predictions were experimentally confirmed by Parkin et al. and Yusha et al. [21, 22]. Remarkably, by 2005 TMR read heads were introduced into commercial hard disk drives.

3 Experimental Methods

3.1 Junction Fabrication

The most common method to form tunnel junctions involves the sequential deposition, usually by thermal evaporation through shadow masks, of a bottom electrode, organic layer(s), and a top electrode orthogonal to the bottom electrode, resulting in a crossbar structure. Another approach involves deposition of continuous films and subsequent photolithography to determine the junction size. Szulczewski et al. described a procedure that combined sputtering, thermal evaporation, and photolithography [23]. The process yielded 40% of nonshorted junctions and the standard deviation of TMR values in working junctions was $\pm 2\%$ of the average value. A microphotograph of a finished device is shown in Fig. 3. The principal advantages of these fabrication methods is the generation of a four-terminal device. In such devices the contact electrical resistance of the leads is canceled. Another important aspect is that the junction resistance must be at least an order of magnitude greater than the lead resistance. If not, micromagnetics simulations have shown that current crowding can occur, and the measured TMR will be an overestimate of the true value [24]. A few other approaches have been attempted and yielded two-terminal devices. For example, electromigration of a narrow constriction in a metal wire can lead to a nanoscale gap [25]. The formation of nanopore junctions has also been utilized [26]. Since both of the approaches requires access to high-fidelity electron-beam lithography tools, they are somewhat specialized, and not routinely practiced. One study reported the fabrication of micron- to millimeter-size junctions by soft-contact printing techniques; however, the *I*/*V* measurements on the final structures showed considerable hysteresis, probably because the Co electrodes were oxidized [27]. Regardless of what approach is used to create a sandwich structure, one important criterion is that the ferromagnetic electrodes must have different coercive fields, that is, the magnetic energy to switch the magnetization direction. This is accomplished by using magnetic "hard" (e.g., Co) and "soft" (e.g., $Ni_{80}Fe_{20}$) layers. Since the coercivity

Fig. 3 Microphotograph of some junctions prepared by the methods described in [23]

is an intrinsic parameter that depends on many properties, such as film thickness, roughness, deposition conditions, etc., one should measure a hysteresis loop, ideally as a function of temperature.

3.2 Tunneling Criteria

Before reviewing the experimental results, it is germane to outline the criteria expected for tunneling behavior. First, in the tunneling limit there is the expectation that current/voltage (*I*/*V*) curves will be nonlinear, due to the exponential dependence of tunneling current on the barrier width and height. Specifically, the differential conductance should scale quadratically with voltage [28]. Second, if the area of the device is varied (and the thickness is held constant), the resistance-area product should remain constant. If so, this is a good indication of uniform conduction and the absence of "hot-spots" in locally thin regions of the barrier. Third, the resistance should change very little with temperature. In fact one of the most accepted criteria to support a tunneling model is a slight increase (~10–20%) in the junction resistance with decreasing temperature [29]. If the resistance decreases with decreasing temperature, then metallic shorts are likely between the electrodes. Finally, if one of the ferromagnetic electrodes is replaced with a superconductor, then an energy gap (characteristic of a superconductor) should appear in a differential conductance measurement. The examples shown below have been selected because they satisfy these criteria, and represent the first generation of experiments that observed TMR in molecular junctions.

3.3 Spectroscopic Characterization

After the criteria for tunneling are met, there remains another challenge. How is the buried molecule/metal interface characterized? One of the best methods is to use inelastic tunneling spectroscopy (IETS). Since IETS is nicely described in the chapter by Hipps in this series, it will not be discussed in detail here. On the other hand, it is possible to probe the interface at various stages of the top-metal deposition with spectroscopic methods. This approach has provided tremendous insight into the electronic structure of the metal/molecule interface [30]. However, very few photoemission studies have been reported for ferromagnetic metals deposited onto organic molecules [31–38]. Since there are several examples of spin-polarized tunneling through Alq_3 barrier layers with Co electrodes in Sect. 4, we next provide a brief characterization of the Alq_3/Co interface.

Xu et al. used X-ray photoelectron spectroscopy (XPS), ultraviolet photoelectron spectroscopy (UPS), and X-ray magnetic circular dichroism (XMCD) spectroscopy to characterize the electronic and magnetic structure of Co deposited onto Alq_3 films under high-vacuum conditions [35]. Figure 4 shows the N(1*s*), O(1*s*),

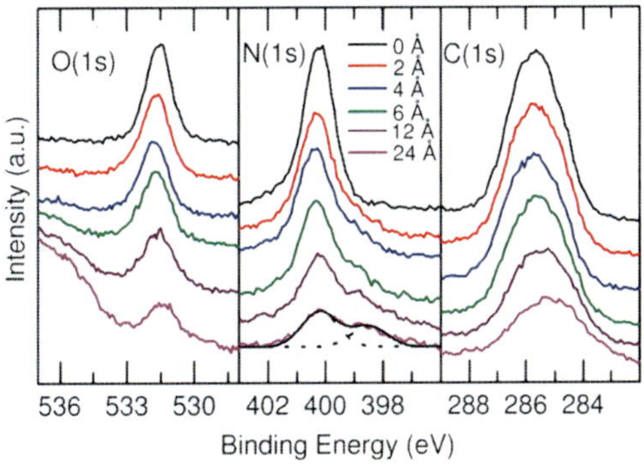

Fig. 4 X-ray photoelectron spectra of Alq_3 core levels before and after 2, 4, 6, 12, and 24 Å thick deposition of Co. Taken from [35] with permission

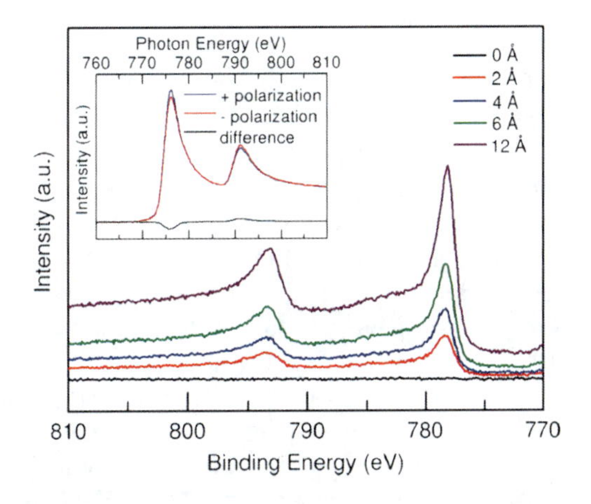

Fig. 5 X-ray photoelectron spectra of the $Co(2p_{3/2})$ core levels before and after a 2, 4, 6, 12, and 24 Å thick cobalt deposition on Alq_3. The *inset* shows the XAS and XMCD spectra recorded at 300 K for a 20-Å deposition of cobalt on Alq_3. Taken from [35] with permission

and C(1*s*) core-level signals from a 10-nm Alq_3 film before and after deposition of 2, 4, 6, 12, and 24 Å of Co. Initially, the pristine Alq_3 film has a single N(1*s*) peak at 400.3 eV. After deposition of 2 Å Co, a low binding-energy peak at 398.5 eV appears. A shift in 1.8 eV to lower binding energy has been seen previously for the initial stages of Ca, Mg, and Al deposition on Alq_3 [39, 40]. With further Co deposition, the intensity of the peak at 398.5 eV increases, and the peak intensity at 400.3 eV decreases. The N(1*s*) spectrum after the 24-Å Co deposition can be deconvoluted with two Gaussian peaks, to indicate clearly two chemical states. The O(1*s*) core level shifts from 531.7 to 532.1 eV after a 4-Å Co deposition. The broad background that appears on the high binding-energy side is due to a Co Auger transition. There is slight shoulder developing on the low binding-energy side for the C(1*s*) peak that becomes obvious after the 24-Å Co deposition. This may suggest decomposition of the quinolate ring. There is no shift in the Al(2*p*) core-level binding energy at any Co coverage (data not shown). Due to the attenuation effect in XPS, the core-level spectra shown after the 24-Å Co deposition suggest that there is a chemical interaction between Co and Alq_3. After a 50-Å Co deposition (data not shown), all the core-level signals from Alq_3 are almost completely attenuated, which implies limited diffusion of Co into Alq_3. The $Co(2p_{3/2})$ region

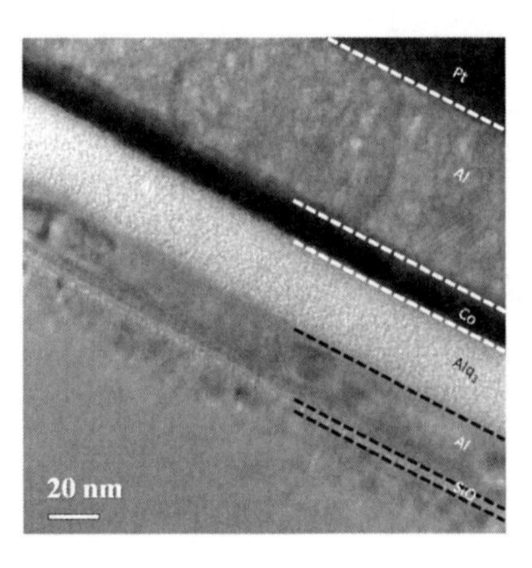

Fig. 6 Cross-sectional TEM image of a 10-nm Co film deposited onto a 30-nm Alq_3 film. Taken from [35] with permission

before and after Co deposition is shown in Fig. 5. The binding energy of the Co ($2p_{3/2}$) peak is found at 778.5 eV after a 2-Å Co deposition, and shifts to 778.0 eV after a 12-Å Co deposition. The binding energy of metallic Co is 778.0 eV, while the binding energy of the Co^{2+} is known to be near 780 eV [41]. Since the intensity of the Co signal increases and the Alq_3 core levels decrease when increasing the cobalt thickness, it is an indication that most of the Co is accumulating within the near-surface region, and not penetrating deep into the Alq_3 layer. Moreover, the Co ($2p_{3/2}$) binding energy is consistent with metallic Co if the film thickness exceeds 10 Å. The inset of Fig. 5 shows an XMCD spectrum (the difference spectrum) measured at 300 K for 20 Å of Co deposited onto Alq_3: it shows a clear signature for ferromagnetism. The decrease in the Alq_3 core-level signals and the increase in the Co ($2p_{3/2}$) suggest that there is minimal diffusion of Co into the Alq_3 films. This is directly observed in the TEM image shown in Fig. 6.

4 Review of Seminal Spin-Polarized Tunneling Studies Through Molecules

4.1 Single Molecule and Alkanethiol Monolayers

In 2004 Pasupathy et al. were the first to report tunneling through a single molecule using ferromagnetic electrodes [42]. Using the electromigration technique on Ni wires with different coercivity, Pasupathy et al. were able to isolate single C_{60} molecules in the gap. However, the yield of working devices was rather low, and only 36 out of 1,200 junctions exhibited TMR behavior. At 1.5 K one of the junctions showed a TMR of −80% and another −38% at a 10-mV bias. The sign and magnitude is opposite and larger, respectively, than the Jullière model prediction of +21%, if the spin-polarization of Ni is assumed to be 31% [43]. The authors also observed the Kondo effect, which is characteristic of a coupling between a localized spin and conduction electrons. The experimental evidence for the Kondo effect was twofold: (1) a splitting in the zero-bias conductance at low temperature when an external magnetic field is applied and (2) a logarithmic decrease in the zero-bias conductance when the temperature is increased. The authors proposed that the Kondo resonance was responsible for the deviation from Jullière model.

Later in 2004 Petta et al. used nanopore geometry to study tunneling through a self-assembled monolayer (SAM) composed of octanethiol molecules on Ni electrodes [44]. The nanopores had a diameter in the 5–10 nm range, into which octanethiol molecules were allowed to self-assemble from a 1 mM solution in ethanol for 48 h. Approximately 100–400 molecules would be in such a pore, assuming a packing density of 5 molecules/nm^2 [45]. To avoid damage to the molecules, the top Ni electrode was deposited at 0.1 Å/s to a total thickness of 3 nm, with the substrate held at 77 K. In some junctions a 1-nm Ti layer was deposited before the Ni. The authors reported no substantial difference in the TMR curves of junctions with or without the Ti. Control experiments verified that TMR was only observed in junctions containing the SAM. At 4.2 K, without an external magnetic field, the *I/V* curves were nearly symmetric around zero bias. Using the Simmons tunneling model, the authors fit the *I/V* curves and obtained a barrier height of ~1.5 eV, which is similar to the barrier height for alkanethiol SAMs formed in nanopore junctions with Au electrodes [46]. When an external magnetic field was −0.6 to 0.6 T and back at 4.2 K, the junction resistance changed between two levels, which signals the TMR phenomenon. There was a direct correlation between the TMR value and the magnitude of the junction resistance. In general, large TMR values were measured for junctions with the highest resistance. For example, the TMR was ~1%, 3%, and 7 % at 10 mV bias when the junction resistance was ~1, 10, and 100 MΩ, respectively. Regardless of resistance, all the junctions showed a strong temperature and bias dependence. When the bias was increased from 10 to 50 mV, the TMR dropped to almost zero with the temperature fixed at 4.2 K. In addition, when the bias was fixed and the temperature was increased from 5 to 30 K, the TMR nearly vanished. Finally, Petta et al. reported evidence for telegraph noise in some of the junctions, that is, two-level resistance changes. Taken together, the experimental evidence suggested that the SAMs contain defects at between one and ten molecule sites in the junctions that might be responsible for the fluctuations.

In 2006 Wang and Richter measured TMR and inelastic tunneling electron spectroscopy (IETS) for Co/SAM/Ni nanopore structures [47]. A nonlinear *I/V* curve measured at 4.2 K was symmetrical about zero bias. Wang and Richter noted that above 50 K the *I/V* curves showed a slight temperature dependence, but below 50 K they were independent of temperature. An IETS spectrum at 4.2 K revealed vibration excitation of the Ni–S, C–S, and C–C stretching mode and a CH_2 deformation mode. The same peaks were observed after application of small magnetic field up to 0.6 T. The TMR at 4.2 K did not show distinct resistance changes between two states. Nonetheless Wang and Richter estimated a 9% TMR value between the maximum and minimum valves of the resistance. The TMR decreased when the junction bias coincided with the first vibration excitation (Ni–S), so they speculated that excitation of the molecular vibration maybe the main cause of the TMR decrease instead of magnon excitations.

4.2 Alq_3 and Rubrene Based Tunnel Barriers

In 2007 Santos showed the first reproducible and robust set of experiments to demonstrate TMR through a thin layer Alq_3 at room temperature [48]. In these experiments up to 72 junctions can be fabricated at once with different junction area and Alq_3 thickness. As a result, the bottom and top electrodes are the same and help to reduce systematic variations that are inevitable in different depositions. Figure 7 shows the electrical behavior of Co(8 nm)/Al_2O_3(0.6 nm)/Alq_3(1.6 nm)/$Ni_{80}Fe_{20}$(10 nm) as a function of temperature. The *I*/*V* curve is symmetrical about zero bias and was fit to the Brinkman model [28] and yielded a barrier height of 0.47 eV. The corresponding G(V) curves at 300, 77, and 4.2 K in Fig. 7 show the expected parabolic dependence as a function of bias and no zero bias anomalies. The inset in Fig. 7 shows the junction resistance scales exponentially with increasing Alq_3 thickness. These data suggest that the junctions are pinhole free and exhibit proper tunneling characteristics. In Fig. 8 the TMR is plotted as a function of external magnetic field and bias. In Fig. 8a the TMR curves were measured at a 10-mV bias, as the external magnetic field was scanned back and forth from −100 to +100 Oe. The abrupt switching of the resistance at the coercive fields of the electrodes signals is observed, and the maximum TMR increased slightly upon cooling the junctions from 300 to 4.2 K. In Fig. 8b the bias dependence of the TMR is shown. Upon increasing the bias from 10 to 100 mV, the TMR value drops from ~7% to ~4% at 4.2 K. The inset in Fig. 8b shows a cross-section TEM image of the junctions, and confirms that the Alq_3 is continuous, as is implied from the *I*/*V* curves. In addition to the TMR measurements, Santos et al. measured

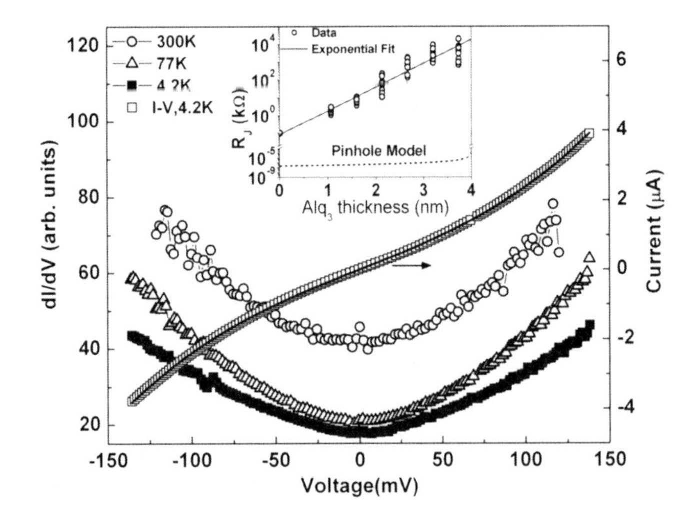

Fig. 7 *I*/*V* characteristics for an 8-nm Co/0.6-nm Al_2O_3/1.6-nm Alq_3/10-nm $Ni_{80}Fe_{20}$ junction. The fit to the *I*/*V* curve is shown as the line through the data points. The *inset* shows the exponential dependence of the junction resistance (R_J) vs Alq_3 thickness, for a total of 72 junctions made in a single deposition. Taken from [48] with permission

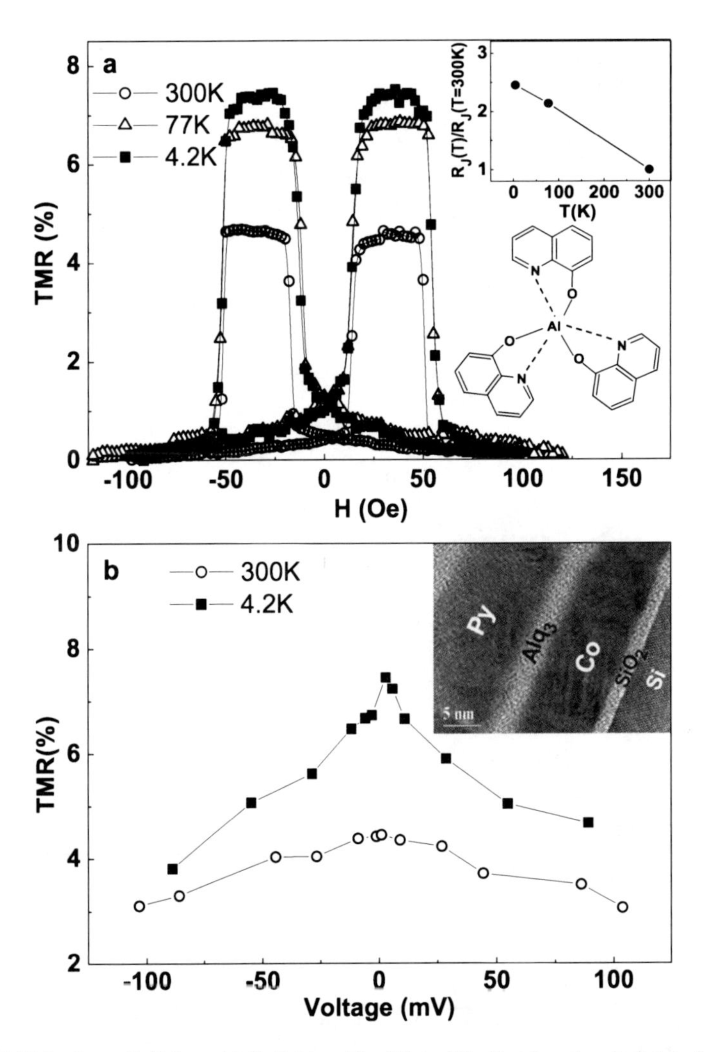

Fig. 8 TMR for 8 nm Co/0.6 nm Al_2O_3/1.6 nm Alq_3/10 nm $Ni_{80}Fe_{20}$ junction. In (**a**) the TMR was measured at a 10-mV bias. The *inset* shows the temperature dependence of the junction resistance and the chemical structure of the Alq_3 molecule. In (**b**) the bias dependence of the TMR is shown. The *inset* in (**b**) is a cross-sectional high-resolution TEM image of the junction showing the continuous barrier. Taken from [48] with permission

the spin-polarization of the top electrodes using the Tedrow–Meservey technique. Figure 9 shows the conductance vs bias for Alq_3 junctions at 0.4 K, using a superconducting ~4-nm Al bottom electrode with and without an Al_2O_3 layer. The G(V) data were fit to extract the spin polarization and the barrier height. For the junction shown in Fig. 9 with the Al_2O_3 barrier, the spin-polarization of the Co electrode was determined to be 27%. In other junctions (data not shown) the spin-polarization was 30% and 38%, respectively, for Fe and $Ni_{80}Fe_{20}$ top electrodes. In contrast, when no Al_2O_3 barrier was present on the superconducting

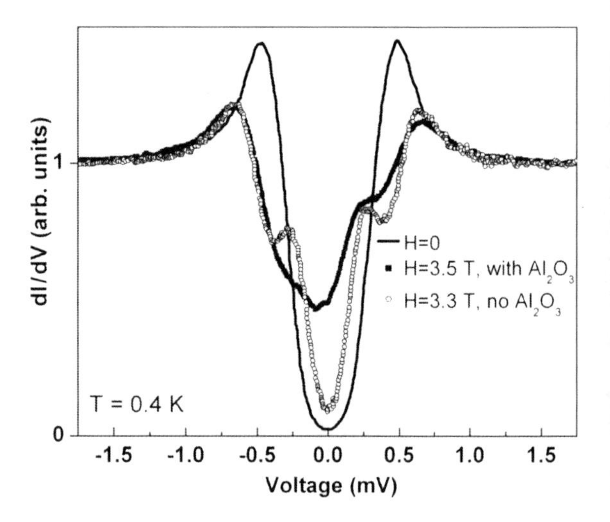

Fig. 9 Conductance of a 3.8 nm Al/Al_2O_3/8 nm Alq_3/8 nm Co junction (*solid squares*) and 3.7 nm Al/3.7 nm Alq_3/3 nm Co/6 nm $Ni_{80}Fe_{20}$ junction (*open circles*) with and without a 3.5-T magnetic field. Taken from [48] with permission

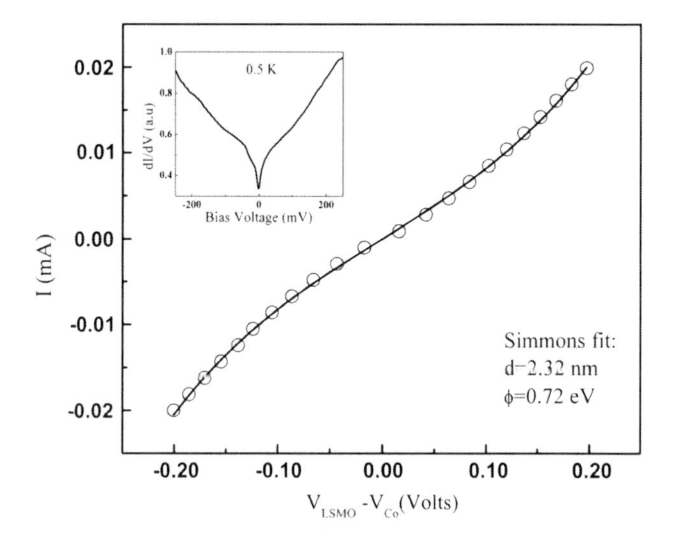

Fig. 10 Current–voltage data (*circles*) for a 15-nm TPP junction at 11 K together with a fit to the data (*line*). The *inset* shows the conductance vs voltage at 0.5 K. Taken from [50] with permission

Al electrode, the polarization of the Co top electrode dropped to 6% and the barrier height increased to 1.8 eV. Santos et al. rationalized the increase in the barrier height in the absence of the Al_2O_3 layer to the surface dipole layer that is known to exist at Al/Alq_3 interfaces [49]. One of the key findings of the Santos study was the observation of a positive TMR value both with and without the Al_2O_3 layer, as expected based on the known sign of the electrode polarization [17].

Shortly after Santos et al. published their results, Xu et al. fabricated organic spin valves using $La_{0.67}Sr_{0.23}Mn_3O$ (LSMO) and Co as the bottom and top electrodes, respectively, as the spin-injecting and spin-detecting ferromagnetic metal with tetraphenylporphyrin (TPP) and Alq_3 thin films as the spin-conducting layers

[50]. Figure 10 shows a typical *I/V* curve for a 15 nm thick film of TPP at 11 K; the inset shows the conductance vs voltage *V* of the same junction measured at 0.5 K. There are two characteristics to highlight. First, the *I/V* curve is nonlinear, suggestive of tunneling. Consequently, the curves were fit to the Simmons model to extract the effective barrier width and height [51]. The fit (shown as the thin line through the data points) yields a barrier height of 0.67 eV and a width of 2.5 nm. Second, the *I/V* curves are symmetrical under positive and negative bias, which indicates a similar barrier height at the Co/TPP and LSMO/TPP interfaces. This latter point is consistent with the fact that the work functions of Co and LSMO are about 4.8 eV [52]. The first ionization potential of TPP molecules in films is known to be 5.4 eV [53]. Consequently, the highest occupied molecular orbital of should be located ~0.6 eV below the Fermi energy of LSMO and Co, which is in very good agreement with fits to the Simmons model. The *G*(*V*) measurement at 0.5 K shows the expected voltage dependence for tunneling in a "ferromagnet/insulator/ ferromagnet" junction [29]. The linear contribution to *G*(*V*) observed at low biases can be qualitatively understood in terms of magnon-assisted tunneling [54]. Basically, electrons that tunnel from the negatively biased ferromagnetic electrode arrive in the positively biased ferromagnetic electrode with energy above the Fermi energy in that electrode. These "hot" electrons can then lose energy by emitting a magnon. As a result, the process leads to the linear contribution to *G*(*V*) that saturates beyond the bias voltage corresponding to the maximum magnon energy in the magnon-emitting electrode.

In Fig. 11 the change in the junction resistance (at ~2 mV) is plotted as an external magnetic field which goes from 1,500 to −1,500 Oe (circles) and is then reversed from −1,500 to 1,500 Oe (triangles) with the sample held at 80 K. When the magnetic field is between ~20 and ~200 Oe, the magnetization direction of the LSMO layer becomes antiparallel to the Co layer. Using the conventional definition of MR, defined as $\Delta R/R_{ap} = (R_{ap} - R_p)/R_{ap}$, where R_{ap} is the junction resistance in the antiparallel configuration and R_p is the junction resistance in the parallel configuration, respectively, then the device exhibits an "inverse" MR. The inset shows the junction resistance as a function of temperature (with no magnetic field). As the temperature is decreased from 80 to 11 K, the junction resistance increases by almost a factor of two [29]. Consequently, the data in Figs. 10 and 11 are convincing evidence for tunneling.

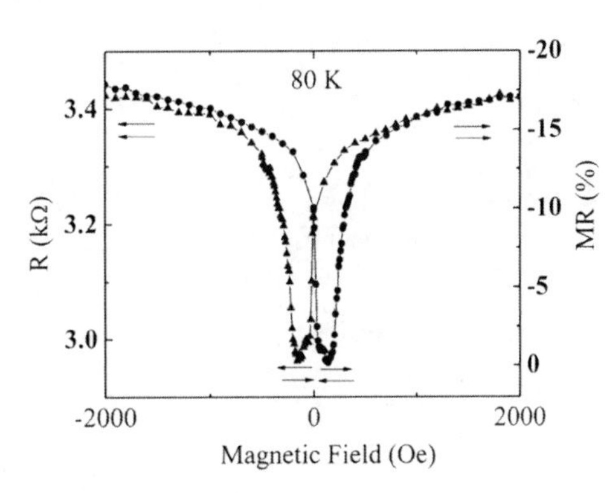

Fig. 11 Magnetoresistance curves of an LSMO/20 nm TPP/5 nm Co junction measured at 80 K. The *arrows* indicate the relative magnetization orientation of LSMO (*bottom arrow*) and Co (*top arrow*) electrodes. Taken from [50] with permission

Figure 12 shows the temperature-dependent MR for TPP and Alq_3, and the inset shows that the MR also decays rapidly with increasing bias voltage. The decrease in TMR can be largely attributed to the loss of spin polarization in LSMO with increasing temperature. The MR drops to half its maximum value at about 70 mV. There is a slight, yet reproducible asymmetry in the bias-dependence. The MR decays more slowly when the bias is negative. Qualitatively, De Tersa et al. have seen the same trend [55]. They observed an inverse magnetoresistance in magnetic tunnel junctions composed of "35 nm LSMO/2.5 nm $SrTiO_3$/30 nm Co/5 nm Au." Based on the relative density of states in the Co d band and LSMO, they concluded that a higher tunneling probably exists, when majority electrons in the LSMO are transferred into the Co minority states when the bias voltage is less than zero. However, recent results by Barraud et al. indicate that positive MR is observed in nanoscale junctions of "LMSO/Alq_3/Co" [56].

In 2008 Shin et al. used IETS and transmission electron microscopy (TEM) to characterize the chemical integrity and morphology of rubrene ($C_{40}H_{24}$) layers after deposition of an Fe top electrode [57]. The IETS spectra were consistent with the known IR- and Raman-active normal modes, which led the authors to conclude there were no chemical reactions with Fe. Cross-sectional TEM images showed continuous rubrene layers between the bottom Co layer and top Fe layer, with no evidence for small particle formation. Similar to the study by Santos et al., they found that the presence of an Al_2O_3 layer had a profound effect on the tunneling

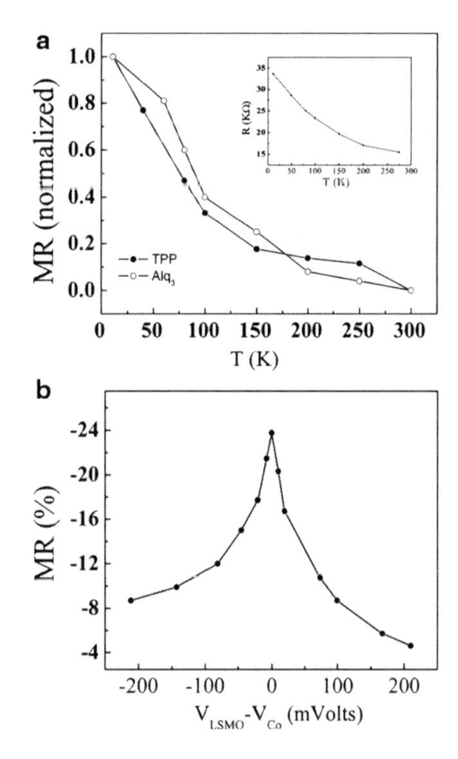

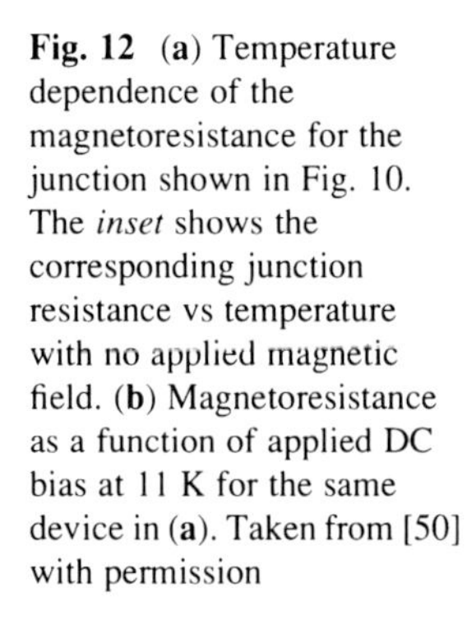
Fig. 12 (**a**) Temperature dependence of the magnetoresistance for the junction shown in Fig. 10. The *inset* shows the corresponding junction resistance vs temperature with no applied magnetic field. (**b**) Magnetoresistance as a function of applied DC bias at 11 K for the same device in (**a**). Taken from [50] with permission

behavior of rubrene films. Specifically, rubrene layers grown on Al_2O_3 showed only a slight temperature-dependent $G(V)$, whereas the rubrene films grown directly on Co showed a much more pronounced temperature-dependent $G(V)$. In Fig. 13 the normalized $G(V)$ for these junctions is plotted against $T^{-0.25}$. Such a plot can help establish whether the transport occurs through Mott's variable-range-hopping model [58]. It is evident that $G(V)$ is proportional to $T^{-0.25}$ at higher temperature (~120–300 K) for the rubrene films grown directly on Co. The Al_2O_3 also greatly influenced the TMR. At 4.2 K the TMR value was 16%, and dropped to 6% at room temperature. In contrast, no TMR was observed in rubrene junctions without the Al_2O_3 layer. Finally, Fig. 14 shows the loss of spin-polarization as a function of increasing rubrene thickness. The decay can be described by an exponential function. The authors ascribe the characteristic decay constant, 13 nm, to be an estimate of the spin-diffusion length.

In 2009 Schoonus et al. proposed a model to support experimental evidence for sequential tunneling in Alq_3 junctions [59]. A schematic depiction of an Alq_3 molecule in a junction between two ferromagnetic electrodes is shown in Fig. 15. The model assumes that, as the total Alq_3 layer thickness (d) increases, the total tunneling current (J) is the sum of a direct current (J_{direct}) and a two-step current ($J_{\text{2-step}}$):

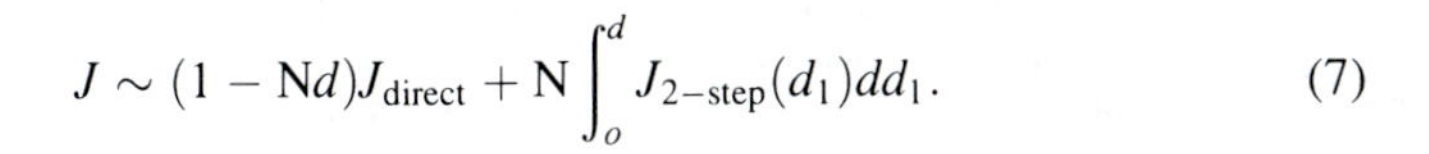

$$J \sim (1 - \mathrm{N}d)J_{\mathrm{direct}} + \mathrm{N}\int_{o}^{d} J_{2-\mathrm{step}}(d_1)dd_1. \tag{7}$$

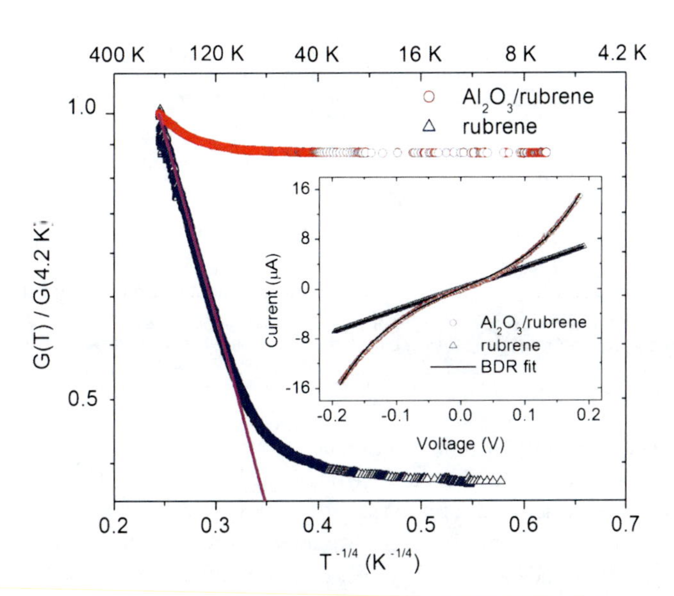

Fig. 13 Temperature dependence of the tunneling conductance for a 3.8 nm Al/6 nm rubrene/ 15 nm Co junction and 3.8 nm Al/0.5 nm Al_2O_3/5.5 nm rubrene/15 nm Co junction. The *solid line* is a fit the VHR model to the data at high temperatures for the rubrene-only barrier. The *inset* shows *I*/*V* curves for junctions at 4.2 K. Taken from [57] with permission

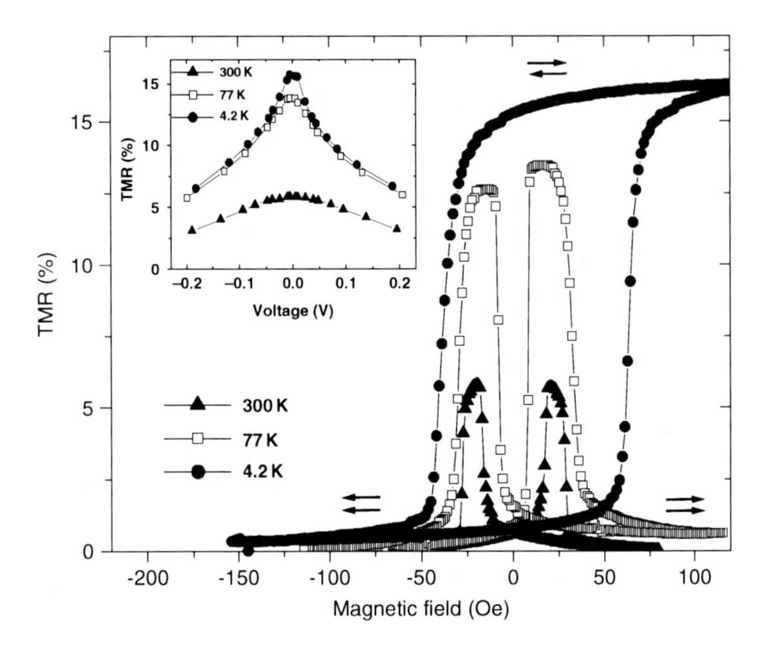

Fig. 14 TMR vs applied magnetic field for an 8 nm Co/0.5 nm Al_2O_3/4.6 nm rubrene/10 nm Fe/1.5 nm Co junction. The *arrows* indicate the magnetic configuration of the Co and Fe electrodes at various applied fields. The *inset* shows the bias dependence of the TMR. Taken from [57] with permission

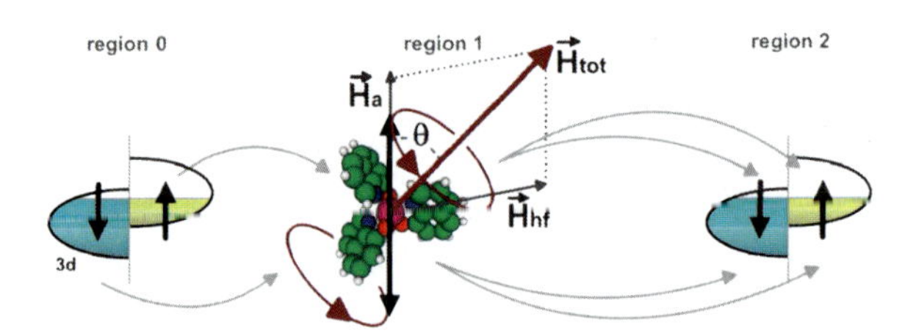

Fig. 15 Schematic diagram illustrating spin procession around the sum of the hyperfine field H_{hf} and the applied magnetic field H_{a} on an Alq_3 molecule, only showing the downstream tunneling from or to majority and minority 3*d* spin bands of the ferromagnetic electrodes. Taken from [59] with permission

The direct tunneling current is assumed to be proportional to the exponential decay of the wave function, $J \sim \exp[-\kappa d]$, where $k \equiv 4\pi\,(2mU)^{1/2}$, U is the barrier height, and m is the electron mass. Equation (7) also assumes that there is a homogenous linear distribution of N molecular sites, which can be occupied at a certain position, labeled as d_1. The two-step tunneling current arises from conservation of current into and out of the intermediate site (n_1, d_1). These two processes are depicted in Fig. 16a. The interesting prediction of the model is a transition in the junction resistance (which is proportional to $1/J$) with increasing layer thickness, as shown schematically in Fig. 16d. Tunnel junctions were made by

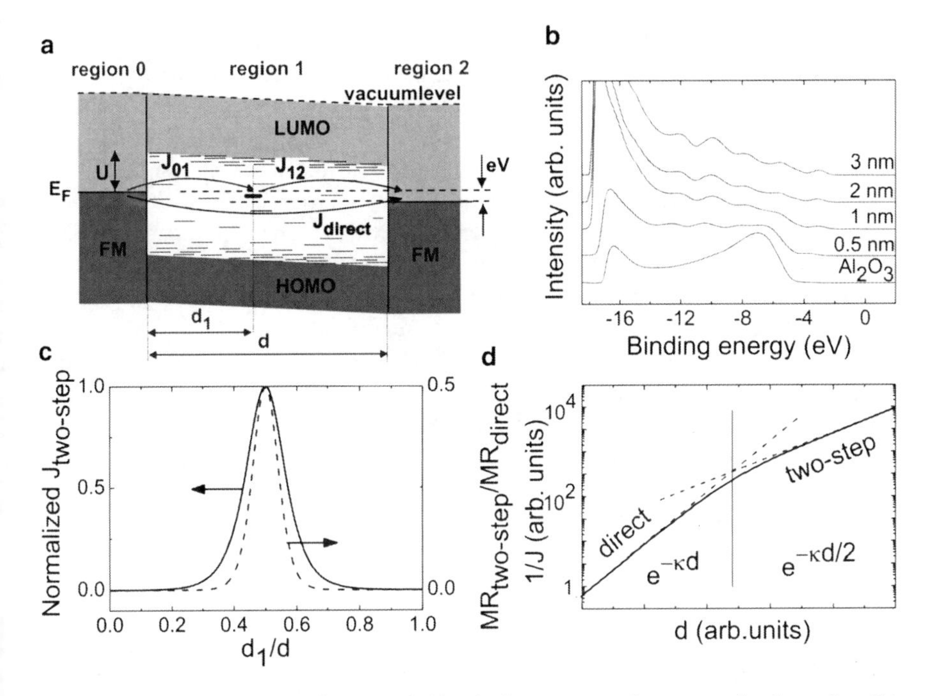

Fig. 16 (**a**) Schematic band diagram of Alq_3 in between two ferromagnetic electrodes. (**b**) Valence band spectra for an increasing Alq_3 layer on Al_2O_3, where the characteristic occupied molecular orbitals are seen in the 3-nm film. Calculation of (**c**) junction resistance and TMR for two-step tunneling as a function of d_1. (**d**) $1/J$ as a function of d. Taken from [59] with permission

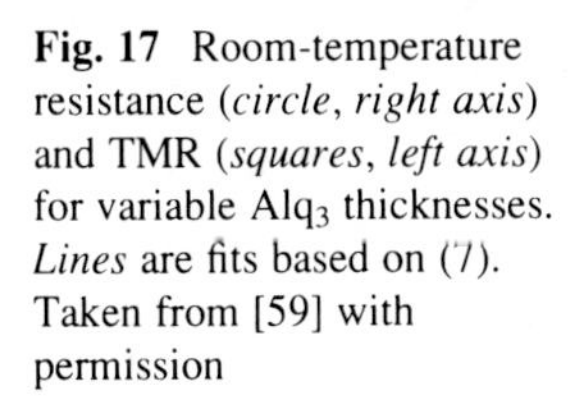
Fig. 17 Room-temperature resistance (*circle, right axis*) and TMR (*squares, left axis*) for variable Alq_3 thicknesses. *Lines* are fits based on (7). Taken from [59] with permission

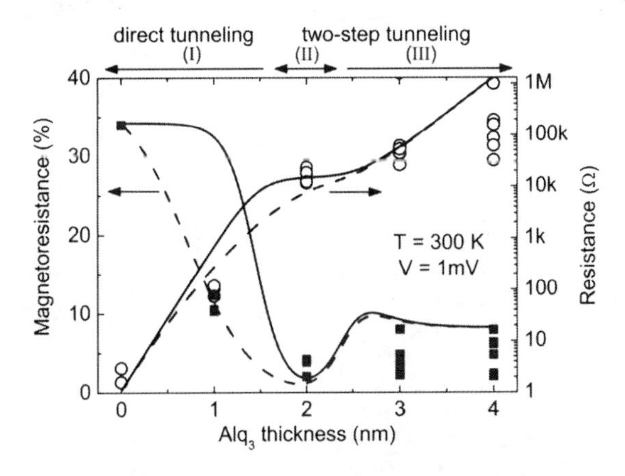

depositing a 2-nm Ta adhesion layer onto a glass substrate, followed by a 2-nm CoFeB layer, and then a 1.2-nm Al layer. The Al layer was completely oxidized by an in situ plasma treatment, followed by deposition of Alq_3 at 110 K and by a 20-nm Co layer. The junction resistance and TMR at 300 K are shown in Fig. 17 for different Alq_3 layer thicknesses. The first trend to point out is the rapid decrease in the TMR when 1 and 2 nm of Alq_3 were deposited onto the Al_2O_3 layer. In this thickness regime, the direct tunneling current dominates, and the decay in TMR can be ascribed to the exponential decay of the spin-polarized wave function with increased d. For junctions with Alq_3 films greater than 2 nm, a small increase in the TMR was observed, with a concomitant increase in the junction resistance. The authors believed that the transition in the junction resistance and TMR values signaled the onset of the multiple-step tunneling predicted by (7). The solid line and dashed lines in Fig. 17 are the results of fitting the data to (7). The details of the fitting procedure and extracted parameters can be found in the original publication [59].

In 2009 Szulczewski et al. were the first to report junctions using an MgO spin-filter in a tunnel junction with Alq_3 [23]. Figure 18 shows the electrical and magneto-resistive characteristics of the MgO/Alq_3. The MgO/Alq_3 series show TMR values of up to 16% at room temperature, for $t = 0$, which falls to 12–13% at zero bias upon inserting the Alq_3, and remains unchanged with increasing barrier thickness, as shown in Fig. 18b. The I/V curves are quite asymmetric for the MgO barriers. Maximum TMR actually occurs at a negative bias of 100 mV. The magnetoresistance changes sign at a positive bias of about 250 mV for the 2-nm

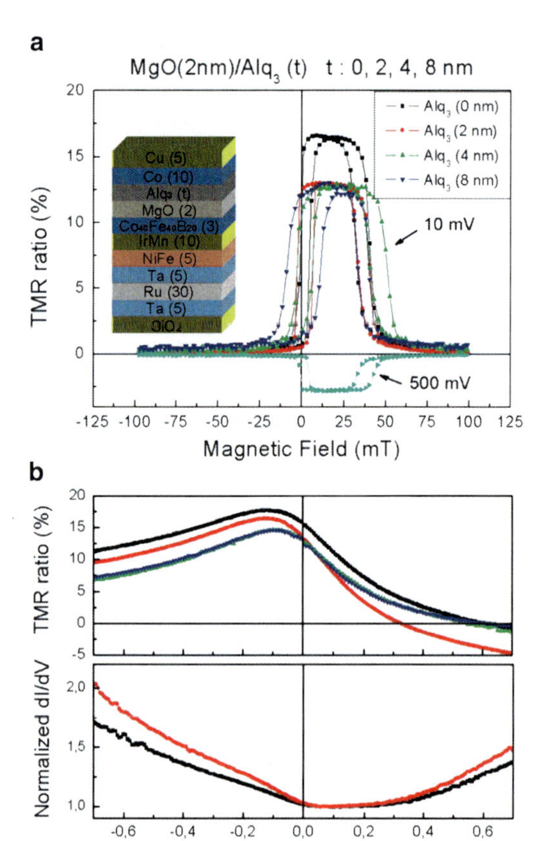

Fig. 18 Characteristics of Alq_3 junctions with a CoFeB/MgO spin injection layer. The resistance switching curve is plotted as a function of applied field for different thicknesses or bias in the *top panel*, and the TMR ratio is plotted as function of bias, together with the dI/dV curve for $t = 0$ and 2 nm in the *bottom panel*. The curve for $t = 2$ nm at 500 mV is included in the *top panel*. Taken from [23] with permission

Alq_3 tunnel barrier, reaching a maximum negative value of −5%. Positive bias corresponds to a flow of electrons from the CoFeB pinned layer into the stack. The top panel in Fig. 18 shows the resistance/magnetic field (*R*/*H*) curves, measured at different bias or barrier thickness. The TMR is 13% for $t = 0$, and it falls to 8–9% for 1 nm of Alq_3. These results are consistent with those of Santos et al. [48], in the 2–4 nm thick Alq_3 films, where the transport is by tunelling. However, in the 4–8 nm thick films elastic tunneling is unlikely, and transport is either by multistep tunneling or hopping, as observed by Schoonus et al. [59]. Initially, the resistance-area (RA) product of the junctions increases exponentially with increasing barrier thickness, and then tends to saturate, as shown in Fig. 19. Assuming that the tunelling resistance of the Alq_3 tunnel barrier increases exponentially with thickness as $\exp(t/t_0)$, and that the resistance in the hopping regime increases as t, the total resistance-area product of the composite barrier is the sum of the resistances of the parallel tunelling and hopping channels:

$$\mathrm{RA}(t) = \left[1/(2\mathrm{RA}(0) + \rho r t) + 1/(2\mathrm{RA}(0)\exp(t/t_0)\right]^{-1} \tag{8}$$

where RA(0) is the resistance-area product of the MgO barrier. The characteristic thickness deduced from the fit in Fig. 19 is $t_0 = 0.8$ nm, and the resistivity of the organic film in the hopping regime is $\rho = 2.0\ \mathrm{k}\Omega\ \mathrm{m}$, which agrees with a literature value for amorphous films of this material [60]. The data of Fig. 19 show no significant reduction of magnetoresistance in the hopping regime, at least out to 8 nm, which suggests that approximately three-quarters of the spin polarization is preserved in the rubrene layers. The asymmetric bias-dependence of the TMR with MgO (Fig. 18) and the sign change at positive bias in the tunelling regime are a feature that reflects the asymmetry of the top and bottom interfaces, which could be chemical or magnetic in nature. The magnetoresistance depends on the overlap of the ↑ and ↓ Fermi-surface cross-sections for the two ferromagnetic electrodes. The voltage at the sign change, +250 mV, is related to the bottom electrode, and may correspond approximately to the exchange splitting of the energy of the tunneling electrons in crystallized CoFeB, which has a much lower Curie temperature than cobalt.

A very comprehensive set of electrical/magnetic measurements on "LSMO/LAO/rubrene/Fe" spin-valves was reported by Yoo et al. in 2009 [61]. These authors varied the thickness of the rubrene layer between 5 and 50 nm, and were able to distinguish between the tunneling and hopping regimes discussed above. In the TMR limit, the TMR was about 12% at 10 K, and monotonically decreased as the temperature was increased; above 250 K no TMR was observed. This trend in the MR with increasing temperature has been observed in most spin-valves using LSMO as the bottom electrode. When the rubrene layer was increased to 20 and 30 nm, the MR at 10 K decreased to ~6% and ~2%, respectively. For rubrene layers thicker than 40 nm no GMR was observed, which implies that the spin-diffusion length in rubrene at 10 K is 10–20 nm. This value is consistent with the 13 nm spin-diffusion length for rubrene at 0.4 K estimated by Santos et al. from the decay of the

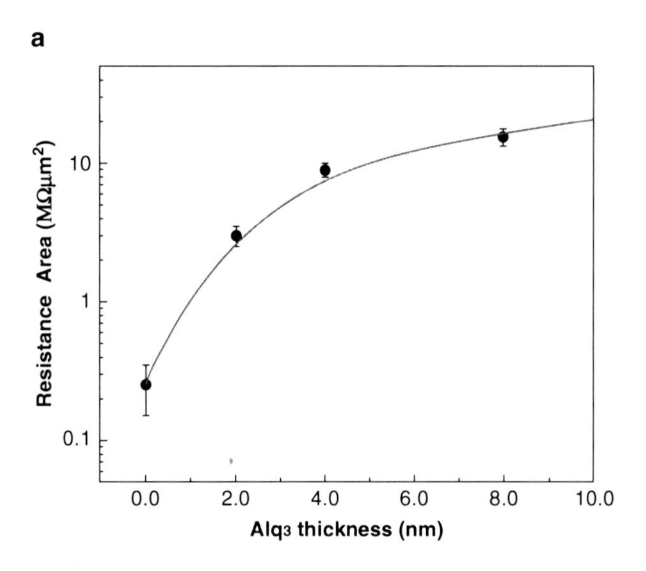

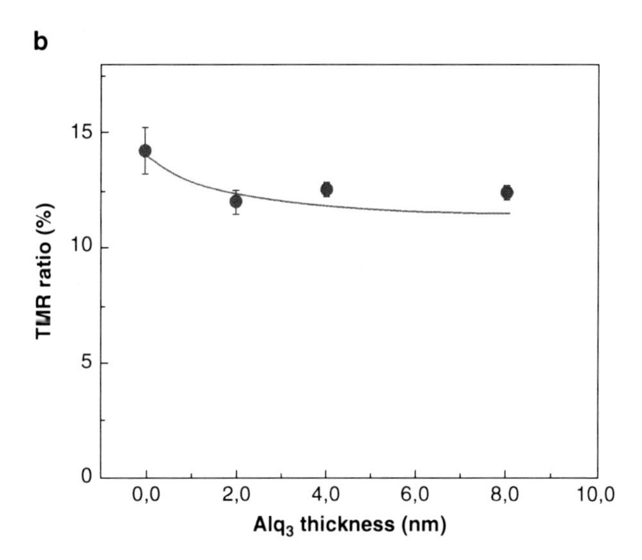

Fig. 19 Resistance-area product (**a**) and magnetoresistance (**b**) plotted as a function of Alq_3 barrier thickness in MgO/Alq_3 magnetic tunnel junctions. The *solid line* is the fit to a (8), with parallel tunelling and hopping channels. The *line* in (**b**) is a guide to the eye. Taken from [23] with permission

spin-polarization [48]. A similar study was able to distinguish between the tunneling and injection regimes of rubrene spin valves. In 2010 Lin et al. fabricated spin valves onto oxidized Si wafer by depositing a 15-nm Co film, followed by 2.5 nm of Al [62]. The Al film was then oxidized ex situ in an oxygen plasma.

Rubrene films were grown on the oxidized surface, followed by 15 nm of Fe. When the rubrene films were less than 10 nm thick, the *I*/*V* curves were typical for tunnel junctions; specifically, they showed a parabolic *G*(*V*) behavior and very little temperature dependence. At 100 K the TMR was −6% when the rubrene layer was 5 nm thick. In contrast, when the rubrene layer was 15 nm thick, the *I*/*V* curves were nonlinear and temperature-dependent. No TMR was observed when the rubrene films were greater than 15 nm.

In 2010 Barraud et al. fabricated nanoscale junctions in Alq_3 films grown on LSMO films by a nanoindentation technique. Cobalt was deposited into the nanopore to complete the sandwich structure. At 2 K the junction resistance was found to increase exponentially with increasing Alq_3 thickness, which demonstrates that tunneling is the primary transport mechanism. One of the most interesting aspects of this work was the observation of positive TMR. In several other reports, using square-millimeter sized "LSMO/Alq_3/Co" junctions, the sign of the MR had usually been negative. There is one notable exception, where both positive and negative MR in "LSMO/Alq_3/Co" junctions was measured in some samples by the same group [63]. In order to reconcile the positive sign, Barraud et al. proposed a spin-dependent interfacial molecular hybridization model. Essentially, the model assumes that the polarization of the LSMO/Alq_3 interface inverts due to strong coupling between the Alq_3 HOMO and one spin channel in the LSMO.

4.3 Spin-Polarized Scanning Tunneling Microscopy/Spectroscopy

In contrast to all the examples discussed above, scanning tunneling microscopy (STM) and spectroscopy allows one to measure spin-currents through molecules without the need to deposit top metal contact. Furthermore, STM experiments are done under UHV conditions on single-crystal surfaces, so exquisite control of the sample environment is available, even though the difficulty of the experiment increases. One of the first spin-polarized STM studies to visualize ferromagnetic coupling between a molecule and magnetic substrate was reported by Iovichi et al. [64]. In this study, Co was deposited onto a Cu(111) surface to produce triangular-shaped islands about two atomic layers thick; then Co phthalocyanine (CoPc) was evaporated on top. STM images at 4.6 K were taken with Co-coated tips, and showed that the CoPc molecules adsorbed to the top and the edges of the Co islands. Two different types of magnetic islands were identified in spin-dependent G(V) plots vs tip-sample bias (in the absence of CoPc molecules). For one type of Co island, a strong peak in G(V) occurred at −0.28 V, which was defined as an "antiparallel" (↑↓) configuration. A second type of Co island was defined as the "parallel" (↑↑) configuration, because a second G(V) peak was observed at a −0.6-V sample-tip bias. Note that this definition of the parallel and antiparallel configurations does not imply a relative orientation of the tip and island magnetization. After CoPc deposition on the islands, spin-polarized G(V) spectra were recorded over the center of molecules, i.e., directly over the Co^{2+} site. For CoPc molecules on the ↑↑ islands there was a broad peak near −0.19 V, which was much weaker than for CoPc adsorbed on the ↑↓ islands. This suggested that the spin-polarized tunneling current through the CoPc was linked to the magnetization of the island; this was confirmed by averaging the results of several different molecules and tips. Using first-principles density-functional theory calculations of the spin-dependent conductance, the authors suggested a ferromagnetic coupling between the local spins in the molecule and the substrate.

Another spin-polarized STM study on CoPc found evidence for strong molecule-substrate hybridization of orbitals. Brede et al. reported submolecular resolution STM images of single CoPc molecules adsorbed onto a two monolayer-thick Fe film on a W(100) surface [65]. They combined spin- and energy-dependent tunneling data to visualize variations in the spin-polarized current through different regions of a single CoPc molecule. The authors were able to simulate accurately the experimental spin-polarized STM results with the aid of state-of-the-art ab initio calculations.

5 Concluding Remarks

This review has focused on spin-dependent tunneling through molecules. In general, the experimental data reported so far suggest two common mechanisms of spin-transport. In one limit, the wave function of the two electrodes overlap in the barrier and primarily lead to nonresonant incoherent tunneling through the molecules. In this regime, evidence has been found for inelastic excitation of molecular vibrations. In another limit, where the wave functions of the ferromagnetic electrodes are too far apart to overlap, one can observe multistep tunneling from molecule to molecule. It appears that the chemical, magnetic, and structural details of the interfaces are key to how much of the spin-polarization is preserved. Unfortunately, it has been difficult to elucidate a direct cause-and-effect scenario, since the details of device fabrication vary slightly from lab to lab. However, one clear observation has emerged. The presence of an amorphous Al_2O_3 or crystalline MgO tunneling barrier dramatically improves spin-polarized tunneling in Alq_3 barriers. The role of a tunnel barrier may function to increase the interfacial spin-dependent resistance in a similar way, to solve the conductivity mismatch at the metal/inorganic semiconductor interface [66, 67]. The studies highlighted in Sect. 4 are likely to motivate future studies that will reveal more insight into spin-injection and spin-ejection across ferromagnetic metal/molecule interfaces [68]. In particular, we are beginning to see more photo-emission and magnetometry studies aimed at probing the electronic/magnetic structure of such interfaces. Several important questions still remain unanswered. For example, what is/are the mechanism(s) for spin relaxation? What factors determine the sign of the TMR? What are the roles of hyperfine and spin-orbit coupling? Given the intense activity in this field over the past few years, it is likely that we will begin to find the answers to these questions soon.

Acknowledgment The author would like to thank US National Science Foundation and the Science Foundation of Ireland for funding and the assistance of former/current students and colleagues including Mr. J. Brauer, Prof. W. Butler, Prof. A. Caruso, Prof. J. M. D. Coey, Prof. A. Gupta, Prof. P. LeClair, Prof. G. Mankey, Prof. J. Moodera, Prof. R. Schad, Mr. H. Tokuch, and Dr. W. Xu.

References

1. Burstein E, Lundqvist S (1969) Tunneling phenomena in solids. Plenum, New York
2. Slonczewski JC (1996) Current-driven excitation of magnetic multilayers. J Magn Magn Mater 159:L1–L7
3. Berger L (1996) Emission of spin waves by a magnetic multilayer traversed by a current. Phys Rev B 54:9353–9358
4. Myers EB, Ralph DC, Katine JA, Louie RN, Buhrman RA (1999) Current-induced switching of domains in magnetic multilayer devices. Science 285:867–870
5. http://www.IDTekEx.com
6. Baibich MN, Broto JM, Fert A, Nguyen Van Dau F, Petroff F, Etienne P, Creuzet G, Friederich A, Chazelas J (1988) Giant magnetoresistance of (001)Fe/(001)Cr magnetic superlattices. Phys Rev Lett 61:2472–2475
7. Nguyen TD, Sheng Y, Rybicki J, Veeraraghavan G, Wohlgenannt M (2007) Magnetoresistance in π-conjugated organic sandwich devices with varying hyperfine and spin-orbit coupling strengths, and vary dopant concentration. J Mater Chem 17:1995–2001
8. Bloom FL, Wagemans W, Kemerink M, Koopmans B (2007) Separating positive and negative magnetoresistance in organic semiconductor devices. Phys Rev Lett 99:257201
9. Golfe N, Desai P, Shakya P (2008) Separating the roles of electrons and holes in the organic magnetoresistance of aluminum tris(8-hydroxyquinoline) organic light emitting diodes. J Appl Phys 104:084703
10. Hu B, Yan L, Shao M (2009) Magnetic-field effects in organic semiconducting materials and devices. Adv Mater 21:1500–1516
11. Martin JL, Bergeson JD, Prigodin VN, Epstein AJ (2010) Magnetoresistance for organic semiconductors: small molecule, oligomer, conjugated polymer, and non-conjugated polymer. Syn Met 160:291–296
12. Wagemans W, Koopmans B (2010) Spin transport and magnetoresistance in organic semiconductors. Phys Status Solidi B:1–13
13. Cottet A, Kontos T, Sahoo S, Man HT, Choi MS, Belzig W, Bruder C, Morpurgo AF, Schonenberger C (2006) Nanospintronics with carbon nanotubes. Semiconductor Sci Technol 21:S78–S95
14. Tombros N, Jozsa C, Popinciuc M, Jonkman HT, van Wees BJ (2007) Electronic spin transport and spin precession in single graphene layers at room temperature. Nature 448:571–574
15. Kartsovnik MV (2004) High magnetic fields: a toll for studying electronic properties of layered organic metals. Chem Rev 104:5737–5782
16. O' Handley RC (2000) Modern magnetic materials: principles and applications. Wiley, New York
17. Meservey R, Tedrow PM (1994) Spin-polarized electron tunneling. Phys Rep 238:173–243
18. Maki K (1964) Pauli paramagnetism and the superconducting state II. Prog Theor Phys 32:29–36
19. Jullière M (1975) Tunneling between ferromagnetic films. Phys Lett 54A:225–226
20. Butler WH, Zhang XG, Schulthess TC, MacLaren JM (2001) Spin-dependent tunneling conductance of Fe/MgO/Fe sandwiches. Phys Rev B 63:054416

21. Parkin SSP, Kaiser C, Panchula A, Rice PM, Hughes B, Samant M, Yang SH (2004) Giant tunneling magnetoresistance at room temperature with MgO(100) tunneling barriers. Nat Mater 3:862–867
22. Yuasa S, Nagahama T, Fukushima A, Suzuki Y, Ando K (2004) Giant room-temperature magnetoresistance in single-crystal Fe/MgO/Fe magnetic tunnel junctions. Nat Mater 3:868–871
23. Szulczewski G, Tokuc H, Oguz K, Coey JMD (2009) Magnetoresistance in magnetic tunnel junctions with an organic barrier and a MgO spin filter. Appl Phys Lett 95:202506
24. Moodera JS, Nassar J, Mathon G (1999) Spin-tunneling in ferromagnetic junctions. Annu Rev Mater Sci 29:381–432
25. Park H, Lim AKL, Alivisatos P, Park J, McEuen PL (1999) Fabrication of metallic electrodes with nanometer separation by electromigration. Appl Phys Lett 75:301–330
26. Ralls KS, Buhrman RA, Tiberio RC (1989) Fabrication of thin-film metal nanobridges. Appl Phys Lett 55:2459–2461
27. Fan X, Rogow DL, Swanson CH, Tripathi A, Oliver SRJ (2007) Contact printed Co/insulator/Co molecular junctions. Appl Phys Lett 90:163114
28. Brinkman WF, Dynes RC, Rowell JM (1970) Tunneling conductance of asymmetrical barriers. J Appl Phys 41:1915–1921
29. Akerman JJ, Escudero R, Leighton C, Kim S, Rabson DA, Slaughter JM, Schuller IK (2002) Criteria for ferromagnet–insulator-ferromagnet tunneling. J Magn Magn Mater 240:86–91
30. Hwang J, Wan A, Kahn A (2009) Energetics of metal-organic interfaces: new experiments and assessment of the field. Mater Sci Eng R 64:1–31
31. Popinciuc M, Jonkman HT, van Wees BJ (2006) Energy level alignment symmetry at Co/pentacene/Co interfaces. J Appl Phys 100:093714
32. Popinciuc M, Jonkman, van Wees BJ (2007) Energy level alignment at Co/AlOx/pentacene interfaces. J Appl Phys 101:093701
33. Zhan YQ, de Jong MP, Li FH, Dediu V, Fahlman M, Salaneck WR (2008) Energy level alignment and chemical interaction at Alq_3/Co interfaces for organic spintronics devices. Phys Rev B 78:045208
34. Aristov VY, Molodtsova OV, Ossipyan YA, Doyle BP, Nannarone S, Knupfer M (2009) Ferromagnetic cobalt and iron top contacts on an organic semiconductor: evidence for a reacted interface. Org Electron 10:8–11
35. Xu W, Brauer J, Szulczewski G, Driver MS, Caruso AN (2009) Electronic, magnetic, and physical structure of cobalt deposited on aluminum tris(8-hydroxy qunioline). Appl Phys Lett 94:233302
36. Borgatti F, Bergenti I, Bona F, Dediu V, Fondacaro A, Huotari S, Monaco G, MacLaren DA, Chapman JN, Panaccione G (2010) Understanding the role of tunneling barriers in organic spin valves by hard X-ray photoelectron spectroscopy. Appl Phys Lett 96:043306
37. Wang YZ, Qi DC, Chen S, Mao HY, Wee ATS, Gao XY (2010) Tuning the electron injection barrier between Co and C_{60} using Alq_3 buffer layer. J Appl Phys 108:103719
38. Chan YL, Hung YJ, Wang CH, Lin YC, Chiu CY, Lai TL, Chang HT, Lee CH, Hsu YJ, Wei DH (2010) Magnetic response of an ultrathin cobalt film in contact with an organic pentacene layer. Phys Rev Lett 104:177204
39. Shen C, Kahn A, Schwartz J (2001) Chemical and electrical properties of interfaces between magnesium and aluminum and tris-(8-hydroxy quinoline) aluminum. J Appl Phys 89:449–460
40. Pi TW, Liu Ch, Hwang J (2006) Surface electronic structure of Ca-deposited tris (8-hydroxyquinolato) aluminum studied by synchrotron radiation photoemission. J Appl Phys 99:123712
41. Moulder JF, Stickle WF, Sobol PE, Bomden KD (1995) Handbook of X-ray photoelectron spectroscopy. Eden Prairie, Minn
42. An P, Bialczak RC, Martinek J, Grose JE, Donev LAK, McEuen PL, Ralph DC (2004) The Kondo effect in the presence of ferromagnetism. Science 306:86–89
43. Monsma DJ, Parkin SSP (2000) Spin polarization of tunneling current from ferromagnet/Al_2O_3 interfaces using copper-doped aluminum superconducting films. Appl Phys Lett 77:720–722

44. Petta JR, Slater SK, Ralph DC (2004) Spin-dependent transport in molecular junctions. Phys Rev Lett 93:136601
45. Widrig CA, Alves CA, Porter MD (1991) Scanning tunneling microscopy of ethanthiolate and n-octadecanethiolate monolayers spontaneously adsorbed at gold surfaces. J Am Chem Soc 113:2805–2810
46. Wang W, Lee T, Reed MA (2003) Mechanism of electron conduction in self-assembled monolayer devices. Phys Rev B 68:035416
47. Wang W, Richter CA (2006) Spin-polarized inelastic electron tunneling spectroscopy of a molecular magnetic tunnel junction. Appl Phys Lett 89:153105
48. Santos TS, Lee JS, Migdal P, Lekshmi IC, Satpati B, Moodera JS (2007) Room temperature tunnel magnetoresistance and spin-polarized tunneling through an organic semiconductor barrier. Phys Rev Lett 98:016601
49. Baldo MA, Forrest SR (2001) Interface-limited injection in amorphous organic semiconductors. Phys Rev B64:085201–085218
50. Xu W, Szulczewski GJ, LeClair P, Navarrete I, Schad R, Miao G, Guo H, Gupta A (2007) Tunneling magnetoresistance observed in $La_{0.67}Sr_{0.33}MnO_3$/organic/ molecule/Co junctions. Appl Phys Lett 90:072506
51. Simmons JG (1963) Generalized theory for the electric tunnel effect between similar electrodes separated by a thin insulating film. J Appl Phys 34:1793
52. Dediu V, Hueso LE, Bergenti I, Riminucci A, Gorgatti F, Graziosi P, Newby C, Casoli F, de Jong MP, Taliani C, Zhan Y (2008) Room-temperature spintronics effects in Alq_3-based hybrid devices. Phys Rev B 78:115203
53. Ishii H, Sugiyama K, Ito E, Seki K (1999) Energy level alignment and interfacial electronic structure at organic/metal and organic/organic interfaces. Adv Mater 11:605–625
54. Zhang A, Levy PM, Marley AC, Parkin SSP (1997) Quenching of magnetoresistance by hot electrons in magnetic tunnel junctions. Phys Rev Lett 79:3744–3747
55. De Teresa JM, Barthléméy A, Fert A, Contour JP, Lyonnet R, Montaigne P, Vaures A (1999) Inverse tunnel magnetoresistance in $Co/SrTiO_3/La_{0.7}Sr_{0.3}MnO_3$: new idea on spin polarized tunneling. Phys Rev Lett 82:4288–4291
56. Barraud C, Seneor P, Mattana R, Stéphane F, Bouzehouane K, Deranlot C, Graziosi P, Hueso L, Bergenti I, Dediu V, Petroff F, Fert A (2010) Unravelling the role of the interface for spin injection into organic semiconductors. Nat Phys 6:615–620
57. Shim JH, Raman KV, Park YJ, Santos TS, Miao GX, Satpati B, Moodera JS (2008) Large spin diffusion length in an amorphous organic semiconductor. Phys Rev Lett 100:226603
58. Mott NF (1987) Conduction in non-crystalline solids. Oxford, New York
59. Schoonus JJHM, Lumens PGE, Wagemans W, Kohlhepp JT, Bobbert PA, Swagten HJM, Koopmans B (2009) Magnetoresistance in hybrid organic spin valves at the onset of multiple-step tunneling. Phys Rev Lett 103:146601
60. Mahapatro AK, Agrawal R, Ghosh S (2004) Electric-field-induced conductance transistion in 8-hydroxyquinoline aluminum (Alq_3). J Appl Phys 96:3583
61. Yoo JW, Jang HW, Prigodin VN, Kao C, Eom CB, Epstein AJ (2009) Giant magnetoresistance in ferromagnet/organic semiconductor/ferromagnet heterojunctions. Phys Rev B 80:205207
62. Lin R, Wang F, Rybicki J, Wohlgenannt M, Hutchinson KA (2010) Distinguishing between tunneling and injection regimes of ferromagnetic/organic semiconductor/ferromagnet junctions. Phys Rev B 81:195214
63. Vinzelberg H, Schumann J, Elefant D, Gangineni, RB, Thomas J, Bücher B (2008) Low temperature tunneling magnetoresistance on $(La,Sr)MnO_3$/Co junctions with organic spacer layers. J Appl Phys 103:0937220
64. Iacovita C, Rastei MV, Heinrich BW, Brumme T, Kortus J, Limot L, Bucher JP (2008) Visualizing the spin of individual cobalt-phthalocyanine molecules. Phys Rev Lett 101:116602
65. Brede J, Atodiresei N, Kuck S, Lazic P, Cacuic V, Morikawa Y, Hoffmann G, Blugel S, Wiesendanger R (2010) Spin- and energy-dependent tunneling through a single molecule with intramolecular spatial resolution. Phys Rev Lett 105:047204

66. Rashba EI (2000) Theory of electrical spin injection: tunnel contacts as a solution of the conductivity mismatch problem. Phys Rev B 62:R16267–R16270
67. Fert A, Jaffrès H (2001) Conditions for efficient spin injection from a ferromagnetic metal into a semiconductor. Phys Rev B 64:184420
68. Szulczewski G, Sanvito S, Coey JMD (2009) A spin of their own. Nat Mater 8:693–695

Index